Georgia in Transition

Schriften zur Internationalen Entwicklungs- und Umweltforschung

Herausgegeben vom

Zentrum für internationale Entwicklungs- und Umweltforschung

der Justus-Liebig-Universität Gießen

Band 26

PETER LANG

Frankfurt am Main · Berlin · Bern · Bruxelles · New York · Oxford · Wien

Lorenz King/Giorgi Khubua (eds.)

Georgia in Transition

Experiences and Perspectives

PETER LANG
Internationaler Verlag der Wissenschaften

Bibliographic Information published by the Deutsche Nationalbibliothek
The Deutsche Nationalbibliothek lists this publication in the Deutsche Nationalbibliografie; detailed bibliographic data is available in the internet at <http://www.d-nb.de>.

ISSN 1615-312X
ISBN 978-3-631-58977-9

Printed in Germany 1 2 3 4 5 7

www.peterlang.de

Foreword of the Editor

The Center for International Development and Environmental Research (Zentrum für internationale Entwicklungs- und Umweltforschung, ZEU) was founded in 1998 as an interdisciplinary and interdepartmental research institute of the Justus-Liebig-University, Giessen. Presently, faculty members from the following disciplines are associated with ZEU: Agricultural and Nutritional Sciences, Economics, Environmental Safety, Geography, Law, Statistics and Econometrics as well as Social and Cultural Sciences. Upon ZEU's foundation, the following five sections were established:

- The use of natural resources and environmental preservation,
- Nutrition security,
- Human capacity in socio-economic systems,
- Institutional foundation of regional development processes,
- Transformation of economy and society.

These sections also indicate the issues which research is currently focused on. A restriction of ZEU's activities to specific geographic regions is not considered.

The series 'Schriften zur Internationalen Entwicklungs- und Umweltforschung' (Publications on International Development and Environmental Research) aims to present the results of ZEU's research activities and the state of discussions in the fields of our research to a broader public. The authors appreciate suggestions and critical remarks, in order to reconsider their own positions.

The volumes of the series will appear intermittently and may be purchased in bookstores or directly from the publisher.

Giessen, January 2009

Professor Dr. Ingrid-Ute Leonhäuser
Chairwoman of the Board

Preface

Georgia is a small country the size of Bavaria, located in the southern Caucasus. It has an area of 69,700 km² and a population of 4.4 million persons, around 84% of which are Georgians. Embedded into the Greater Caucasus Mountains to the north, the Lesser Caucasus Mountains to the south, and the Black Sea to the west, Georgia's landscape and climate conditions are very diverse. Being a traditional agricultural country, Georgia was the main exporter of agricultural products such as fruits, vegetables, and wine during Soviet times. Moreover, the living standard of the Georgian population was one of the highest of all Soviet countries during this time. The situation changed enormously with the breakdown of the Soviet system and the beginning of the transition process. Like all other former Soviet countries Georgia had to face the challenges of radical political and economic reforms.

In April 1991 Georgia declared its independence. The following transition period was characterized by civil war and political unrest. Some of the problems could not be solved yet. After a sharp economic decline during the first years of transition, Georgia was able to recover after the financial crisis in Russia at the end of the 1990s. Since then the GDP has grown annually by up to 10% and more. Especially the living conditions in urban areas have improved in recent years. Europe is particularly interested in the stability of the region. The second-largest oil pipeline in the world, as well as important transportation routes run through Georgia from Asia towards the West and are consequently an important lifeline of the European economy.

Generally, the transformation of social systems in post-socialist countries is described as a process which is set in motion by political will and political actions with the aim of replacing an old social order with a new one. This change of system includes political and socio-economic levels. Hence, reshaping is a time-intensive process which does not only include the change of formal institutions but also requires the adaptation of market-economic and democratic ways of thinking and behaviour. In view of the difficulties connected with simultaneous changes in various areas as well as with the temporal discrepancy between long-term cycles of economic development and the short-term time horizons of voters, there has even been talk of the "impossibility of holistic reforms" or the "dilemma of simultaneity". Still, the experience of post-socialist transformation has shown that countries with a higher degree of democratization also made better progress with regard to economic reforms and have thus proved that these two aims may actually be reached simultaneously.

For ten years now there has been a lively interdisciplinary exchange of knowledge between the Ivane Javakhishvili State University in Tbilisi / Republic of Georgia and Justus-Liebig-University Giessen (JLU), which has been and is mainly initiated and coordinated by the Center for International Development and Environmental Research (ZEU) at JLU. An official cooperation agreement between both universities exists since 2005. Within its framework, a regular exchange of scientists and students takes place. If one wanted to take stock of the joint research activities so far, it would become apparent how much the transformation processes taking place in Georgia have influenced the cooperative projects with regard to their orientation, progression and results. Instead of such an account, this book, which is supposed to express the multidisciplinarity of transformation, has come into being.

First of all, it portrays the development of Georgia in a difficult period of change and is divided into three main parts. The first part deals with the jurisprudential and political aspects of transformation. By way of depicting the reform processes in civil areas and those pertaining to civil law, we will go into the reshaping of Georgia's legal system. Furthermore, the section covers the problems concerning statehood and democratization in Georgia and describes the course former elections, the most important means of expressing social will, have taken. In addition, the possibilities and limits of conflict resolution in Abkhazia and South Ossetia are highlighted, as well as the processes of integrating ethnic minorities in Georgia.

The second part of the anthology deals with the challenges arising from the renunciation of the centralist economic system in favour of a market oriented economy. In this context, the question is raised which direction the Georgian economy should orientate itself towards, not least because of the sanctions imposed on Georgian products by Russia. In addition, macro-economic and political reform processes are investigated, as for example decentralization aspirations and the tax reform. Another contribution covers the investment climate in Georgia. Due to the traditionally important role agriculture plays in the Georgian economy, the economic situation of Georgia's agricultural economy is examined.

Land use change and environment are the main topics of the book's third part. The changes in agriculture and land use after the breakdown of the Soviet Union are presented and analysed. Environmental problems are one consequence of these changes. The problems are known but the political will and the financial means for melioration are missing. Especially in a irrigation district south of Tbilisi the uptake of heavy metals by food crops is a danger to people. The ecological state of the river Kura, the longest river in Georgia, has also been studied. Like most countries, Georgia is affected by climate change. The trends in air temperature and precipitation in different regions of Georgia during the 20th century are analysed and the mosaic character of climate change in Georgia is presented.

As editors we would like to thank all authors for their contributions, all reviewers for their suggestions and comments, Kerstin Kötschau, Nino Chkoidze and Michael Schäfer for their editorial work, Sabine Viehmann and Dr. Matthias Höher for the technical transformation of the manuscript, Arnd Federspiel for linguistic editing, and the *VolkswagenStiftung* as well as the *Deutscher Akademischer Austausch Dienst* (DAAD) for financial support of the projects realized so far.

Giessen, January 2009

Prof. Dr. Lorenz King
(Cooperation Delegate of
Justus Liebig University)

Prof. Dr. George Khubua
(Rector of Ivane Javakhishvili
State University)

Content

THE DEVELOPMENT OF CIVIL LAW IN GEORGIA 1
Lado Chanturia

THE TRAJECTORY OF GEORGIAN ADMINISTRATIVE LAW - IMPRESSIONS OF AN ADVISOR 19
Gerd Winter

ELECTIONS IN GEORGIA 45
Marie-Carin von Gumppenberg

STATENESS AND DEMOCRATISATION IN GEORGIA.............................. 59
Nino Chkoidze

CONFLICT RESOLUTION IN GEORGIA: A NEVER-ENDING STORY? 77
Marietta S. König

ETHNIC MINORITIES DURING THE TRANSITIONAL PERIOD 97
Mamuka Komakhia

CHALLENGES OF DECENTRALISATION IN THE REPUBLIC OF GEORGIA 115
Tanja Kirn, Elguja Khokrishvili

INVESTMENT CLIMATE: CHALLENGES OF TRANSITION 131
George Tomaradze

THE EUROPEAN VECTOR OF ECONOMIC REFORMS IN POST-REVOLUTION GEORGIA 149
Vladimer Papava

THE NATIONAL ECONOMIC MODEL AND GLOBALIZATION (THE CASE OF GEORGIA) 165
Teimuraz Beridze

IMPACT OF RUSSIAN SANCTIONS ON THE GEORGIAN ECONOMY 175
Eric Livny, Mack Ott, Karine Torosyan

TAX REFORM IN GEORGIA - WHICH ALTERNATIVES SHOULD BE PURSUED? 195
Elguja Khokrishvili, David Narmania

AGRICULTURE IN GEORGIA- COMMERCIAL SECTOR OR SOCIAL SAFETY NET? 221
Kerstin Kötschau, Eka Sepashvili, Nino Narimanidze

AGRICULTURE AND LAND USE CHANGE IN GEORGIA 241
Alexander Didebulidze, Tengiz Urushadze

UPTAKE OF HEAVY METALS BY FOOD CROPS FROM HIGHLY POLLUTED KASTANOZEMS IN AN IRRIGATION DISTRICT SOUTH OF TBILISI, EASTERN GEORGIA 265
Peter Felix-Henningsen, Diedrich Steffens, Tengiz Urushadze, Eliso King-Narimanidze & Besso Kalandadze

CLIMATE CHANGE IN GEORGIA DURING THE 20TH CENTURY 285
Michael Schaefer, Mariam Elizbarashvili, Lorenz King & Ramaz Meskhia

INVESTIGATION OF THE ECOLOGICAL STATE OF GEORGIA'S MAIN RIVER KURA IN THE FRAMEWORK OF THE NATO - OSCE PROJECT: SOUTH CAUCASUS RIVER MONITORING 301
Nodar Kekelidze, Teimuraz Jakhutashvili, Eremia Tulashvili, David Kekelidze & Lela Mtsariashvili

I. LAW AND POLITICS

THE DEVELOPMENT OF CIVIL LAW IN GEORGIA

LADO CHANTURIA

1 How the Civil Code was Created

On June 26, 1997 the Parliament of Georgia unanimously passed the first Civil Code[1] of independent Georgia. Many years of assiduous work carried out at Tbilisi State University under the leadership of Professor *Sergo Jorbenadze* with immense support from the Ministry of Justice was finally completed after a one-year period of diligent consideration by the Parliament[2]. This culminated in this unprecedented event in the history of the country. All participants in this great creative/historic process had every reason to be proud. For the first time in the history of Georgia, a civil law code sharing almost all achievements of developed western countries and enhancing the legal foundation of a new society based on private property and freedom of contract had been adopted.

The Civil Code did not just serve as the basic law for drafters of the Civil Code. It also signified the beginning of a new epoch which had been the dream of Georgian jurists of many generations[3]. It was intended to connect Georgia to the cultural world of the West, thereby facilitating the infiltration of developed legal culture into Georgia. This aim was actually achieved to a great extent[4]. Nowadays western developed countries do not doubt that Georgia is a western-oriented country.

1 Bulletin of Parliament, #31, July 24, 1997, p.1

2 Professor *Besarion Zoidze* with his inherent diligence and honesty recounts the history of the creation of the Civil Code of Georgia in his book "Reception of European Private Law in Georgia", Tbilisi, 2005. His work "Historical Background of the Reform of Civil Law in Georgia" published in the journal "Law", 1996, #5-6, p.54-61 is also worth mentioning.

3 *Sergo Jorbenadze*, Principal Problems of the Future Civil Code of the Republic of Georgia. In the book: "Law Reform in Georgia", Tbilisi State University, 1994, p.139.

4 This is obviously supported by the fact that many young Georgian people have been educated in western countries since the adoption of the Civil Code .

I. Law and Politics

The significance of the Civil Code for any country acknowledging private property, entrepreneurial freedom and freedom of contract cannot be overemphasised. It is not accidental that it is often referred to as the constitution of economic life. However, its function is of greater significance than merely the regulation of economic life. This served as a precondition for the fact that the governmental commission for the elaboration of the Civil Code was established immediately upon the restoration of the state of Georgia in 1991. The commission was always fortunate, since all the Ministers of Justice of those years unanimously and undoubtedly acknowledged the great significance of the Civil Code for the country[5]. Therefore, the group elaborating the Civil Code was always encouraged.

The necessity of the creation of a civil code was acknowledged in other post-Soviet republics as well. Russia took the initiative and the Centre of Private Law which was established within the framework of the Parliamentary Assembly of the Commonwealth of Independent States (CIS) under the chairmanship of Professor *Alexander Makovskij*, started working on a model Civil Code for the CIS. Georgia was also invited to take part in the elaboration of this Code[6].

During the same period, the basis for a cooperation with German colleagues, under the leadership of the great patriot of Georgia and especially dear friend of the Georgians, Professor of the University of Bremen, *Rolf Knieper*, was established.[7] As a result of this cooperation, the new law was practically established and developed in Georgia.[8]

5 The Commission elaborating the Civil Code worked in collaboration with three Ministers of Justice: *Johny Khetsuriani, Kote Kemularia* and *Tedo Ninidze*. Work on the code commenced under the reign of *Johny Khetsuriani* and was completed under the reign of *Tedo Ninidze*. In 1996, the latter submitted the final project of the Code first to the government and then to the Parliament. *Mr. Ninidze* recalled the reaction of the President of Georgia of those times, *Eduard Shevardnadze,* when the Minister of Justice submitted to him the Civil Code consisting of approximately 600 pages. "I assume that by this we have finally bid a farewell to the Soviet law," was *Shevardnadze's* evaluation.

6 In the 90s of the past century there were many politicians as well as scientists in Georgia who supported the adoption of a CIS model code. In regard to this discussion see: *Lado Chanturia*, Perspektiven der Privatrechtsreform in Georgien, in Wege zu neuem Recht (Boguslavskij, Knieper Hrsg. Berlin. 1998, S. 35 ff; Die Verhütung von Chaos in der Rechtsschöpfung – eine wichtige Aufgabe der Rechtsreform, in Wege zu neuem Recht (*Boguslavskij/Knieper* Hrsg. Berlin. 1998, S. 157 ff.

7 *Rolf Knieper* made an outstanding contribution not only in the adoption of the Civil Code or other laws. His merit is invaluable since he provided many young Georgian lawyers with the opportunity of further education in Germany. He was rewarded for this contribution, first he became an honorable doctor of Tbilisi State University (the first lawyer holding the title of honorable doctor), and later he became a freeman of Georgia (2001).

A choice should have been made between the participation of the group elaborating the CIS model code, composed mainly of Soviet specialists in civil law, and cooperation with Germans brought up in the traditions of the country of classic civil law. The opinion of *Sergo Jorbenadze* proved to be decisive in making the choice. As usual he made the right choice. This has repeatedly been confirmed during my collaboration with colleagues in many post-Soviet countries.[9]

Separate parts of the Civil Code prepared by members of the commission and written in Georgian were translated into German. On the basis of their expertise, German colleagues were able to elaborate a new version of the Code, every single sentence and article of which was continuously revised during the course of four years. It was a very intricate work which was further complicated by the fact that Georgian law was completely different from that of many institutes. Georgian legal language was also unequipped for such institutes, therefore *Sergo Jorbenadze* used to call the process of refinement of separate norms of the Code the "Sisyphean toil". Traces of these difficulties are more obvious today than was the case then. This is quite understandable given that Georgian private law has developed considerably over a ten year period. Legal studies and court practice have manifested many problems and events which the authors of the Code could not possibly have predicted or anticipated then. The scope of knowledge and expertise of Georgian specialists of civil law was enriched by new contributions from western-educated legal scholars whose claims to separate points of the code were often grounded and reasoned. The task of legal researches is to identify and critically evaluate legal norms, suggest means whereby these norms, etc. can be perfected. However, the Civil Code is a creation of the epoch in which it was developed and therefore it is quite natural that it reflects merits and demerits of that epoch. One consideration should be stressed: The Civil Code was not drafted as a law of the transitional epoch. It was created as a law intended to survive for many decades. So-called conjecture norms usually adopted with the aim of achieving particular purposes and

[8] The activities of German lawyers in Georgia should really constitute the subject matter of a separate research which would be an honorable duty for Georgian lawyers. I hope this will be the case.

[9] This can be proved by the fact that the relicts of Soviet law which were abolished in Georgia as far back as 10 years ago are still subject to discussions in many post-Soviet countries. In particular, the forms of property, obsolete institutions of operative management and economic management, the division of law into private and public law, the division of legal entities into legal persons of private and public law, etc.

expressing interests of particular groups are alien to the Civil Code. This is a genuine achievement of the Code and it should be preserved in the future.[10]

Work on the text of the draft Code was carried on in Parliament.[11] Once again, every sentence and clause of the Code constituted the basis for discussion during one year.[12] Unanimity existing between the governing political power and representatives of the opposition during the process of working on the code deserve special acknowledgement. It is probably the most unprecedented occasion in the history of the Georgian Parliament where the Members of Parliament so diligently worked on the text of the law, on every clause of it.[13]

2 The Influence of the Civil Code on the Development of Civil Law

Despite the fact that ten years is not quite a significant period for such a fundamental law as the Civil Code, this Code has already established revolutionary novelties in our society and left indelible traces.[14] Citing several examples should be enough to illustrate changes initiated by the Civil Code.

[10] Due to this fact, the necessity of releasing the Civil Code from relicts of the Soviet Civil Code such as operative management, economic management, etc. so widely spread in other post-Soviet codes is not on the agenda.

[11] The present President of Georgia, Mr. *Michael Saakashvili,* was the chairman of the Legal and Judicial Committee of the Parliament and the present Chairwoman of the Parliament, Ms. *Nino Burjanadze,* was his deputy. The latter supervised the group working on the final text of the Code for one year.

[12] As a result, differences exist between the text of the draft code submitted to Parliament and the final text of the Code.

[13] It had an impact on the voting results: the Civil Code was passed unanimously. In addition to the text of the draft code, all amendments made as a result of consideration, as well as voting results are kept in my personal files.

[14] According to the results of one of the surveys carried out in 1999, the Civil Code was deemed to be the most important law of the 20th century ranked after the Constitution of Georgia. Newspaper "Banks and Finances", 1999, #24

Thanks to the Civil Code, Georgian jurists of new generations, some of whom were educated in western countries, have returned to Georgia.[15] They were provided with the opportunity to get familiar with civil law in accordance with the new code of their own country. Many of them acquired further knowledge abroad.[16] The jurists of this generation feel very confident while participating in discussions and expressing their qualified opinions on problematic issues which constitute the subject of discussion of lawyers representing the system of developed law.[17]

The Civil Code provided the basis for the development of Georgian civil law. Numerous publications (considering the scope of Georgia) prepared and published in regard to the many institutes of the Civil Code obviously demonstrate this fact.[18] Six-volume commentaries to the Civil Code should be specially acknowledged since these were prepared and published for the first time in the history of Georgian law and despite the flaws (rather obsolescence) still deserve the attention of Georgian lawyers.[19]

More than once, the Civil Code of Georgia was the subject of legal research carried out abroad.[20] The critical attitude of some authors to the Code by no means diminishes its great importance. Foreign authors have also devoted their works to the Georgian Civil

15 Jurists who (studied)/(were educated) in Germany are distinguished from the point of view of quantity as well as quality. This is due not only to the expertise of Georgian jurists, but to the highly developed legal studies and legal education in Germany. I had the opportunity to witness the zeal and eagerness demonstrated by young Georgian lawyers working in the libraries of leading research institutions in Germany.

16 The support of foreign organizations assisting Georgian jurists through grants and scholarships to facilitate their education abroad is really worth mentioning. I would restrict myself to mentioning only some of them: DAAD, OSI, Erasmus, Masky, etc.

17 The presentation of *Lika Kinkladze* made at the second German-Georgian symposium on corporate law held in Tbilisi can serve as one of the numerous proofs of this. See: Sorgfaltspflicht und Verantwortlichkeit in der Aktiengesellschaft nach georgischem und deutschem Recht. Tagungsband. II. Deutsch-georgisches Symposium zum Gesellchaftsrecht. 2003. Tbilisi, 2003.

18 e.g. *Besarion Zoidze*, Property Law, *Roman Shengelia*, Family Law, *Zurab Akhvlediani*, Law of Obligations, etc.

19 Commentaries to the Civil Code of Georgia in six books, published by "Samartali".

20 I would restrict myself to citing only some of them: *Kupraschwili Teimuras*, Schweigen beim Vertragsabschluss. Eine rechtsvergleichende Untersuchung unter Berücksichtigung von Deutschland und Georgien. WVB. Berlin. 2006; *Kereselidze David*, Der allgemeine Teil des Georgischen Zivilgesetzbuches von 1997. Eine rechtsvergleichende Untersuchung, Peter Lang Verlag. 2004 usw.

Code.[21] The Georgian Civil Code has also been mentioned in scientific collections published abroad.[22]

Numerous manuals and textbooks have been written in accordance with the Civil Code[23] enabling large-scale teaching of the Civil Code at the university. This is vitally important since law does not simply consist in the norms or regulations of law. First of all, law is the product of legal education. Modern law can only function in an educated society, i.e. where society members are well aware of their own rights and obligations. Hence, legal education is of the greatest importance. At the same time, the peculiar role of educated lawyers in contemporary states based on the rule of law is obviously evident. It can be said that states governed by the rule of law actually exist due to the presence of well-educated lawyers. The norms of law can be interpreted and employed maliciously and in order to avoid such abuse of interpretation and usage, it is essential that the education of lawyers is founded on proper and developed methodologies, enhanced by humanistic values. The Civil Code has already provided foundations for such approaches, although much more work is required. The methodology of teaching law most obviously requires that further developments and advancements take place.

The Civil Code has had great influence on court practice. Through the publication[24] of court decisions, the Civil Code can be accessed by lawyers and the community and

[21] I assume citing of several examples would be enough: Iris Muth, Neues Recht zwischen Tradition und Systemwechsel : Beratung des georgischen Arbeitsrechts in der Praxis. Baden-Baden : Nomos, 2004; *Mario Pellegrino*, Postkommunismus und Zivilrecht. Das Obligationenrecht Georgiens. 1996.

[22] *Lado Chanturia,* Das neue Zivilgesetzbuch Georgiens: Verhältnis zum deutschen Bürgerlichen Gesetzbuch. In Aufbruch nach Europa. 75 Jahre Max-Planck-Institut für Privatrecht. Herausgegeben von Jürgen Basedow, Ulrich Drobnig, Reinhard Ellger, Klaus J. Hopt, Hein Kötz, Rainer Kulms, Ernst-Joachim Mestmäcker. Mohr Siebeck.2001. S. 893-904

[23] Professor *Zurab Akhvlediani*, a devoted member of the editorial group elaborating the code, was one of the first to draw up a manual on the basis of the Civil Code. *Zurab Akhvlediani* attended every meeting of the group. He was an exceptional person, whose early and unexpected death caused great sorrow and suffering to each of us. His "Law of Obligation" continues to be a handbook in the Law of Obligations for Georgian students.

[24] Decisions of the Supreme Court of Georgia have been published since 1999. Georgia was the first post-Soviet country which made concrete court decisions available to the public. Not only those normative resolutions adopted by the Supreme Court, but also judgments pronounced in regard to the concrete case create judicial law (Richterrecht) and demonstrate how in fact the specific norm of the law is implemented.

also applied practically. Statements of the law are simply literary texts[25] unless they are relevantly implemented, while court practice provides information on the application of norms of law.[26] The connection between legal researches and court practice is really necessary since it ensures that outcomes of legal research are not only interesting but also applicable in practice. [27]

Some judgments pronounced on the basis of the Civil Code are interesting from the point of view of legal application methods. They have served as the subject matter of separate researches within the context of preparation and adoption of court decisions.[28] With foreign judges, some problematic issues of Georgian court practice have become the subject of discussion and publication.[29]

The Civil Code provided foundations for the launch of international contacts, collaboration with foreign colleagues and fulfilment of joint legal researches and their further development. Numerous projects existing in various fields today are closely related to the adoption and implementation of the Civil Code.[30] Particular reference should be made to German scientific works which obviously prevail in scientific papers of Georgian specialists of law. German civil law still remains an important source for the development of researches of the civil law of Georgia. In turn, the German science of law is supported by European law. Hence, German law contributes to the popularity

[25] Since texts are an essential part of the functioning of jurisprudence, jurisprudence is considered to be a form of hermeneutic science. See *George Khubua*, Theory of Law, Tbilisi, 2004, p. 27

[26] It is important not to confuse law with sociology of law which is mainly concerned with law in action. See: *George Khubua*, Theory of Law, Tbilisi, 2004, p.21.

[27] Despite the negative image presented of judges and the judicial authority prevalent in the post-Soviet countries, court judgments must necessarily remain the subject of scientific researches and criticism.

[28] In this regard see: *Hein Bölling, Lado Chanturia*, Die Arbeitsmethode des Richters im Zivilprozess an Hand von Fällen, Lösungen und Entscheidungen zum georgischen Recht. "Jisaia' Tbilisi, 2003

[29] *Hein Bölling*, Recognition as an Unworthy Heir and the Deprivation of the Right to a Compulsory Portion according to the Civil Code of Georgia (and Some Comments on the Method of Interpretation of Law), in: Georgian Law Review 6/2003. p. 517-525.

[30] In regard to positive and wide experience existing in respect of the adoption of the Civil Code and its implementation see: *Knieper Rolf*, Juristische Zusammenarbeit: Universalität und Kontext. Wiesbaden. 2004.

of not only German, but European law in Georgia.[31] This aspect of German-Georgian academic cooperation in the sphere of law is extremely important.

3 The Influence of Civil Law on the Development of Private Relationships

The Civil Code is firstly the legal act of private law. It is especially important and necessary for any society in which private property, freedom of contract, creation of private corporations, private entrepreneurial (commercial) activity and the like was forbidden over a 70 year period. The historical and social mission and function of the Civil Code was to restore the society which diverted from the path of true historical development to a normal state. The Civil Code was intended to establish private conditions for those freedoms and rights that were declared and legally consolidated by the Constitution. In my opinion, the Civil Code successfully accomplished this mission.[32]

The Civil Code, the first in the post-Soviet region, did not permit the conversion of property into forms which in many CIS countries are related to different regimes of property law, thus making the fundamental principle of equity of the participants of market relations doubtful. The unified concept of property is undoubtedly a great achievement of the Georgian civil law.[33] The past ten years have obviously proven that this decision, which had a lot of critics, was the right one. The classification of property according to forms belongs to the sphere of economic science, but is legally of no significance as regards the definition of the concept of property (at least it should not

[31] On May 24 of this year, a conference on the "Europeanization of Georgian Law" will be held at the Max-Planck Institute of Hamburg based on the initiative of the director of the institute Professor Hopt. Georgian lawyers whose scientific or practical activities are related to Germany will participate in this conference.

[32] According to the new Civil Code, tens of thousands of new titles to movable and immovable property have been created, thousands of agreements have been executed; legal entities have been incorporated, etc. All the aforementioned illustrate that the Civil Code is really implemented and applied to regulate social relationships.

[33] See: Besarion Zoidze, Reception of European Law in Georgia. p. 251-260.

be).[34] Along with property, other proprietary rights have been widely applied, among which especially should be mentioned the *usufruct* as a form of use of property subject to payment as well as free of charge,[35] *hypothecation,* etc. The institution of the apartment owners association made an attempt to establish a pragmatic and rational organization of the owners of privatized state apartments. The fact that in this sphere the law was not implemented as intended could be attributed to low levels of legal awareness and unimaginably low quality of self-organization.[36]

As opposed to the CIS model civil code, the Georgian Civil Code details regulated legal status of legal persons. First of all, it refers to the division of legal entities into legal persons of private and public law, representing a revolutionary novelty in the whole post-Soviet area.[37] The code defines and regulates in detail legal entities of private law, i.e. legal entities' establishment of the rights of any persons of private law.[38] By introducing the concept of legal persons of public law, the code determined the state's legal framework, within which the state is entitled to establish legal persons.[39] All legal

[34] Some Russian specialists of the civil law share this opinion although the Russian Civil Code is a product of the diversity of the forms of property. See: *Суханов Е.А.* Проблемы правового регулирования отношений публичной собственности и новый Гражданский кодекс // Гражданский кодекс России: Проблемы. Теория. Практика. М., 1998. С. 205-227.

[35] It was specially spread as an institute of transfer of state property into free of charge use. In this regard, the Georgian law on transfer of state-owned building-constructions with the right of use which was passed by Parliament on April 14, 2006, is worth mentioning.

[36] With the aim of changing this situation, on July 11, 2007 the Parliament of Georgia passed the law on Apartment Owners' Association and accordingly, articles 210-232 were withdrawn from the Civil Code.

[37] This kind of classification of legal entities is extraneous to the law of post-Soviet countries causing chaos in matters of regulation of legal entities. The declaration of monopoly as regards the concept of legal entity by specialists of civil law makes all the legal entities created by the State in different spheres "outlawed". Some Russian scientists had foreseen this problem and an attempt to introduce the concept of legal persons of public law to the Russian law was made although this attempt has not succeeded so far. See: *Чиркин В.Е.* Юридическое лицо публичного права. Норма. М.2007. In regard to this problem also see my speech made at the international conference held in Alma-Ata (Kazakhstan) on May 17, 2007. *Ладо Чантурия*, Гражданское право и юридические лица публичного права: особенности правового регулирования. В кн.: Гражданское право в системе права. Материалы международной научно-практической конференции в рамках ежегодных цивилистических чтений в Алматы 17-18 мая 2007 г. Алматы. 2007. С. 240-251

[38] Non-profit legal entities are regulated in the Civil Code, while commercial legal entities are subject to the law on entrepreneurs.

[39] The law of Georgia on legal entities of public law is unprecedented for post-Soviet countries. It plays a significant and very positive role in the system of Georgian law.

persons of public law, and primarily the state, are equal participants in civil relationships. Thus, the Georgian Civil Code rejected the widely spread opinion in Soviet (as well as in post-Soviet) law to the effect that the state is the particular subject of civil law. The state is a common subject of civil law. If it performs some distinguishing functions such as lawmaking or the right of expropriation of property, this characterises the state first of all as a legal person of public law. If the state exercises this right and deprives the owner of its property, it acts not in a civil framework, but within the frames of public law. This competence of the state is not determined by the Civil Code but by the public law – constitution and other statutes of public law.

Contract law can be distinguished from the point of view of influence on social relationships, since it represents the precondition for circulation of private law, development of economic life and widening of personal initiative. If we just imagine the number and kind of agreements made by people every day, we can easily ascertain what role is attributed to the principle of freedom of contract sustained by the Civil Code in our country. Free property circulation, freedom of property disposal, undertaking of obligations and securing thereof – all this is accomplished within the framework of contract law. [40]

Freedom of property and contract facilitated unimpeded participation of real estate, first of all of land, in civil circulation. Although the introduction of private property on land was a political decision, legal grounds for its implementation were determined by the Civil Code. Freedom of contract is widely used in banking, insurance and other spheres bearing economic significance.

Several aforementioned examples should be sufficient to illustrate the influence of the Civil Code on the legal development of private relations. Expectations related to the Code were justified to a large extent although in individual cases it became necessary to update the Civil Code. Some of them will be referred to below.

4 Amendments to the Civil Code

A number of amendments have been made to the Civil Code over the past ten years. Despite the fact that some of these present some doubts, each amendment reflects the ongoing processes in our society, which are characterised by clear dynamism. Such

[40] On novelties sustained in this sphere by the Civil Code see: Besarion Zoidze, Reception of European Law in Georgia. p. 262-291

legislative amendments are concomitant to any law involving the Civil Code and should not evoke surprise or indignation. At every stage of development, a country faces new tasks and priorities and lawmakers need to realize this. Besides, the modern state is involved in the globalization process which, in turn, has an impact on national law. This influence is especially evident in those spheres where internationally acknowledged rules are applied (corporate law, obligations security, etc). Only those amendments which have changed initial norms of the Civil Code will be referred to below.

4.1 Intellectual Property Law

The first important amendment to the Civil Code was made in 1999 and addressed intellectual property rights. The adoption of separate laws on copyright and neighbouring rights[41] induced such amendments, which comprise of full withdrawal of norms regulating copyright property and personal non-property rights as well as neighbouring rights from the Civil Code. [42]

Notwithstanding the fact that the norms of the law on copyright and neighbouring rights were practically identical to the norms of copyright law included in the Civil Code, the adoption of separate laws was a political decision and represented one of those preconditions for the entry of Georgia into the World Trade Organization. [43] Following the adoption of separate laws, the issue of withdrawal of the norms of copyright law from the Civil Code was naturally placed on the agenda, since from the point of view of legislative techniques existence of parallel norms is impermissible. [44] This is the very circumstance explaining the excerption of the norms of copyright law from the Civil Code and not their noncompliance with contemporary demands and standards of current copyright law.

[41] Legislative bulletin of Georgia, 1999, #28, article 35

[42] Legislative bulletin of Georgia, 1999, #26, article 134

[43] Sandro Jorbenadze, Author's Rights in life. Tbilisi, 2006, p. 3

[44] This is one of the important peculiarities of the Georgian Civil Code, distinguishing it from civil codes of other CIS states. As regards the codes of these countries, it is characteristic to include the norms which are further repeated in separate laws, thus often causing non-compliance between these norms. This causes inconvenience and misunderstanding in law enforcement. For more details with regard to this problem see: *Осакве Кристофер,* Сравнительный анализ гражданских кодексов России и Казахстана 1994 г.: Биопсия экономических конституций двух постсоветских республик. Алматы. 1998.

As a result, although the fifth book of the Civil Code is still called Intellectual Property Law, it is limited to those norms which, apart from copyright law, refer to laws regulating industrial property (Articles 1100-1105).

4.2 Right of Pledge on Movable Things and on Rights

The concept of the Right of Pledge was completely changed by the law adopted on June 30, 2005. Consequently, articles 254-285 were edited and assumed new wordings.[45]

The amendment clarified that movable things or immaterial property could be used as security for monetary as well as nonmonetary demand.[46] Within the context of nonmonetary demands, usage of the institute of security interest was so unnatural that the legislator had to provide a clearer definition although it caused even greater confusion. The following norm of the law should serve as an illustration: "Pledge securing nonmonetary demand can be effective provided it is expressed in monetary form" (part 4 of article 254).

The editorial nature of the amendments is evidenced by the fact that possessory pledges and registered pledges similar to those present in the code prior to the amendments are presented as separate forms although the rules of origination thereof remain the same in essence.

The rule relating to the pawning of possessions can be considered a novelty (article 260), although it is surprising that in this respect only pawnshops are considered separately. This should obviously be attributed to the influence of the Russian Civil Code. [47]

The structure of counter-pleas which can be applied towards a pledgee by a pledgor who is not a personal debtor with respect to the claim has also changed. Previously, such counter-pleas addressed the deduction of monetary demands or appealing

[45] Terminological inaccuracies and defects are evident in the new edition. E.g. the terms "Movable thing" and "Mortgaged property" are ambiguous (art. 254, part 1), etc.

[46] The old edition included only the term "demand" and did not clarify whether this demand was monetary or nonmonetary. (Article 254, Part 1)

[47] The rules of pawning are envisaged by Article 358 of the Civil Code of Russia, although it provides more detailed regulations in respect of pawning relationships than the Civil Code of Georgia does.

demands (old edition of article 259 of the Civil Code), but at present a counter-plea that has been rejected by the personal debtor with respect to the claim after the creation of the pledge, may be applied to the pledgee (article 263).

The final amendments with respect to security interest were made to the Civil Code on December 29, 2006. Pursuant to this amendment, the rules relating to the securing of mechanical vehicles changed. In particular, a notarised document registered with the legal entity of public law – the Service Agency of the Ministry of Internal Affairs of Georgia - is required for the creation of security interests over cars or other vehicles (Part 4, Article 258). Therefore, the place of registration of security interest over vehicles has changed from the Public Register to the Ministry of Internal Affairs.

4.3 Non-commercial Legal Persons

The most important and numerous amendments have been made to parts of the Civil Code determining the legal status of non-commercial legal persons.

Initially, amendments dealt with the rule of registration of non-commercial legal persons – unions and foundations. Namely, in accordance with the amendments of May 25, 2006, the registration of unions and foundations was no longer undertaken by the court or Ministry of Justices but was completely transferred to the tax body.[48] In so doing, the state and tax registration of legal entities was equated and registration carried with it greater tax-legal implications than it used to have. Although the principles of registration, accuracy and publicity have remained the same, it is obvious that focus has been dedicated to tax control and those significant aspects of private relationships traditionally related to the institute of registration.

The amendments of December 8, 2006 are even more revolutionary, since in accordance with these amendments, legal-organizational forms of non-commercial legal entities such as unions and foundations were altogether abolished and the

[48] Part 1 of Article 31 of the new edition as of May 25, 2006

unified, although more abstract concept of "non-entrepreneurial legal entities" was introduced. [49]

The Civil Code made a choice of extreme liberalization in relation to the regulation of non-commercial legal entities. The Code does not imperatively determine either charter requisites or the structure and authorities of managing bodies or rights of members. These issues are regulated within the competence of a charter. Such liberalization might not be expedient in a country of low legal culture having no traditions of civil society. In such conditions, the task of legislation is to provide detailed legal regulations not only in order to clarify game directives, but also to ensure that they are comprehensive to participants of the relationship. Accomplished changes have so far completely failed to meet/fulfil this requirement.

The greatest flaw inherent in this legislative reform is that the law on united non-profit legal entities no longer exists. This means that as many forms of legal entities will exist and that as many charters will be drawn up. On the one hand, this will entail complications in the legal reforms within this sphere and on the other hand, legal research of these organizations will be impeded. In fact, the whole legislative reform has been dedicated to the registration of non-profit legal entities, to legislative determination of registration requisites.

Apart from the aforementioned very disputable provisions, legislative changes have introduced completely new, previously unknown concepts such as: "Organizations managing on the collective basis of property rights" (point "b" of part 3 of article 29). What is implied under "management on collective basis" should be clarified. Should it be understood as an organization incorporated on the basis of membership or on the basis of joint (united) property? The reason for the singling out of the national sports federations should also be specified.

Hence, amendments made to the Civil Code in respect of non-commercial legal entities deserve critical evaluation. This part of the Civil Code requires reconsideration.

[49] This amendment has obviously been influenced by American law, using the expression "Nonprofit-Organisation": (Klaus J. Hopt, Dieter Reuter, Stiftungsrecht in Europa: Eine Einführung. Heymanns. 2001. S. 2 ff) for the determination of non-profit legal entities. Every state in the USA has a law on non-profit corporations, which is based on the model law elaborated in 1964 by the Institute of Justice of America and the Lawyers' Association of America. Model Nonprofit Corporation Act: Thomas von Hippel, Begriffsbildung und Problemkreise der Nonprofit-Organisationen. In: Nonprofit-Organisationen im Recht, Wirtschaft und Gesellschaft (Hrsg. Hopt/von Hippel/Walz). Mohr Siebeck. Tübingen. 2005. S. 36 ff.

4.4 Transfer of Title to Immovable Property

One further amendment made to the Civil Code which should also be criticized relates to the rule governing the transfer of a title to immovable property. Prior to the amendments of December 29, 2006, notarization of the agreement on acquisition of immovable things and registration of the purchaser in the Public Register was an obligatory procedure. This rule is envisaged in the legislation of many countries and apart from providing protection for the rights of purchasers of immovable property, guarantees legal security and stability. Notarization of some kinds of transactions is aimed at strengthening preventive control which can help to avoid making mistakes which might cause further problems for the parties to the agreement. Notaries are obliged to provide qualified legal consultations while notarizing transactions.

Following the amendments of December 29, 2006, a written agreement between the parties and registration of the title to the property as stated in the agreement in the public register in the name of the purchaser was sufficient (article 183). This novelty greatly endangered the legal security which had existed before in the market of immovable property and which at least had slightly protected the interests of the parties. It can easily be anticipated that this amendment will cause many problems. Therefore it is quite natural that citizens react carefully to such deals and still prefer to notarise agreements related to immovable property.

The rule of notarization of agreements has been abolished in regard to registered pledges as well as mortgages, which will also undoubtedly cause a great number of misunderstandings and problems.

The legislators have to reconsider these issues and restore the norms approved in contemporary law which serve to protect the interests and safety of the parties involved in the legal relationship.

4.5 Apartment Owners Association

On July 11, 2007, a law on the apartment owners association was adopted, abolishing those articles of the Civil Code (namely, articles 210-232) which had regulated the legal status of the apartment owners' association. In fact, the relevant norms of the Civil Code served as the basis for the new law. Besides, the norms regulating relationships with local self-governing bodies supplemented the law, thus representing more public-legal norms (article 29).

It should be mentioned that the earlier apartment owners association should have been regulated by a separate law. In 1993-1994, a bill was prepared with this purpose but failed in its attempt to become law. Later it was decided that the norms of this bill would be included in the Civil Code. As a result, the new law represents the realization of the initial legislative intent.

On the whole, it can be said that the new law is quite refined and takes account of many issues which were previously not reflected in the Civil Code. In this case, it is also important to ensure practical implementation of the rules stipulated by the law.

5 Outlines of the Path to be Passed

From the point of view of history, ten years is not a considerable period indeed, but for those people who have devoted their lives to the Georgian Civil Code, begun living with the Civil Code, started families, acquired property, concluded agreements, were educated, etc. it has been quite an important time. Many things considered as commonplace and trivial today were absolutely the contrary at the time when work commenced on the Civil Code. Therefore, new circumstances have evoked new requirements and interests, which is also absolutely natural. The Civil Code of Georgia is not ready to decelerate on its progress but, on the contrary, to provide reliable legal support.

Notwithstanding the resilience and flexibility of the Civil Code, considerations regarding the preparation of a new code still exist. Unless no other alternative exists and the new code is passed, it is vital to observe and complete the processes going on in the European Union in respect of codification of the civil law. Provided Europe is in need of a common civil code, it could easily be adopted in Georgia.

The future of the Civil Code must relate not only to infinite amendments but also to its deeper implementation, more profound investigation and greater awareness of its separate institutes. To achieve this aim, much more needs to be done than has been done so far. Numerous institutes of the Civil Code have not constituted the subject matter of serious scientific researches for the time being. Neither have those issues of application of such institutes been studied in practice. New commentaries to the Civil Code should be prepared to take into account contemporary levels and traditions of preparing such commentaries, paying more attention to court practices which reflect achievements of the Georgian Civil Law. Juridical journals of civil direction must be established. Georgia lacks the tradition of specialized juridical periodicals which presents a serious obstacle and deprives representatives of different spheres of the opportunity to carry on professional or scientific dialogues. It is desirable to elaborate the methodology of teaching civil law. A complicated legislative act such as the Civil Code undoubtedly requires this sort of handbook. Lack of methodology is evident in court judgments, in the process of university legal education, as well as in the sphere of the training of separate legal professions.

Despite the fact that court decisions, especially those decisions of the Supreme Court, are published in Georgia, no professional and scientific discussions in regard to these decisions have been carried out in Georgia. For the time being, court practice has not been established in researches of Georgian specialists of civil law, without which the development of the science of law is unimaginable. From all that has been mentioned above it can be inferred that it is vital to determine new tasks and approaches for scientific researches in the sphere of civil law.

THE TRAJECTORY OF GEORGIAN ADMINISTRATIVE LAW - IMPRESSIONS OF AN ADVISOR[1]

GERD WINTER

1 Tasks in Transition

According to the socialist concept of identity, state and society are not separated. Rather the state apparatus is identical with the people, and even in its relative separation, it still is only its organ.[2] The state is at the same time an economic actor, a planning and regulatory body and a supplier of services and facilities.[3] In this conception, administrative law is not a law covering the legal relationships between the state as a legal body and natural or legal personalities, but organizational law that determines the state apparatus' decision-making structures, powers and regulatory means, which establish obligations and responsibilities for citizens and economic actors alike.[4]

However, the real conditions in 'socialist' societies developed differently. The state apparatus, steered by the party, became detached from society, even more than its competitor, the state of bourgeois society, which formally rests in the separation of state and society. Without legitimacy the extensive administrative penetration of society paralyzed social creativity. In the late phase of 'real life socialism', this paralysis could

1 My knowledge of the Georgian development rests in many consultancies concerned with the reform of administrative law, especially the work on the Administrative Codex and the Rules of the Administrative Courts.

2 See for a particularly clear notion of the identity theory and particularly influential in the GDR, Karl Polak, *Zur Dialektik in der Staatslehre*, Berlin 1963 and id., *Reden und Aufsätze*, Berlin 1968. Cf. Uwe-Jens Heuer, *Marxismus und Demokratie*, Baden-Baden 1989, pp. 290 ff., pp. 359 ff.

3 Heuer, op. cit., pp. 374 f. mentions a similar spectrum of state functions, i.e. the development of productive forces, the control of the national income, distribution and the appropriation of nature.

4 For the example of the environmental law of the GDR mentioned in the regional cultural law, Autorenkollektiv, *Landeskulturrecht*, Berlin 1986, pp. 47 ff., pp. 58 ff.

be cushioned somewhat, because economic law allowed economic actors relative autonomy and granted them a kind of individual rights[5], and because administrative law introduced some forms of a quasi rule of law, such as the intra-administrative right to lodge complaints.[6] Overall, however, the notion of a people's democracy prevailed; the proponents of a more direct responsibility for businesses and local authorities, of political pluralism and a socialist rule of law met no response for a long time[7], and when they were listened to the regime was already discredited.

With the fall of the Iron Curtain, these structures were modeled after a totally different conception – the Western conception of representative democracy and the rule of law. This endeavor required to establish legal relationships between the state and its citizens aimed at framing societal freedom, providing infrastructure and ensuring redistribution of wealth.

Legal development was faced with two fundamental tasks: the construction of the new legal edifice and the creation of instruments enabling and directing the move from the old into the new dwelling. How this has been accomplished shall be illustrated with reference to the Georgian legal system. Among the transition countries Georgia is a suitable example, due to two reasons: the country consistently cast its transition in legal forms, and in this respect it has been the fastest and for a long time most advanced among other Eastern European and Central Asian states of the Community of Independent States (CIS). This, however, does not imply that today Georgia is a *de facto* social market society and a constitutional state. But to the extent that law can support their accomplishment, legal structures are readily available.[8]

[5] For the GDR, Autorenkollektiv, *Marxistisch-leninistische Staats- und Rechtstheorie: Lehrbuch*, Berlin 1975, pp. 480 ff.; Autorenkollektiv unter Leitung von U.-J. Heuer, *Wirtschaftsrecht*, Berlin 1985, pp. 65 ff. Subjective rights are understood as organizational means of the state and show no resemblances with fundamental rights.

[6] For the GDR, W. Bernet, 'Verwirklichung der sozialistischen Gesetzlichkeit in der staatlichen Leitung durch Rechtsmittel', *Staat und Recht* 1980/81, pp. 13 ff.

[7] In the GDR, e.g. Heuer, op. cit. and U.-J. Heuer, G. Klinger, W. Panzer, G. Pflicke, *Sozialistisches Wirtschaftsrecht-Instrument der Wirtschaftsführung*, Berlin 1971.

[8] In comparison to Georgia and the mainstream of transition countries, the German Democratic Republic is an untypical case, as the self-initiated *Wende* was rather quickly taken out of the hands of the citizens. Rapidly, a new system was introduced with the result of economic wealth but socio-psychological degradation. In all other transition countries matters were reverse: transformation in poverty but in proud self-determination.

The reasons why Georgia chose law as the frame for its transition are partly contingent, as they are rooted in the involved actors[9], but they are also of a more structural nature. Traditionally, Georgia is a self-confident civil society. During the Soviet regime it was considered a region difficult to govern. This renders understandable the country's quick try to establish legal forms that would enable societal intercourse without state intervention, i.e. especially civil law. A further factor was the relationship with Russia. Apart from two years in the early phase of the Soviet era, Georgia had been occupied by foreign powers since the 13th century, i.e. by Persians and Turks, and since about 1800 by the Russians. In order to capitalize on and secure the opportunity for independence after the break-down of the Soviet Union, a political system with constitutional authority had to be established quickly. Democratic and constitutional institutions could provide Western recognition, e.g. membership within the European Council, and hence safeguard against potential Russian ambitions.

2 The Instruments of Transition

The most important engine of transition is a proactive and democratically legitimate government. After the civil war had established the territory and the fundamental political orientation of Georgia, the layout of its political system had to be considered. As in most transition countries, a presidential system was adopted.[10] In the first one and a half decades of transition it was rather the parliament and its staff that fostered the process of legal reform. Since the so-called Rose Revolution in late 2003 the president and his apparatus has become dominant in law reform and enforcement. However, the revolutionary potential of moving society ahead was gambled away by precipitate actions towards economic growth that often disrespected the rule of law and threw the regime into a legitimacy crisis.

One of the most important legal means of transition was a legal framework for the privatization of state property, especially companies, dwellings and land. In many transition countries, privatization was implemented *ad hoc* by dint of executive acts and contracts, with the effect that assets were concentrated in the hands of very few – people who were often neither the technically, economically nor socially most capable.

9 In Georgia, a group of young lawyers emerged quite early, which established a reform network in the legally relevant institutions and made use of foreign, especially German, aid.

10 Constitution of the Republic of Georgia of 17 October 1995.

In Georgia the Act on the Privatization of State Property of 1997 provided the framework for calls for tenders, auctions and private lease/sale (yet, it came too late as many assets had already been privatized in a rather opaque manner). Additionally, the Act on the Privatization of Agriculturally Used Estates was established in 2005.

The reduction of state regulation further fostered the transition from a state to a market economy. Almost any economic activity had been subject to often manifold authorization in the past. Now the principle of freedom of trade had to be introduced. This was meant to be achieved by reforming the Licensing Act, which freed many trades from the requirement to get a license to set up a business – excluded were only a small number of socially problematic industries[11].

A further tool of transition is the fundamental reorientation of administrative infringement codes. Such codes used to be the core of administrative law in the states of the Soviet Union. They contain a plethora (sometimes up to 400) of prohibitions covering almost all sectors of economic, social and cultural life, they provide sanctions ranging from fines to forced labour and arrest, they establish general requirements of prosecuting violations (such as causality, error in fact and law, intent and negligence, the liability of enterprises, etc.), and they open up procedures of complaint in the courts. This somewhat authoritarian approach of administrative law had to be cut back to the minimum necessary in a liberal society in order to make way for the new administrative law which shall facilitate rather than control societal freedom.

3 The New Legal Structure

The essential forms of economic transactions had to be established, such as different types of contracts, rules for liability, property rights, security interests, intellectual property, types of companies, competition rules and so forth, and they are all available now. However, many of these forms of intercourse have remained underused, as many transactions still follow the pattern of the bazaar, i.e. they are *ad hoc* and cash-based. Nonetheless, large investments, which require a variety of legal forms of intercourse, have increased over time.

[11] Insurance companies, banks, weapons production and sale, aviation and water transport, stock markets, lotteries, pharmaceutical industry and sale, technical test of motor vehicles, building industry, audit firms, private educational establishments.

The market with its forms of intercourse under private law does, as is generally known, not work without a degree of administrative intervention. Even if the latter is reduced as much as possible, the necessity remains to implement political goals, provide for infrastructure and redistribute wealth through law making and law enforcement. For such functions organizational structures, financial resources and staff is needed, which also have to be established by legal norms. This is primarily the task of administrative law.

In the following, I will outline the genesis of modern administrative law in Georgia. In doing so, I will focus on general administrative law and the law of administrative courts. Organizational and sectoral administrative laws shall only be mentioned in passing, such as:

- concerning organizational law: the Act on Local Self-Administration of 2005, whose autonomy model is qualified by the fact that local communities depend upon the state allocation of funds; the Act on the Structure of the Executive Power, which determines the authorities of the state administration and their competences, and the Act on the Types and Generation of Normative Acts;
- concerning sectoral laws: the Police Law; the already mentioned Administrative Infringement Code, the Law on Licensing; the Public Procurement Law with a Code of Practice adapted to EU law; the Act on the Certification of Products and Services, the general Environmental Act entailing the instruments of environmental protection and establishing fundamental obligations for the handling of environmental media, yet without already creating concrete legal relationships; the Act on Environmental Permits, which is heavily influenced by the German Federal Immission Control Act; the Water Act and the Act on the Safety and Quality of Foodstuffs.

4 The Main Problems of a New Administrative Law

The issues administrative law has to address are various, as are the resulting legislative aims. Many problems are not limited to transition countries, as they emerge in other states as well, but of course they also have to be solved in transition countries. Among these general problems are:

- The tendency to concentrate power within the bureaucracy: the subjective legal position of the individual and the introduction of liability serve as correctives in cases of harm inflicted by public authorities;
- The inaction of the administration: action can, for instance, be triggered by setting terms and establishing obligations to act;

- The lacking competence of staff: competence can, among others, be achieved by qualification requirements in the law of the public services and the right to a hearing in administrative proceedings;
- Schematic and monocratic decision-making: Decision-making can be made more responsive by the duty of administrative bodies to carefully investigate the circumstances of the case, the duty to give reasons for the decision, and the possibility of cooperative forms, such as the contract under administrative law;
- The insistence on already established solutions: more flexibility can be reached by the possibility of attaching subsequent conditions to an administrtaive act as well as to withdraw it if a new situation so requires;
- The keeping secret of administrative information and autocratic decision-making: more openness can be introduced by the right of access to information and the introduction of rights to hearing and participation serve as correctives.

Beside these widely known problems and their solutions, corruption poses a particular problem for Georgia and other transition countries. Corruption had already been rampant in the Soviet system, but only in postsocialism it turned into a ubiquitous phenomenon strangulating economic development. Doubtlessly, corruption is also widespread in 'the West', and even seems to increase with the recent tendency to informalization in administrative law, yet it has not reached the level prevalent in 'the East'.

The reasons for corruption and the means to fight it in the public sector are numerous and cannot be discussed here. Suffice it to name the socio-cultural core of corruption: the perception that a public office is a good like any other good on the market. Accordingly, the office has a value that can be 'sold'. It is the task of the rule of law to constitute public office as a *res extra commercium*, to institute state functions independent of those of the market.

Many legal areas are called to ensure this, among them administrative law. Every candidate norm of administrative law should hence be checked not only in terms of its substantial expediency but also as to whether it provides a potential source for hidden, extra income. If, for instance, the introduction of a recurring safety inspection for motorcars is discussed, it has not only to be asked whether this is suitable to ensure road safety, but also whether this new obligation can be 'traded' for bribes. Until a few years ago, in Georgia the badge which certified that a vehicle was safe was available without check for about 100 US$, depending on the type of motorcar. The driving license was available without any lesson and test for a slightly amount.[12]

The drafting of new legal norms therefore requires the consideration of how a provision can best be immunized against bribery. This aim can, among others, be achieved by limiting those activities which need administrative authorization, by fettering administrative discretion, by more precisely phrased provisions that bind the administration, by setting deadlines for decisions, by the right of access to information and by establishing the opportunity to sue for particular administrative acts.

Such legal framing of administrative functions must be flanked by economic means, like the increase of wages of public officials to a sufficient standard; for after the end of the Soviet Union public servants have increasingly lost direct gratifications (such as free holidays in public facilities, public housing, etc.) and their income suffered by inflation and increasing prices.

One measure of course depends upon the other: an administration strongly bound by administrative law would cost less to the individual client; this raises the willingness to pay taxes and thus increases the state revenue which then allows the state to better remunerate its civil servants.

After the Rose Revolution of 2003, the new government fought corruption decidedly. Thousands of civil servants, e.g. policemen, were urged to ask for their dismissal in exchange for the promise to be able to reapply. These positions were now better salaried. In this way, small corruption could be eradicated to a large extent. The entire approach, however, had been immensely harsh and unfair, and it was problematic under constitutional law. Although the Georgian constitution does not protect 'the rights acquired in the public service' like the German Basic Law[13], still the freedom of occupational choice[14] was infringed by unfounded dismissals and layoffs based on extorted consent.

Corruption was also battled at the level of politicians and high officials. In order to tackle the issue of sufficient evidence in regular prosecution under criminal law, two ways of procedural simplification were adopted: the 'life style' approach and plea-bargaining.

[12] Evidence from an interview with a Georgian attorney.

[13] Article 33(3) German Basic Law.

[14] Article 30(1)(4) Georgian Constitution.

'Life style' approach means: if the prosecutor has sufficient suspicion that a person has acquired her wealth (e.g., real estate) in an illegal manner, he can ask the administrative chambers of Rayon courts to confiscate these assets; by dint of shifting the burden of proof, the defendant has then to prove that the property in question has been acquired legally (e.g., by heritage). This procedure has been successfully applied several times, but it met judges in the administrative courts who objected to it as unconstitutional or who applied it only reluctantly, rendering it insignificant in the fight against corruption; so much so that it will probably be abolished altogether in the near future.

Plea-bargaining means that an indictment under criminal law can be refrained from or withdrawn, if the defendant admits to the charge and commits to pay a fine. Such proceedings have turned out to be the core instrument in the fight against corruption. They have generated enormous resources for the state budget and for secret or private cash funds as well. The main reason for this development is that plea-bargaining mainly takes place between the prosecutor and the defendant, allowing the prosecutor to push through state interests in raising funds and save prosecution costs. Although the deal requires approval by the court this is no corrective because judges are often pressed by public prosecutors to provide agreement.[15] This lack of independent supervision has fostered the perversion of the instrument. The prosecution is hardly forced to specify its indictment; some indications often suffice to bring a defendant to the heel.

In consequence, the cancer – the assumption of an office as a tradable commodity - is in a way doubled by plea-bargaining rather than being abated. Not only the office of the accused but also the office of the prosecutor is turned into a commodity. Through plea-bargaining, the office of the prosecutor - objective prosecution according to the standards of criminal law alone - becomes an object of barter, if the prosecutor trades guilt and justice for expediency and state revenue. In addition, other inappropriate factors can influence prosecution practice. This might explain why members of the previous but hardly those of the current regime have been prosecuted because of corruption, and why those who were charged were often picked out arbitrarily.

The judiciary had also been affected by bribery and hence became an object of the fight against corruption. The informal rules of such affairs had previously looked like

[15] This is different in the country of origin of plea bargaining, the USA, where the judge is independent and influential. That this fact was ignored in Georgia is an example for how problematic the transfer of a legal concept can be.

this: a prosecutor offers the reporting judge an amount of money or is encouraged by the latter to do so. The judge discusses the offer with his or her colleagues, in larger cases also with the court president, and agrees to share the offered sum. Is the offer insufficient, the judge can approach the defendant. If the latter agrees to a sum, and this leads to a ruling in his favor, the sum accepted by the judge has to be paid back. Communication in this matter avoids the direct pattern of money for verdict, as bribery is illegal; it rather takes a roundabout way; judges then point to the high workload of the court and hence the potential time span of the proceedings, or more directly to a car in need of repair or to the high study costs of their children.[16]

Fatal to the development of a modern economy, this system was battled with political actions permanently on the brink of collapse. Already during the transition period after 1988 it was decided that all judges should be dismissed at the end of 1999. This meant that in 1999 all re- or newly employed judges had to pass an exam in the newly introduced law. The exam questions were guarded like a state secret. About the best 40 percent of those passing the examination were employed. At the same time, wages were raised to a level guaranteeing a sufficient standard of living for the judges. Although the success cannot be evaluated systematically, according to both the impression of many experts and my own view, based on a variety of intensive contacts with new judges, highly qualified lawyers who in their majority do not accept bribery and fulfill their function with the necessary objectivity have been chosen for the higher courts.

In the dynamic of the Rose Revolution of 2003/2004, the incremental success of the first years after 1999 was seen as deficient. New activities were undertaken. Still perceived as corrupt, many judges were urged to 'voluntarily' apply for their dismissal in return for safe retirement. Some judges, who refused this procedure as they objected to its implicit accusation, were sacked by dint of disciplinary proceedings. One case has gained particular notoriety. Five judges of the Supreme and Appellate Court were blamed for failures, which were either of a technical nature or belonged to the usual discretion in interpreting legal norms. These accusations disregarded the fact that disciplinary proceedings have to respect judicial independence and can only be applied in cases of severe organizational misconduct or significant violations in applying the law. In this vein, the accusations would have justified an amiable or adhortative conversation with the Court President at best. To the contrary the insistence on taking drastic measures for (if at all) negligible misbehaviour caused the impression that the

[16] Evidence from interviews with Georgian attorneys and judges.

authorities misused disciplinary proceedings in order to avoid regular anti-corruption proceedings. In addition, there is indication that the accusation had a second and even more important goal: the expulsion or domestication of self-conscious and politically independent judges.

In that respect the outlined case is indeed a turning point for a new development in judicial policy. With such drastic measures, corruption among judges is probably largely contained. Instead, judges seem to become politically more careful and more adaptable, especially in cases in which state interests are at stake. Furthermore, the qualification of some judges, who succeeded their dismissed colleagues, can be doubted. It would be detrimental, if in such a manner the old dependency on bribery would be replaced with a new dependency on priorities of governmental policy.

5 The Project of Reforming Georgian Administrative Law

In Georgia a General Administrative Code and a Law on Administrative Court Procedure came into effect on 1 January 2000. The General Administrative Code (GAC) encompasses provisions on the organization of the administration, access to administrative information, administrative procedures, the shape and withdrawal of administrative acts, administrative enforcement, contracts under administrative law, state liability, complaint procedures and procedures of rule-making. The Law on Administrative Courts Procedure (LACP) refers fundamentally to the Code of Civil Procedure and adds only those provisions that are specific for disputes under administrative law.

The notion to comprehensively codify administrative law, instead of leaving it open to judicial casuistry, was initiated by the European Council, which demanded of Georgia to introduce means of controlling the administration before being granted accession to the Council of Europe. As Georgia anticipated that its accession would increase Western solidarity and foster stability in the region, the drafting took place under considerable time pressure from the very beginning; the strain even increased towards the autumn of 1999, as parliamentary elections were pending and the old parliamentary majority favored the envisaged reform of the administrative law. Thence, both laws came into existence after hardly more than two years of elaboration. This comparatively short preparatory phase has to be taken into account when assessing the text in detail; it has doubtlessly caused many gaps, redundancies, inaccuracies and some incomprehensiveness within both legal texts.

During the drafting phase of both laws, a number of Western countries offered advice from the perspective of their system of administrative law. The project took off with a seminar in The Hague in October 1997, organized by the European Council and bringing together the high representatives of Georgia and Armenia, who had similar plans at the time, and consultants from the Netherlands, France, Poland, the United States and Germany. The seminar addressed conceptual issues of an administrative law reform, such as how the new legislation was related to the traditional Administrative Infringement Codes, whether an inner-administrative complaint procedure should be mandatory before court review, etc. A second seminar took place in May 1998 in Strasbourg, yet this time without Armenia.

The actual drafting of the GAC was the subject of a project by the GTZ (Deutsche Gesellschaft für Technische Zusammenarbeit) in collaboration with the Center for International Legal Cooperation (CILC) in Leiden and the United States Agency for International Development (USAID).

The main work rested with two Georgian lawyers. One, about 40 years old, had already worked as a judge, the other, about 25 years old, served as an assistant to the parliamentary law committee. Both had studied law in Georgia and had only scarce knowledge of Western administrative law. One only spoke Georgian and Russian, the other also some passable English. They knew the Georgian social and political conditions very well and were experienced in the fight with real life socialism. Favored by an immense grasp, they were both able to comprehend Western administrative law codifications very quickly. They partly worked with Dutch consultants in Groningen and in The Hague and with me in Bremen.

Both Georgian lawyers studied and evaluated the English versions of legal texts of Dutch and German and less so of American and French administrative law, discussed the individual articles to be drafted with their consultants and then formulated them in the Georgian language. The outcome was translated into English and discussed at a number of seminars with about 20 Georgian lawyers and in changing composition with two Dutch, one American consultant and me, as German adviser. However, these seminars were primarily meant to increase the interest of the Georgian lawyers in the codification project. For deeper substantial suggestions, the project was all too new. In the following time, I more and more became the principal advisor. I went through the drafted texts in detail and discussed them again with both authors.

The drafting of the LACP went somewhat differently. Much time had been spent on advice given by Dutch consultants to draft a complete set of Rules for Administrative Courts, which entailed no reference to the Georgian Code of Civil Procedure and neglected the fact that separate administrative courts were not intended for Georgia. Following my advice, a simpler solution was then chosen; it left the project by amending the Code of Civil Procedure. As the parliamentary schedule left almost no time for further complex consultations the resulting relatively short text (with its 35 articles) was in large parts drafted by myself.

After they had come into effect, both laws were intensely reviewed and applied in court practice. However, little is known as to what extent the administrative bodies have become aware that they must act within new forms and have to obey new requirements (e.g., the hearing and justification of administrative acts, the rule against bias, the protection of legitimate expectations in cases of the annulation and withdrawal of administrative acts, the suspensive effect of complaints, etc.).

The judicial practice and younger university lecturers have identified many deficiencies within the first version and demanded some amendments. From time to time, I was consulted. I also made suggestions of my own but tried to ensure that the overall systematics is not impaired by *ad hoc* ideas. Between 2000 and 2007, a total of eight amendments to the GAC and ten amendments to the LACP were adopted.

The provisional character of the first version and the number of amendments have now caused both laws to be fundamentally reworked.

6 Experiences from Consultation

I would like to add some generalisable experiences from this and other consultations I have been involved in. Western legal consultants when offering advice are in a way pushed into an external perspective on the law they are familiar with. The consultancy situation is a challenge to develop a deeper understanding of what shall be transplanted. It is not sufficient to know the shape of the law from literature, jurisdiction and practice; one is compelled to comprehend its foundations anew, to re-evaluate it critically and to compare its particularities with other legal systems. Only after such considerations the task at hand can actually be approached, namely to grasp the problems of a country in its transition from old to new social circumstances and to relate law, which emerges in such conditions, to them. Old doctrines about state functions, back to Adam Smith and buried in positivist dogmatics, become significant again and face the question which functions an administration in a transition country should fulfill.

Decades of legal-dogmatic precision work, e.g. in the law of the administrative act, state liability and the access to legal protection, have to be judged against the ruthless yardstick of simplicity and transparency. A helpful experience was that, for some time, the often overly complicated German law had already been pressured by Community law into becoming simpler.[17]

Closer attention has to be paid to current Western developments in the 'philosophy' of public administration. Of foremost interest are approaches called New Public Management (NPM), which shift the thrust of administration from legality and legitimacy to effectiveness and efficiency.[18] I believe it would be fatal in a situation of transition to propagate managerial approaches at the cost of democracy and the rule of law. NPM can only unfold on a stabile political system based on democratic law-making and court supported law-application. Recent governmental practice in Georgia is a case in point: legal rules have often been set aside in order to facilitate major investment projects. This has cost the Georgian government general acceptance, as became manifest in the demonstrations in November 2007.

Finally, as law is mainly created for deficient situations, the main phenomena of administrative deficiencies have to be identified, which administrative law is then meant to battle. Which, it has to be asked, are the society-specific flaws of public administration? Are the particularities of a state characterized by the power of an old caste, which at least succeeds accidentally? Is it indolence and arbitrariness which burdens a society, or is it the systematic exploitation of powers as a corrupt source of income? Above, I have already tried to address these issues.[19]

Consultation is, however, not only characterized by the intellectual challenge it entails; it is also in many ways socially structured. The central condition for success is doubtlessly the fact that counterparts who know their own legal and social system and are open for innovation must be found. Such lawyers are generally scarce, but in transition countries they are even harder to find, as the low wages in the public service forces them to have multiple employments, leaving them with no time for thorough legal study. Often, the best turn to the legal consultation of Western investors or are appointed by international institutions.

[17] An example is the cumbersome construct of norm-concretising administrative guidelines, see ECJ Case C-361/88 (Commission v. Germany), Rep. 1991, I-2567; R. Streinz, Europarecht, Heidelberg (C. F. Müller Verlag) 7th ed. 2005, p. 152:

[18] K. Schedler, I. Proeller, New Public Management, Bern (Haupt Verlag) 3rd ed.2006, p. 54 ff.

[19] See for a more elaborate discussion of finding a way between the generalisability of many legalconcepts and the need to adapt them to the recipient country's tradition R. Knieper, Möglichkeiten und Grenzen der Verpflanzbarkeit von Recht, Rabels Zeitschrift 72 (2008) 88-113.

Foreign donor organizations are then confronted with the question of whether they should pay domestic drafting, in order to secure some qualified cooperation. If they choose to do so the often-noted risk arises that the receiving country and the individual expert do not develop an authentic interest in the project. Economic empirical research has proven that money often demotivates real involvement.[20] According to the logic of exchange, the counterpart acts rational if he comes up with a draft without much investment of time and work and "sells" it perhaps a number of times. Thus, it seems to be more promising for a legislative project to work with extra-economic benefits or delayed gratifications, such as the reputation of being the 'mother'/'father' of a law, which can later be of benefit in the exercise of legal consultation or in access to higher qualified positions. In the case of younger lawyers, the interest in learning and new subject matters also creates motivation.

On the other side, it seems cynical if the foreign consultant receives his or her *per diem* and the needy counterpart is expected to more or less work *pro bono*.

Donor organizations solve this dilemma in different ways. While American organizations rather turn to paid work, the German GTZ tries to follow the rule that Western consultation is paid for by donor countries and the work of staff in the receiving countries is paid for by the latter itself. But also the second concept allows for some flexibility: some nonpersonnel costs (such as office facilities, travel, literature, etc.) can, for instance, be covered by the donor, and individual achievements, such as the successful completion of a draft or legal commentary, can be remunerated moderately. In the cases of the Georgian General Administrative Codex and the Law of Administrative Court Procedure, it has proven valuable to invite experts from Georgia to Germany for some weeks. Here they were able to work undisturbed by their everyday business. At the same time, they received some kind of compensation in form of a modest *per diem* and a cultural program.

A further condition for successful consultation is that experienced translators and interpreters accompany consultants unfamiliar with the language in the receiving country. Their task is particularly difficult, as legal terms are defined in their individual country of origin and have to be linked to each other. This issue is well known from legal comparison, yet it becomes even more severe during consultation as the draft of a law has to establish an altogether new legal context. It is almost inconceivable how many misunderstandings can arise, and it is often puzzling to notice the difference

[20] Cf. B. Frey, Markt und Motivation: Wie ökonomische Anreize die (Arbeits-)Moral verdrängen, München 1997.

between one's understanding and the matters actually addressed.[21] Some interpreters have the ability to come up with sentences of erudite hollowness. This is most often the fault of the consultant himself if he has not introduced the translator to the respective terminology. In general, however, it is indispensable that the consultant and his counterpart communicate directly in one language, even if the common language is not native for either of them.

A further condition is that law-making policy is in general open to innovation. The offices of ministries and parliamentary committees are swamped with well-meaning, yet unread reports of foreign experts, and too often it is virtually only the consultant's supervisor in the donor organization who has indeed read the paper. Ideal conditions exist if the counterpart belongs to a network with some influence on law making in the receiving country. It is also essential that the respective law is part of the legislative schedule for the election period. A minimum requirement should be that its preparation has been arranged in a written cooperation agreement between the receiving state and the donor organization.

If a consultant has access to networks relevant to law making, the temptation grows to follow the networks' tactical moves. The situation arose, for instance, that the adoption of the Georgian GAC offered the opportunity to deal with many individually contested issues, which would disappear in the plethora of problems of the overall legal work, e.g. the public procurement law. From the perspective of the legal system, it does not belong into the Codex but the opportunity to introduce it was tempting.

The professional answer is to uphold the quality of law as the ultimate yardstick. The political network has to take responsibility for cut backs, the consultant should not be involved. Often, it becomes quickly obvious that a tactical move has been unsuccessful. Then, responsibility rightly also rests with the foreign expert. In the mentioned example, the public procurement law had indeed been adopted with the overall legal package, but within months it was rendered obsolete by a separate political discourse and an according special law.

The necessity that donors coordinate themselves should also be mentioned. Often, a number of organizations contribute to a project without this being planned and coordinated from the beginning. For example, recently a meeting of all donor organizations, which offer consultation in administrative matters in Georgia, established that almost all of them focus on the training of administrative judges, although there are

[21] For example, a translator stated that the president of a municipal council had to *sing* the decisions of the council. What was meant?

only 30 to 50 of them in the entire country. No wonder that administrative judges do not or seldom react to seminar invitations. However, no organization had turned to the training of administrative officials, although the new administrative law was primarily addressed to them and implemented completely unfamiliar new standards in their interaction with the citizens.

To turn to another example of lacking donor coordination: an essential innovation of the new administrative law is the right of access to information. Following my advice, the draft contained a chapter that mainly took up and generalized EU standards for the access to environmental information. Only after the first reading in parliament, a totally new text was introduced, based on an US initiative. Due to the rush, the text was out of place, both stylistically and substantially. It introduced its very own terms in stark contrast to the rest of the Codex, it was partly very detailed, provided its own recourse to legal action and cases of liability, and it provided the Georgian administration too quickly with requirements that were too ambitious, e.g. by demanding that every incoming information had to be categorized as public or classified and had to be included in a public registry.

Some partners in the consulted countries use the lack of coordination among the donor organizations to their personal advantage; as said before, if collaboration is paid for, they sometimes 'sell' their work a number of times. Other partners feel pressured and clueless with regard to the often contradicting pieces of foreign advice, which pile up in their offices in the form of heavy reports. Thus, the coordination of consultation and a division of labor among donor organizations is essential. This is, however, hard to achieve as donor organizations compete with each other when it comes to certain projects, and as they pursue different political aims and often represent different politico-legal traditions. It is, hence, of large significance that decision-makers in the receiving country notice the problem on the spot, and supervise and coordinate the different forms of consultation on offer.

7 Homegrown and Foreign Elements in the New Administrative Law

Different legal traditions have found their way into the General Administrative Codex and the Rules of the Administrative Courts in Georgia. However, the adaptation to Georgian needs and the particularity of the legal amalgam has created a genuine Georgian solution.

7.1 The General Administrative Code

The first chapter of the GAC begins with a *short statement of its core contents* (the making and implementation of administrative acts and contracts) and a general definition of its *purpose* (the protection of civil rights while pursuing public interests). The following provision contains definitions of the most important terms, including the administrative agency, the administrative act, the administrative contract, and the administrative procedure.

In the ensuing articles the *principles* of administrative law are set out, including equality, the lawfulness of administration, statutory reservation (i.e. that intrusions in individual rights must be basedon a law), the precept to balance interests while exercising discretion, the principle of proportionality, impartiality, open access to the administration, the right to be heard, the official language. These principles follow a cross section of principles in force in many member states of the European Council. The presentation of the main principles serves the purpose to briefly and comprehensively introduce citizens to the spirit of the code. A disadvantage of such a legal approach is that many principles are not self-explanatory and directly applicable, but need some further elaboration in the following chapters. In the draft of an amendment of 2008, it is made clear which principles are applicable by themselves and which ones need additional, more specific provisions.

The second chapter addresses problems of *administrative organization*, among others *mutual assistance between administrative bodies*. They follow German law and are of particular relevance for Georgia, as the cooperation among authorities causes serious difficulties, not least due to the hitherto unsolved question of who carries the costs for additional work. Provisions concerning the geographical competence of authorities are lacking, but they will be introduced with the new draft.

The third chapter regulates the *access to administrative information* in detail. Model is the American Freedom of Information Act. As already mentioned above, this part of the law is stylistically a failure and does not integrate in the overall systematics, caused by a lack of coordination among consultants. In the new draft, the chapter is phrased in a simpler and clearer fashion but without diminishing the right of access to information itself.

The fourth chapter is dedicated to *administrative acts*. Following continental European tradition, it sets out the form, contents, publication, coming into effect, withdrawal and annulment of administrative acts. The current version of the code does not contain provisions about auxiliary conditions to administrative acts; they have been added to the new draft. The preconditions for the annulment of administrative acts are somewhat more simplified in comparison to the over-complicated model, Article 48(2) of the German Administrative Procedures Law. The fundamental idea was that under

Georgian conditions an independent casuistry would develop. Nevertheless, in a first verdict which considered recourse to the protection of confidence, an approach inspired by the Soviet era was chosen pursuing a very strict state-favoring application. Meanwhile, amendments have corrected such tendencies.

Subject of the fifth chapter is the *administrative contract*. The code provides that such contracts are to be judged under civil law, but aspects of public law and especially constitutional provisions have also to be taken into account. As an administrative contract ranks every contract concluded under the participation of a public authority. This taxonomy abolishes the distinction between contracts under administrative private law and administrative law, which is characteristic for German law but causes much doctrinal complication.[22] The largely unnecessary distinction between settlement contracts and exchange contracts in German law were also considered dispensable. Unfortunately, specific measures aimed at preventing the prevalence of the administration (or else of powerful private parties) in the contractual relationship were lacking. Resounding German law in this respect[23], provisions will be introduced into the code's new version, ensuring the equivalence of services exchanged in the contract, requiring that the money paid in exchange for the public service must be spent for public purposes, and excluding illegimate linkages between services (such as the 'selling' of a license for money). These clauses could also play a significant role in the fight against corruption.

Chapters 6 to 11 concern *administrative procedures*. Not less than five different types of procedures have been introduced. They were heavily influenced by Dutch and German law, although the fifth is a genuine creature.

The *simple procedure* includes provisions about

- the participants of the procedure;
- the right to submit an application, which actually should be a matter of course but which is separately mentioned due to the still prevailing anxious attitude of citizens towards the administration;
- the documents applicants must submit, including the explicit exclusion of administrative powers to demand more documents than required by law – again this is another element against delaying tactics and hence against potential sources for extra income;

22 For the amendments to the new draft, see below.

23 Article 56 of the German Administrative Procedure Act.

- the right of proxy;
- the exclusion of administrative staff in the case of conflicts of interest;
- the maxim of investigation [Untersuchungsmaxime];
- the right to be heard;
- failure to observe a time-limit and restitutio in integrum;
- deadlines of the administration with respect to making decisions (usually one month after making an application).

For the *procedures of collegial bodies*, the law stipulates the decision-making process and the taking of minutes. This would also be the appropriate place for the principle of publicness of all committee meetings, which was introduced in the chapter about the access to information – a requirement which is, however, hardly practicable under current Georgian conditions.

The *formal procedure* with formalized hearing of evidence and contradictory structure represents a court-like administrative decision procedure. It could be made applicable in highly controversial, but individualisable cases like the decision-making about expropriation and compensation. However, it seems that as yet no sectorial law makes reference to this type of procedure.

The *procedure with public participation* is intended for the authorization of large industrial and infrastructural projects with impact on diffuse interests. Its major issue is the notice and comment procedure including details of the public display of documents, terms for comments and the procedure of public hearings. Here too, references to this procedural type are still missing in sectoral law.

The *procedure before an independent body* can be applied to decisions with significant political and economic consequences, e.g. in the privatization of large public enterprises or other public property like frequency ranges or conduction rights. This type of procedure follows the Anglo-American model of independent 'inspectors' or 'administrative law judges'. The provisions are especially meant to allow the appointment of neutral foreign individuals or consultants (e.g. an internationally renowned law firm) as head of the procedure.

Chapter 12 includes the provisions concerning the implementation of administrative acts. Two options were available: the model of Anglo-American law according to which an authority has to obtain a judicial enforcement order in the regular case, or the continental model according to which the administrative body is able to enforce the administrative act itself. In the end, the latter solution was adopted. Following German law, procedural safeguards for those affected were introduced, especially by stipulating that normally an authority has to announce the means of enforcement before actually applying them. The catalogue of sanctions – compulsory fine, execution by substitution, direct coercion, and in the case of monetary claims also the distraint of assets – also corresponds with German law.

Chapter 13 regulates the procedure of inner-administrative complaint. It mainly follows Dutch administrative law. It had to be decided in advance whether the lodging of an appeal should be optional, i.e. whether a party affected has the choice to either appeal to an administrative body or to go to court. The Georgian experts felt that the authorities would hardly change a decision they had made voluntarily and therefore chose the optional solution. This has lead to the practice that almost always courts are appealed to. When the currently beginning training of administrative lawyers shows effect, it will be advisable to introduce the obligatory option due to reasons of institutional economy. The new draft already includes this.

Chapter 14 concerns the issue of state liability. The official liability of employing administrative bodies for their employees follows the German model established in the civil code, to which the GAC refers. In addition, a further claim to liability exists, which equals the German claim to compensation for damage caused by not unlawful burdensome administrative activies; yet, in the interest of protecting the state treasury, the preconditions of liability in Georgia have been phrased more carefully: on the part of the harmed there has to exist both a particular sacrifice for common welfare and a particularly serious damage. Finally, the chapter also contains liability provisions for unlawful state action.

The final chapter includes provisions concerning the procedure for the generation of sublegal norms. This chapter has to be seen in relation to the Law of Normative Acts. While the latter in particular defines the types of normative acts and clarifies questions of hierarchy, legality and the consequences of faults, the GAC specifies the procedure for the enactment of such norms. The model for this chapter is the American rule-making procedure under the Administrative Procedure Act. It is especially intended that the drafts of statutory instruments or administrative regulations have to be publicised and sent to the affected administrative bodies and associations. The public as well as the authorities and associations then have the right to submit their comments. Additionally, the formal procedure of the 8th chapter on public participation of the GAC is found applicable. This means that every time a potentially costly hearing has to take place. In this respect, the new draft introduces some simplifications.[24]

[24] For the corresponding American experience with the so-called ossification of formal proceedings and notions of simplification, see Mashaw, Merrill and Shane, *Administrative Law*, 4th edition 1998, pp. 604-610.

7.2 The Law of the Administrative Court Procedure

Georgia does not possess an independent administrative judiciary. Regular courts decide upon administrative matters. In inferior courts, i.e. the so-called rayon or city courts and (if implemented) magistrates' courts, administrative and civil matters are decided by the same judges, while district courts, courts of appeal and the Supreme Court have particular administrative panels.

The LACP included 35 articles in its first version of 2000. Since then it has been amended ten times. Most changes came as a reaction to legal practice, which revealed shortcomings of the first version but also indicated that LACP has indeed found its way into practice. Some amendments also originated in the political realm. This applies in particular to a number of special proceedings including proceedings of confiscation of illegimate property (the 'life style approach' explained above), the admission for the control of economic activities, the fight against domestic violence, and the hospitalization in psychiatric institutions.

The LACP originally followed an easy legislatory strategy: the Code of Civil Procedure generally applies to proceedings before administrative panels, while the LACP includes only those provisions that differ from the Code of Civil Procedure.

Despite these conceptually clear distinctions, there were points of contention in judicial practice, especially concerning the issue of whether those rules of the Code of Civil Procedure that are not particularly regulated in the LACP have to be applied directly or only in analogy (i.e. by way of interpretative adaptation). In order to provide greater clarity, a parliamentary working group which should draft a more detailed LACP was established in 2007. This reversed the legislatory strategy of the LACP in force: only those rules of the Code of Civil Procedure should find application in administrative proceedings, which were explicitly referred to in the LACP. Furthermore, the working group is instructed to clarify additional substantial issues that have arisen in judicial practice. In the meantime the completed draft encompasses almost 100 articles.

The central question remains which kind of disputes should be heard in administrative proceedings and which ones in civil proceedings. Such issues of distinction can be resolved more easily if both proceedings are dealt with by the same jurisdiction, as is the case in Georgia. Here competing courts are absent, which, in a system of separated civil and administrative jurisdiction, lead to turf wars and attempts to extend competences. Still, even in a unitary system there should be clear criteria for the separation of competences.

Therefore, it also had to be determined in Georgia, which criterion should underlie the term of a dispute under administrative law: the purpose of those legal norms deciding the dispute (the criterion of public interest), their subject matter (criterion of power relations) or their nature (criterion of law specially designed to control administrative bodies). As all of these criteria are not sufficiently selective and as a simple solution was considered necessary, the authors of the first version of the LACP opted for a formal criterion: accordingly, administrative law cases were to concern legal relationships involving an administrative body as one or even two or more parties. While this poses no issues for administrative acts, it leads to a significantly extended application of the LACP in the area of administrative contracts. This implies that, for instance, rental agreements for a flat between the state and individuals fall into the area of administrative law and therefore have to be decided upon by the administrative panels of courts (respectively, if particular panels do not exist, under application of the LACP).

This confronted the administrative panels of courts with a flood of cases in legal areas in which civil courts are usually far more experienced, as the example of tenancy law shows. *Vice versa*, the formal selection criterion also excluded those actors under civil law who fulfill public functions from administrative proceedings.

This is the reason why judges came to demand a substantial criterion for the distincition of civil and administrative law cases. The definition of contracts under administrative law was therefore narrowed down to contracts that were concluded to exercise sovereign authority.[25] At the same time, the definition of authorities was extended to private persons who exercise sovereign authority.[26]

Characteristic for the administrative proceedings are the so-called procedural maxims [*Prozessmaximen*], especially the issue of whether the matter in dispute is at the disposal of the parties or the court (party disposition), and whether it is the parties' or the courts' task to feed facts and provide evidence in the proceedings (principle of contradiction vs principle of investigation).

The first chapter of the LACP explicitly confirms the principle of party disposition, which also applies to civil proceedings, for administrative proceedings, yet curbs it insofar as the administration is bound to legal provisions when making a composition, waiving or accepting a claim. This provision, also present in German law, reduces the opportunity for defendants and third parties to bribe the administration into making a composition, waiving or accepting a claim *contra legem*. Therefore, the court has to examine whether the termination of proceedings follow legal requirements.

[25] Article 2(1)(g) General Administrative Codex (version of 2005).

[26] Article 2(1)(a) General Administrative Codex (version of 2005).

In German Law, a mixture of the contradiction and investigation principles determines the introduction of dispute matter and proofs. While the parties are primarily responsible the court is authorized to complement facts and proofs on its own account. For the Georgian LACP it was to be considered that Article 85(3) of the Georgian constitution prescribes legal protection according to the principles of both equality and the competition of the parties. This is why the contradiction maxim was adopted as a basis.It was however somewhat qualified, as the court is allowed to demand the submission of information (especially administrative files) and order certain proofs (especially expert opinions). I believe it could be argued that constitutional law allows for a more proactive role of the court as a compensation for the disparate starting positions of private individuals and the administrative bodies in a proceeding. Following this argument, the new draft tries to extend the court's room for maneuver.

The second chapter of the LACP contains some provisions concerning jurisdiction. The competence of the first instance was allocated to inferior and district courts. District courts are courts of first instance in more significant cases, i.e. administrative acts and decrees by the president and ministers as well as by the councils and mayors of the large cities.

Chapter 3 states that a judge is excluded if he has been party to an administrative proceeding which has dealt with the same matter before.

Chapter 4 deals with the matter of costs. After a prolonged discussion about the need to simplify access to administrative legal protection, the legislator nevertheless opted for a liability for court costs with the exception of disputes under social law. Following the Code of Civil Procedure, the losing party owes the resulting costs. It facilitates matters that a nonpaid advance on fines does not cause the proceedings to be suspended. In practice, however, parties sometimes take advantage of this fact, as cost payment is often refused and the court has to press its debtors.

Chapter 5 concerns the time allowed for appeal in cases of failure to provide instructions about possible legal remedy. This matter actually belongs into the Code of Civil Procedures, but is missing there.

Chapter 6 includes provisions concerning the parties. It determines that administrative bodies and not their founding entity (the state, local communities) act as parties. The rationale is that in this way administrative bodies learn better from their mistakes, as according to Georgian budgetary law not the state or localcommunitees but the individual administrative agency carries the costs of court proceedings.

Chapter 7 includes particular provisions concerning the introduction of subject matters and the hearing of evidence; it thus concretizes the specific amalgam of proceedings and investigation maxims. Among other things, it is stressed that on the one hand both parties are required to contribute and that on the other hand the court can present matters, but only if the parties are allowed to comment. A stronger obligation to contribute exists for administrative authorities, in so far as they are obliged to submit the files concerning the case. With regard to the secrecy of files worth of protection, there are no *in camera* proceedings as in German law. Rather, the authorities are allowed to withhold information, which has to be justified and can be reviewed by court; additionally, the files kept secret have to be paraphrased. Accordingly, the parties have the right to substantial, unlimited access to legal files, with the exception of drafts of and preparatory documents for judgments.

Chapter 8 determines the types of complaints, temporary legal protection and the contents of rulings. As types of complaints, it stipulates the motion of quashing an administrative act, the motion of compelling an administrative body to issue an administrative act, to provide a service or desist from some activity ('mandamus' in English law) and the motion to declare the existence or non-existence of a legal relationship between the parties. The very controversial issue of the right of standing was solved in a *via media* between approaches oriented towards subjective rights and those oriented towards protected interests.[27] For all types of complaints, it is equally required that the plaintiff has to assert that he is directly and individually affected in a legally protected interest. Furthermore, it is determined that administrative acts can be annulled without full checking, if a deficit in the legal investigation or evaluation has been established.[28] As the willingness of authorities to obey to judgments cannot easily be assumed in Georgia, the law stipulates that the court can decree the act of administration itself.

Moreover, it is determined that motions of quashing administrative acts have a suspensive effect. Administrative acts are then not allowed to be enforced, except in extraordinary circumstances or if an authority has ordered the immediate execution justified by explicit reasoning. Three tiered constellations pose specific problems here. If upon complaint of a third person (e.g. a neighbour) the administrative act (the construction permit) is suspended the beneficiary (the house builder) can ask the court to repeal the suspensive effect. For this he must show that the complaint has a low chance of success or that substantial disadvantage is caused. *Vice versa*, the party burdened by the administrative act can ask the court to establish suspensive effect if the administrative body has ordered immediate execution.

[27] Cf. G. Winter, 'Individualrechtsschutz im deutschen Umweltrecht unter dem Einfluss des Gemein-schaftsrecht', *NVwZ* 1999, pp. 467 ff.

[28] Similarly Article 113(3) of the German Law on Administrative Court Procedure.

According to the LACP, motions for quashing sublegal regulations are also admissible. As in the case of complaints against individual administrative acts, the precondition is the potential violation of legally protected interests. The complaint has no automatic suspensive effect, yet such an effect can be ordered by court following a respective request.

The 9th and last chapter enables the appeal and revision to the Appellate Court and the Supreme Court respectively. Appeal and revision can be asked for regardless of the extent of the gravamen. Here the notion prevailed that legal protection in public law should be provided even for small amounts under dispute. An admission procedure for an appeal does not exist. It seems unsuitable – at least in an early phase of administrative law, such as in Georgia. Revision is only admissible if the matter is of significance for the legal development, if the court of appeal deviates from the jurisprudence of the Supreme Court, or if essential procedural provisions have been violated.

8 Concluding Remarks

With the introduction of the Administrative Codex and the Rules for the Administrative Courts, Georgian administrative law was put on new foundations in the year 2000. On this basis, a multifaceted jurisprudence has developed, confronting issues that are also on the daily agenda of legal systems with a long tradition in administrative law. Judges and other legal practitioners have authored textbooks and commentaries, in most cases with support of the GTZ and sometimes involving my consultation. A commentary of the LACP as well as a commentary and textbook concerning the GAC have been published. Concomitantly, seminars have taken place repeatedly, inviting administrative judges to discuss current legislation and matters of codification.

Thus, Georgia has moved a step forward on its way to the normal life of a society that both empowers its public administration to take measures for the benefit of society, and protects individual rights against arbitrary administrative action or inaction.

However, in this process three issues have been neglected and are now increasingly addressed: the introduction of the public to the new rights of citizens, the education of administrative staff on the new legal provisions, and the teaching and research of administrative law at universities. Here, the more important tasks loom, because only if legal texts and rulings find their way into the heads of individuals, administrative law really enters into force. To the radical skeptics of institution building it might be replied that good texts are simply indispensable. Georgia has progressed in achieving them, and it is on its way to put them into practice.

ELECTIONS IN GEORGIA

MARIE-CARIN VON GUMPPENBERG

1 Elections in Georgia – A Reflection of the Political Culture

Since 1990 Georgia has seen a series of elections: six parliamentary (1990, 1992, 1995, 1999, 2003, 2008) and six presidential elections (1991, 1992, 1995, 2000, 2004, 2008) – if the 1992 elections for the Chairman of Parliament/head of state are included – were held. In addition, three referenda/plebiscites passed (1991, 2003, 2008) and four local elections were conducted (1991, 1998, 2002, 2006).

Presidential Elections

Election Date	Number of Contestants	Election Results
26 May 1991	6	86% for Gamsakhurdia
11 October 1992 (Direct Elections of the Chairman of the Parliament – Head of State)	none	96% for Shevardnadze
5 November 1995	4	74% for Shevardnadze
9 April 2000	6	80% for Shevardnadze
4 January 2004	6	97% for Saakashvili
5 January 2008	7	53% for Saakashvili

Source: www.cec.gov.ge, OSCE/ODIHR election reports

Despite having held numerous elections, Georgia has never experienced that elections facilitated the change of the country's leadership. As the January 2008 PACE Election Observation Report stated, "[S]ince independence, changes of power have been initiated by revolutionary means rather than by elections". Incumbent presidents never left their office as a result of the voters' choice; they were always forced to leave their position, may this have been by violent or non-violent means. Georgia's first president, Zviad Gamsakhurdia, had to flee the country after a coup d'etat in December 1991; Georgia's second president, Eduard Shevardnadze, had to step down after three weeks of demonstrations in November 2003. The current president, Mikheil

Saakashvili, holds his office since January 2004. He was re-elected for another five years in early elections in January 2008.

Constitutionally and de facto, the president enjoys wide ranging powers. He appoints the Prime Minister who – with the consent of the president and upon approval of the parliament – selects the ministers (Constitution Art. 73.1bff). The parliament serves primarily as a legislative body; it has limited oversight functions. It can be dissolved by the president in case of declaring non-confidence to the government (Constitution Art. 80.4) or if it fails to adopt the state budget (Constitution Art. 93.6). At the same time, the parliament can impeach the president if no less than two thirds of the total number of the members of parliament has supported the decision (Constitution Art. 63.3). Since the introduction of the Constitution in 1995, neither the parliament has ever been dissolved nor has the president been impeached by the parliament.

With the aim to secure broad support for their policy in the parliament, all presidents sought to secure a comfortable 2/3 majority of their own presidential party. In most cases they secured their party's victory in parliamentary elections and in case they did not they aligned with other pro-presidential parties. In 1990, Gamsakhurdia's nationalist and anti-communist *Round Table* coalition won the absolute majority of seats in the parliamentary elections. In the 1995 and 1999 elections, Shevardnadze's *Citizen's Union* received a comfortable majority of the seats in the parliament. However, it did not reach the constitutional majority. As a result it had to align with Aslan Abashidze's *Revival Union*. In the 2004 and 2008 elections, Saakashvili's *United National Movement* won the constitutional majority.

The presidential parties were limited in their existence. Gamsakhurdia's and Shevardnadze's parties only played a role while their heads held the position of president. They emerged with the presidents and their existence ceased once the presidents had left their position. After Gamsakhurdia was forced to leave Georgia, his *Round Table* coalition gained only a very limited number of supporters in the October 1992 parliamentary elections, and President Shevardnadze's *Citizen's Union* disintegrated after the ouster of its chairmen from office in 2003.

The success of all presidential parties – including Saakashivili's *National Movement* – depended on their broad public support. And they had this public support because they targeted all social strata, hereby not only focusing on their own supporters, but also on people from different social backgrounds. In the presidential parties, all relations between the president as the head of the party and the party members/affiliates were based on the principle of patronage and loyalty: patronage from the side of the head of the party and loyalty from the side of the party members to their patron. Party members/affiliates, however, were motivated by various reasons: some presidential parties' supporters hoped to gain benefits from joining the presidential party and to avert repressions from the country's ruling forces; other party supporters – mostly representatives of various powerful business groupings – hoped to gain some immunity and political leverage by joining the presidential party.

Parliamentary Elections

Election Date	Number of Seats Threshold	% of Seats in the Parliament
28 October 1990	250 Seats 4% Threshold	62% for Round Table "Free Georgia", 26% for Communist Party
11 October 1992	235 Seats No Threshold 150 Regional Party Lists 85 Single Mandate Constituencies	16% for Peace Bloc, 9% for 11 October Bloc, 6% for National Democratic Party (and many others)
5 November 1995	235 Seats 5% Threshold 150 National Party Lists 85 Single Mandate Constituencies	46% for Citizens' Union of Georgia, 14% for National Democratic Party, 13% for All-Georgian Revival Union
31 October 1999 (1st round) 14 November 1999 (2nd round)	235 Seats 7% Threshold 150 National Party Lists 85 Single Mandate Constituencies	56% for Citizens' Union of Georgia, 25% for Bloc Revival of Georgia
2 November 2003 28 March 2004 (repeat elections of the proportional component)	235 Seats 7% Threshold 150 National Party Lists 85 Single Mandate Constituencies	68% for National Movement–Democrats, 10% for Rightist Opposition
May 2008	150 Seats 75 National Party Lists 75 Single Mandate Constituencies 5% Threshold	79% for National Movement; 11% for United Opposition; 4% for Christian-Democratic Party; 4% for Labour Party; 1% for Republican Party

Sources: OSCE/ODIHR election reports, Wheatley 2005

Political orientations played only a minor role in the presidential parties. Party programmes in the western sense of understanding never existed. The programmes – if they existed at all – were neither initiated bottom-up nor did they enjoy broad public support. The president was the "programme". His views were reflected in the programmatic orientation of the presidential party: Due to Edward Shevardnadze's background, the *Citizen's Union* viewed itself as a centre-left wing party and identified itself with European social democracy; due to Saakashvili's orientation the *United National Movement* considered itself as a centre-right conservative party. As both presidents supported European and Euro-Atlantic choice, so did their parties.

Nevertheless, programmatic orientations never played an important role in Georgian politics. In all presidential elections it was not "programmatic" issues but personalities that were discussed. Electoral campaigns focused on the person of the president rather than on his "programme". If there were "programmatic" messages, they were dominated by promises of benefits for the poorer parts of the population and by promises of immunity for businessmen.

Presidential parties were always confronted with a plurality of other parties that justified their existence through their opposition to the ruling party. Most opposition parties focused their activities on criticizing the presidential parties rather than proposing real policy alternatives. They, however, were never united in their opposition to the presidential party, no matter whether it was the *Communist Party* in 1990, the *Round Table* Coalition in 1992, the *Citizen's Union* in 2003/4 or the *United National Movement* in 2008. On the contrary, they competed with each other. "On the surface, disagreements were about tactics, but in essence the divisions were about the personalities and the ambitions of the different leaders." (Nodia/Pinto Scholtbach 2007: 9) As a result of these struggles between personalities, political parties mushroomed in Georgia. Nowadays (July 2008), there are nearly 200 parties registered with the Ministry of Justice with only about 20 of them being active and submitting financial reports to the Ministry of Justice.

The pluralism of political parties in Georgia cannot be assessed as an indication for the existence of a genuine democracy in Georgia. It is rather a reflection of the existing civic and political freedoms in Georgia, however, never an illustration of real political competition and broad civic participation. While in the early 1990s private businessmen and paramilitary leaders established parties in order to pursue their interests with non-democratic means (bribing, blackmailing, intimidation), starting from the late 1990s, defected government officials and young businessmen formed parties with the aim to counterweigh the government's economic policy (e.g. *Conservative Party, New Rights*). The possibility of the latter, however, to influence the government's decision making was always quite limited for the following reason:

Power was always concentrated in the hands of the president – as primary source – and of his close associates – as secondary sources. This group constituted the key decision-makers. They elaborated the main directions of the government's policy, sought parliamentary endorsement of their decisions and the support of the courts, hereby undermining the constitutionally established system of checks and balances. Political competition was always between the president and his associates on the one hand and various political and business groupings which sought to influence the inner circle's decisions on the other hand. It was never between political parties about programmes and political orientations nor about the votes of the citizens. Elections were always the battlefield of various influence groups with an aim to gain access to the power circle; they were never a battlefield of the political parties to win the support of the people. As a result, political elites always only had limited interest in gaining real popular support of their policy directions.

Against this background it is clear that ruling elites were mainly interested in the election *results*, never in the electoral *processes*. It was all about winning presidential and parliamentary elections with an overwhelming majority of the popular vote – by all means. Especially, "Shevardnadze believed that it would always be possible for him to win elections and that, if necessity arose, they could be rigged. And, while there would be international criticism, it would soon fizzle out." (Kandelaki/Meladze, 2007: 102) However, presidents always underestimated the dynamics of the inner circle and of the various interest groupings.

Both groups' loyalty was limited – and came to an end when presidential policies did not coincide with their own orientations anymore. In the case of the "Rose Revolution" the inner circle and some interest groupings managed to mobilise broad support of the people. And the Georgian people could be easily convinced of the concept of government change through demonstrations, as they had accumulated deep distrust towards elections. "In the public mindset, all elections were unfair and, therefore, a change of government through elections was not considered a viable prospect." (Kandelaki/Meladze, 2007: 104) In their eyes, power change through demonstration was the only legitimate way to decide about a new president.

2 Free, Fair and Transparent? International Election Observation Missions and Their Main Findings

With the aim to promote election integrity and to restore confidence in electoral processes, international election observer missions have observed elections in Georgia on a regular basis since 1999.

One of these observer missions was that of the OSCE/ODIHR. Before deploying an observer mission OSCE/ODIHR sent a needs assessment mission that assessed the conditions and preparations for the election and advised on the level of OSCE/ODIHR involvement in the elections.

Based on their recommendations, *OSCE/ODIHR* deployed *Standard Election Observation Missions* for parliamentary and presidential elections. A standard mission always consisted of a core team based in Tbilisi and long-term observers who were deployed throughout the country approximately six weeks before the elections. The long-term observers were supported by short-term observers who observed Election Day procedures and specifically counted and tabulated the votes in the night of elections.

The OSCE/ODIHR observer missions assessed elections for thier compliance with principles for democratic electoral processes, including commitments agreed to by all the OSCE participating States, as well as national legislation. Observers closely monitored campaign activities, media coverage, the legislative framework and its implementation, the media situation, the work of the election administration and relevant government bodies, and the resolution of election disputes. Monitoring the coverage of the pre-electoral campaign in the media was an important integral part of the observers' work.

The OSCE/ODIHR observer missions were complemented by *monitoring teams* of the *Parliamentary Assembly of the Council of Europe* (PACE) that assessed the electoral campaign and the political climate in the run-up to the election and after the elections.

The Election Day was observed by OSCE/ODIHR and PACE short-term observers together with observers sent by the OSCE Parliamentary Assembly, the European Parliament, and the NATO Parliamentary Assembly. All these organisations together issued joint statements of the preliminary findings and conclusions on the day after the elections. These statements of preliminary findings and conclusions were delivered prior to the completion of the election process – before the satisfaction of all court proceedings and the announcement of the final results.

OSCE/ODIHR as well as PACE always issued final assessments of the elections that included the conduct of the remaining stages of the election process, i.e. the tabulation and announcement of final results and the handling of post-election day complaints or appeals. While OSCE/ODIHR issued final reports – including recommendations for potential improvements – approximately two months after the completion of the election process, PACE did come up with their reports within a weeks-time after the elections.

Irrespective of the timing of their publication, final reports often received only limited attention. They were mainly read by international donors that hoped to find in these reports some justification for their electoral assistance programmes. International attention was paid mainly to the *Statements of the Preliminary Findings and Conclusions* of the international election observation missions that were presented to the press on the day after the elections. These statements always replicated current geopolitical orientations and domestic politics, and as a result their findings did not necessarily reflect the overall assessment of conduct of the elections. Since 1999, international election observation missions have stated the following about elections in Georgia:

"The conduct of the 1999 parliamentary election *represents a step towards Georgia's compliance with OSCE commitments*. Despite some irregularities, it appears that Georgian voters were generally able to express their will."

"... *considerable progress* is necessary for Georgia to fully meet its commitments as a participating State of the OSCE." (2000 Presidential Elections)

"The 2 November 2003 parliamentary elections in Georgia *fell short of a number of international standards*. Delays and confusion over voter lists contributed to a lack of public confidence in the governmental and parliamentary authorities' capacity to manage an effective and transparent election process."

"The 4 January 2004 extraordinary presidential election in Georgia *demonstrated notable progress over previous elections*, and in several respects brought the country closer to meeting OSCE commitments and other international standards for democratic elections."

"The 28 March 2004 parliamentary elections in Georgia *demonstrated commendable progress* in relation to previous elections."

"While the 5 January 2008 presidential election in Georgia *was in essence consistent with most international standards* for democratic elections, significant challenges were revealed which need to be addressed urgently."

"Overall, these elections clearly *offered an opportunity for the Georgian people to choose their representatives from amongst a wide array of choices*." (2008 Parliamentary Elections)

Figure 1: Assessment of Electoral Processes in Georgia

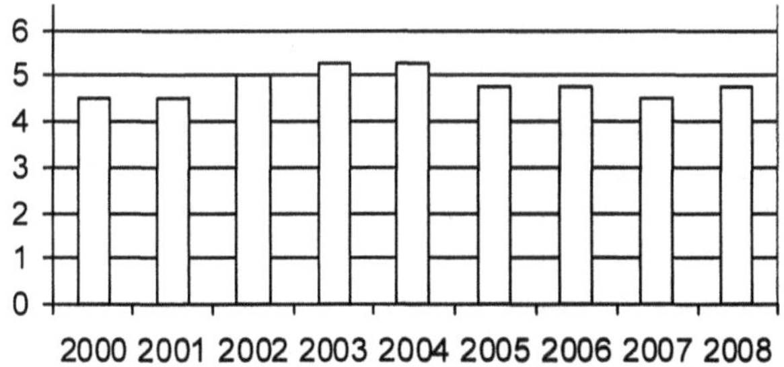

Source: Freedom House: Nations in Transit, Georgia 2008

Most of the above findings are reflected in the Freedom House ratings for Georgia. Freedom House, an independent nongovernmental organization that supports the expansion of democracy and freedom in the world, annually publishes ratings about electoral processes. These ratings are based on a scale of 1 to 7, with 1 representing the highest and 7 the lowest level of democratic progress. As the chart below indicates and as the above mentioned election observer findings state, Georgia saw a deterioration of electoral standards towards 2003/2004, and experienced a significant improvement after 2005 – a trend that, however, was discontinued in 2008.

Electoral standards reached an ever low point in November 2003 when Shevardnadze, having lost the support of his inner circle and of powerful influence groups, tried to win the parliamentary elections by any means. The international observers assessed these parliamentary elections negatively and especially pointed out that "acts of violence against individuals and the property of political parties, as well as instances of intimidation, marred the pre-election period" (International Election Observer Mission: Statement, 3 November 2003), as well as a "widespread and systematic election fraud during and after election day, most obviously in Adjara and Kvemo Kartli regions" (OSCE/ODIHR: Final Election Report Part 1, 24 January 2004). Having observed all these violations themselves, the Georgian people did not have any more confidence in the process' fairness during these parliamentary elections. Subsequently, it was easy for leading opposition figures and powerful influence groups to mobilise the population. After days of demonstrations and declaration of the state of emergency, Shevardnadze had to resign.

3 Key Challenges to Elections in Georgia

Since 2003 election standards improved. However, challenges remained. Among others, the following issues have been repeatedly criticised by international observation missions:

Before almost all elections, the *election code* was amended. Sometimes this happened a few months before the elections, sometimes – as in the case of the 2008 presidential elections – even several weeks (PACE Report Doc. 11502 rev, 23 January 2008) prior to the elections, leaving all election stakeholders with little or no chance to familiarise themselves with the electoral changes.

The electoral code – despite being frequently amended (18 times between January 2004 and May 2008) – was never assessed as being "in compliance with international standards and practices, and enjoying the confidence of all political interests and the electorate" (OSCE/ODIHR: Final Report, 9 July 2000).

Amendments to the elections code never resulted in the eradication of all pertinent inconsistencies, gaps and ambiguities. As a result, the Electoral Law became "far from perfect" and left "a lot of room for interpretation by the Central Electoral Committee" (PACE Report Doc. 9191, 13 September 2001).

There was always a blurred "*distinction between state and party resources*, with the incumbent party apparently enjoying privileged use of state resources" (OSCE/ODIHR: Final Report 1999 Parliamentary Elections). Often the implementation of social welfare programmes was combined with campaigning for the president. E.g. activists of the presidential party distributed vouchers as subsidies from the president during their canvassing in favour of the president (International Election Observation Mission: Statement, 5 January 2008). Having observed these patterns, OSCE/ODIHR voiced concern about the lack of a clear "separation of political party and State on the one hand, and Government and election administration on the other" (OSCE/ODIHR: Final Report, 9 June 2000).

Elections always saw a *heated competition* between the leading political parties and election blocs. Seldom, a constructive issue-based dialogue between the electoral contestants took place. Especially opposition candidates tried to attract the attention of the electorate by verbally attacking the incumbent and by coming forward with allegations of electoral violations, hereby persistently questioning the fairness of the electoral process. They rarely focussed on policy issues.

Occasionally the tone of the campaign went beyond acceptable limits of political competition. All elections saw "*instances of violence and intimidation*" (OSCE Election Observation Mission: Statement, 1 November 1999). Allegations, not always supported by strong evidence, persisted that police and local officials were putting pressure on opposition supporters to desist from campaigning, threatening them with arbitrary arrest or job dismissals. In isolated instances, hand grenade attacks on houses, beatings and even kidnappings were reported.

The *election administration* at central level never experienced institutional stability. Throughout the years it underwent numerous reforms. E.g. in 2003 the Central Election Commission had commissioners appointed on party basis – a provision that was abolished in mid-2005, and re-introduced in late 2007. Besides, the election administration experienced frequent changes of the head (January 2004-May 2008: six heads) which each time resulted in substantial staff changes and in the loss of institutional memory. Only very few heads of the Central Election Commission did enjoy the trust of the opposition and the people. Most of them were accused of being installed by the government and pursuing government interests.

Due to the institutional instability at central level, there were only limited efforts to build up a professional election administration at lower level. As a result, "the election administration at the district and precinct levels often failed to perform all their duties in a uniform, transparent and timely manner." (International Election Observation Mission:

Statement, 3 November 2003) Especially *vote counting and tabulation* failed to meet any standards (OSCE/ODIHR: Post-Election Interim Report, 6-18 January 2008).

Inaccurate *voterș' lists* were a recurrent problem. Despite repeated recommendations to address the issue, little progress was made until the 2006 local elections. Voters' lists were of special concern during the 2003 parliamentary elections when authorities failed to make "genuine efforts to produce accurate and reliable voter registers" (OSCE/ODIHR: Final Report, 28 January 2004). "Errors" denied eligible citizens the right to vote and inflated voters' lists furthered election fraud. Even after the 2003 elections, the review and update of voter registers remained an issue that was only partially addressed by the Georgian authorities.

After the 2008 parliamentary elections the OSCE/ODIHR Election Observation Mission concluded: "The voter registration has overall been improved in Georgia in the last decade. However, many stakeholders continued to express concerns regarding the accuracy of the voter list, claiming that the number of records was inflated due to records of deceased persons in the list, multiple records, and non-exiting buildings as registration addresses." (International Election Observation Mission: Statement, 22 May 2008)

4 Outlook: Raising Political Culture – Starting a Discourse with the People

For years, leading government officials assured the Georgian people and the Western community of their sincere intention to hold free, fair and transparent elections. However, they repeatedly failed to keep their promises. So far, none of the elections held in Georgia met international standards. They were characterised by opposition allegations of unfair practices and election fraud and by international and domestic observers' criticism of serious procedural violations.

Elections were always the battlefield of various influence groups which strove to ensure their access to the inner power circle on the one hand and of the ruling elites which wanted to keep their power position on the other hand. None of these election stakeholders was really interested in gaining the support of the people or forming alliances with the opposition. They were mainly interested in ensuring favourable election *results*, never in conducting fair electoral *processes*. It was about winning presidential and parliamentary elections with an overwhelming majority of the popular vote – by all means. This attitude led to alienation between the ruling elites on the one side and the opposition and the people on the other side. Nowadays (Summer 2008) neither the opposition nor the Georgian people have any confidence in the fairness of the electoral processes.

In Georgia, however, there exists not only distrust towards the ruling elites, but also distrust towards any kind of political elite. There is a broad gap between the country's political leadership, may this be position or opposition, and the Georgian people. Both position and opposition have repeatedly failed to learn from the people about their real needs and to subsequently address the people's problems in their party progammes. Despite repeated incentives of the political parties to gain the support of the majority of the population in elections, they never sought to become deeply rooted in the people or to reflect the people's interests in their programmes.

"In this context", as Wheatley states, "the publicly espoused political platform of parties (...) (is to be) understood as no more than window-dressing, devised to convince the population that the party really cared for them or to show Western governments that Georgia was really a democracy" (Wheatley 2005: 158). Understanding that elections were not about to win the "hearts" of the voters, but to ensure the smooth continuation of the "business" of powerful influence groups, the Georgian people lost their interest in elections as such. They turned to street actions as the only means of influencing the decision making of the political elite. As a result, "street democracy" became the only channel of communication of the people with the political elites.

While people exercise their "street democracy" on a random basis, the political elite communicates to the people on a daily basis through TV and radio. Political parties and politicians reach out to the voters through public debates in which high-ranking representatives of the government and the opposition get involved in direct and heated disputes. These debates are not about issues, but about personalities. Personal attacks on the political counterpart dominate. Another channel that is widely used by the opposition is through events that attract the attention of the media. These may be press conferences at which the opposition protests against specific government actions, physical attacks on government politicians or protest rallies.

None of the parties, however, is really interested in opening this one way street for a two-way traffic by conducting profound research about the people's needs, addressing them in their party programmes, and running an issue-based campaign. As long as these patterns continue to persist in the political discourse, apathy and absenteeism will prevail from the side of the voters and "street democracy" will be the only way of communication of the people's concerns towards the political leadership of the country. In order "to instill broad confidence amongst election stakeholders and the public, a continuing commitment on the part of the Government, as well as from all other political actors (...) (will be) required" (International Election Observation Mission: Statement, 22 May 2008).

5 References

Marie-Carin von Gumppenberg/Udo Steinbach (ed.): Der Kaukasus. Geschichte, Kultur, Polititk. Muenchen 2008

Giorgi Kandelaki/Giorgi Meladze: Enough! Kmara and the Revolution in Georgia. In: Joerg Forbrig/Pavol Demeš (ed.): Reclaiming Democracy. Civil Society and Electoral Change in Central and Eastern Europe. Washington 2007, p. 101-125

OSCE Mission to Georgia/Council of Europe (ed.): Georgia. Local Elections 2006. Lessons learned. Tbilisi 2007

Ghia Nodia: Georgia, in: Freedom House: Nations in Transit 2008, p. 231-250, in: http://www.freedomhouse.hu/images/fdh_galleries/NIT2008/NT-Georgia-final.pdf

Ghia Nodia/Álvaro Pinto Scholtbach (ed.): Political Parties: The Political Landscape of Georgia. Achievements, Challenges and Prospects. Delft 2006

Jonathan Wheatley: Georgia from National Awakening to Rose Revolution. Delayed Transition in the Former Soviet Union. Hampshire/Burlington 2005

Internet sites

www.cec.gov.ge

www.osce.org/odihr-elections

www.coe.int

www.freedomhouse.org

Election Reports

International Election Observation Mission: Georgia — Parliamentary Elections, 21 May 2008. Statement of the Preliminary Findings and Conclusions. Tbilisi 22 May 2008.

International Election Observation Mission: Georgia — Presidential Elections, 4 January 2008. Statement of the Preliminary Findings and Conclusions. Tbilisi 5 January 2008.

International Election Observation Mission: Parliamentary Elections — Georgia, 2 November 2003. Statement of the Preliminary Findings and Conclusions. Tbilisi 3 November 2003.

OSCE Election Observation Mission: Georgia — Parliamentary Elections, 31 October 1999. Preliminary Statement. Tbilisi 1 November 1999.

OSCE/ODIHR Election Observation Mission: Georgia – Extraordinary Presidential Elections, 4 January 2008. Post-election Interim Report 6-18 January.

OSCE/ODIHR Election Observation Mission: Georgia — Parliamentary Elections, 2 November 2003. Final Report, Part 1. Warsaw 28 January 2004.

OSCE/ODIHR Election Observation: Georgia — Parliamentary Elections, 31 October 1999 & 14 November 1999. Final Report. Warsaw 7 February 2000.

OSCE/ODIHR Election Observation: Georgia — Presidential Elections, 9 April 2000. Final Report. Warsaw 9 June 2000.

Parliamentary Assembly of the Council of Europe: Doc. 11502 rev, Honouring of obligations and commitments by Georgia. Report. 23 January 2008.

Parliamentary Assembly of the Council of Europe: Doc. Doc. 9191, Honouring of obligations and commitments by Georgia. Report. 13 September 2001.

STATENESS AND DEMOCRATISATION IN GEORGIA

NINO CHKOIDZE

1 Introduction

The non-violent change of government in November 2003, known as the Rose Revolution, raised hopes for an advance of democratisation and a strengthening of the state capacity in Georgia. The change in regime, produced by massive public protests, powerfully revealed major weaknesses of the Georgian state: inefficient and corrupt public institutions, election frauds, restricted access to political processes, dysfunctional formally democratic institutions (Huber 2004).

In the first four years after its declaration of independence in 1991, Georgia experienced two secessionist wars in South Ossetia and Abkhazia as well as the bloody overthrow of the first Georgian president Zwiad Gamsakhurdia. The country suffered from the sharp economic decline triggered by hyperinflation, a misleading institutional incentive structure, asset-stripping, neglected investments in infrastructure and insufficient provision with public goods. Despite this poor performance, the Georgian political elite under the former Soviet Foreign Minster and the second Georgian President Eduard Shevardnadze was able to maintain the power in every election between 1992 and 2003 by means of keeping a kind of fragile stability, which was flexible enough to adapt to changing circumstances by fluid alliances, altering institutional structures and incoherent policies (Christophe 2004). The massively fraudulent parliamentary elections of 2003 played the role of a catalyst for a clearly needed change. As a consequence of a series of mass protests President Shevardnadze was forced to resign. With the overwhelming majority of votes, the new presidential and parliamentary elections in early 2004 brought to power the charismatic leader of the November protests, Mikheil Saakashvili, and established the parliament dominated by the parties connected with him. Although some observers assessed the events of 2003 rather sceptically and qualified it as an intra-elite struggle for power (Hale 2005), as a change induced to some extent internationally (Küchholz 2005) or even as a coup d'etat (Fairbanks 2004), there were high expectations that the new government would implement urgently needed reforms. Especially the unfinished state-building or weak stateness was seen as a main impediment for the sustainable democratisation (Linz/Stepan 1996).

This paper examines the changes in post-Rose Revolution Georgia in terms of the state functions and democratisation. In the first chapter the concept, which delivers the indicators for their assessment, will be presented. Afterwards follows the description of each function. The conclusion summarizes the main findings.

2 Stateness and Democracy – Definitions and Characteristics

Since 1990 Georgia has been considered as an example of a state hardly capable of accomplishing its main functions: the epithets "failing" (Halbach 2006a), "hybrid" (Koehler/Zürcher 2004) or "weak" (Huber 2004) have been applied to characterise the underperformance of the Georgian state. According to Max Weber, the crucial prerequisite for stateness is a government executing its legitimate monopoly of power, i.e. a government which is able to enforce its political decisions, if needed even by taking recourse to the means of coercion (Weber 1980). The modern understanding of state, however, goes beyond this minimum requirement of the sovereignty and includes the further dimensions of statehood like constitutional and administrative order, rule of law, welfare or social state (Schneckener 2006; Milliken/Krause 2003; Grimm 1996). According to this perception, stateness is defined by three core functions: *security, welfare* and *legitimacy/rule of law* (Schneckener 2006).

Schneckener (2006) formulates a set of indicators for each of the three state functions, by means of which the degree of eroding stateness can be assessed. The level of *security* provision is a crucial indicator for the classification. The central argument for this is that without a reasonable level of secure environment no sustainable development in other areas can be achieved. The essence of the security function is the control of a territory through state monopoly power. This comprises the ability of a government to control resources and the existence of a national army or police for reconciliation of local conflicts and disarming of violent private actors. However, the monopoly power shall not be abused and become a threat to the majority of citizens or for certain population groups. The indicators of the security functions are: degree of control over the entire national territory, degree of control of external borders, ongoing or repeated violent conflicts (such as separatism), the number and political relevance of non-governmental violent actors, actual situation of the state security sector, crime rate, degree of threat executed by governmental bodies towards the physical safety of citizens (torture, deportation etc.).

Welfare function focuses on state transfer payments and public services as well as on distribution mechanisms of economic resources, which are, as a rule, funded by tax revenues (customs duties, taxes, fees, charges, etc.). This function includes the entire state activities in the fields of social and economic, employment, education, health and environmental policies as well as the establishment and maintenance of public infrastructure. The welfare function of the state can be measured by the following indicator: degree of participation of certain population groups in economic resources, continuing economic and/or monetary crises, amount of tax and custom revenues, amount and distribution of government expenditure, amount of external debts, inequality level (such as low state redistribution, urban-rural inequality), unemployment or labour force participation rate, level of human development (HDI development), state of social security systems, state of infrastructure, education and health care.

Legitimacy/rule of law function covers the forms of political participation and decision-making procedures, stability of political institutions and the quality of the judiciary system and public administration. To this area belong the questions concerning the form of political order and regime type. The indicators are: the extent of political freedoms (including expression, freedom of assembly), political participation rights (including active / passive voting rights, competition for offices), dealing with the political opposition; extent of election fraud, degree of political participation of certain population groups (such as minorities), existence of serious human rights violations (torture), acceptance of the regime or of the political order, degree of judicial independence, state of public administration, the extent of corruption and clientelism. Such a normative understanding of the stateness function implies that the states with authoritarian or semi-authoritarian regimes are considered as fragile, contrary to the assumption of being "strong" in terms of the military potential of some of them (Schneckener 2006). Some aspects of stateness are produced selectively, as some state resources are being invested in certain core state functions in order to achieve the regime stability. The concept indicates that certain regimes jeopardize statehood and that one should distinguish between regime stability and state stability. Thus, this model indicates that only states with a certain degree (relatively high compared with the neighbouring states in the region) of democratic structures can be consolidated, i.e. be a sound state.

The indicators of *legitimacy/rule of law* largely correspond with the defining elements of democracy within the approaches of the political regime type research, which became especially in demand in the course of the democratisation waves of the last decades (Pickel/Müller 2006). Alongside other regions, the post-Soviet area, except the Baltic States, experienced the emergence of so-called hybrid political systems, i.e. of regimes that cannot be definitely classified as either fully democratic or authoritarian (Karl 1995; Diamond 2002; Ottaway 2003). In his well-known concept of polyarchy, Robert Dahl defines democracy as "contestation open to participation" (Dahl 1971), i.e. as a system featuring two interdependent dimensions: political participation and political competition. In search of the proper identification of the proliferated formal democracies this positive and empirically easily measurable definition has been extended by the additional dimension, which specifies the constitutional order and the rule of law (Merkel / Croissant 2000). In this regard the "defective democracies" are political systems which combine the existence of free electoral regimes with various traits of authoritarianism such as limited function logic of institutions safeguarding fundamental political participation rights and civil liberties, restrictions in check and balances or constraints in the power of the legitimate authorities (ibid).

The concept of stateness encompassing the above mentioned three functions has been criticised for being too much oriented to the criteria of OECD world standards and for assessing not only the elementary (Weberian) state function, but also its *governance* performance. Alongside the *security* function, the policy-outcomes within the *welfare* function as well as the institutional architecture assessed by the *legitimacy/rule of law* indicators are being considered as the defining characteristics of stateness. Given the fact that in two thirds of today's world's nation-states the deficits in the fulfilment of the three dimensions can be observed, state fragility or limited statehood seems to be rather a norm than a deviation (Risse/Lehmkuhl 2007). Nevertheless, this concept should be understood as a relative benchmark. In particular cases, the state function characteristics can be weighted against each other (Schneckener 2006). Following Merkel's concept, the "embedded democracy", i.e. a fully fledged democratic system, requires the effective power of the elected authorities to be guaranteed. Although this characteristic can be interpreted as the stateness feature, the concept develops a framework which primarily aims to classify a wide range of the mixed regime types. The question of stateness, even in its most basic meaning in Weberian tradition, is not the research object. Schneckener's indicators, on the other hand, seem to be a more appropriate tool to address the questions formulated above.

3 Assessment of State Functions in Georgia before and after the Rose Revolution

3.1 Security

After the collapse of the Soviet Union independent Georgia faced severe difficulties in terms of safeguarding its territorial sovereignty. After the armed conflicts between 1991 and 1993 two breakaway regions, South Ossetia and Abkhazia, have functioned as virtually independent states (Huber 2004). The areas of both secessionist regions constitute about 15 percent of Georgia's territory. In the rest of the country the minorities, Azerbaijanis in the southern Kvemo Kartli region and Armenians in the Samtskhe–Javakheti region, represent 16.3 percent of the population (Coppieters

2005). They are rather poorly integrated in the political and economic life of the country. Some mountainous regions, like the northern part of Swaneti or Pankisi Gorge, where, after the start of the second Chechen war in 1999, large numbers of Chechens lived, also fell beyond Tbilisi's control (Lynch 2006). Although Adjara, the south-western Black Sea region, governed by the authoritarian leader Aslan Abashidze until 2004, did not strive for secession, it refused to transfer taxes and duties to the central budget and retained far-reaching fiscal, economic and political autonomy. The reunification of the Georgian multiethnic state became the Shevardnadze regime's main challenge. The negotiations to solve the "frozen conflicts", however, were mostly deadlocked due to the difficult confidence-building process and the controversial interference from the side of the Russian Federation.

The separatist conflicts were closely related to the limited capability of the Georgian state to properly control the country's external borders. This entailed considerable negative impacts on the country's security (Fritz 2006). Russia, accusing Georgia in supporting Chechen rebels located in Pankisi Gorge, launched military operations on Georgian territory, thereby undermining the latter's sovereignty (Jawad 2006). The so-called *no-go areas* near secessionist regions like Kodori valley or the Gali district have been considered as safe havens for illegal activities (ibid.). The lack of demarcation of the former internal Soviet borders with Russia, Armenia and Azerbaijan further complicates an efficient border management (Huber 2004).

The secessionist conflicts in Georgia occurred against the backdrop of the acute political instabilities. After the coup against the Gamsakhurdia regime, the country fell into chaos and anarchy marked by open looting and the loss of the state monopoly on violence (Jones 1993). The erosion of state security structures between 1991 and 1994 was accompanied by the increased influence of the non-state violent actors. Besides the secession forces in the conflict regions some Georgian partisan formations also belonged to those actors. Given the lack of efficient state security organs, their formations were often even charged with military and rudimentary state security functions (Halbach 2006a). After Shevardnadze's entrance into Georgian political life the non-state violent actors were largely deprived of power. A certain extent of stability was achieved. However, the widespread corruption and insider deals, which run like red threads through the entire Georgian state sector, pervaded the security sector institutions too (Koyama 2002 citied in Halbach 2006a). The Ministry of Defence offered a wide range of opportunities to earn money by selling the genuine public good, like protection, in exchange for bribes (Fairbanks 1995; Huber 2004). The absence of a national security strategy, the lack of professional experience and reform willingness, and fiscal shortages led to the emergence of a venal police as well as to poorly equipped, hardly trained and low-paid army forces and border guards (Fritz 2006).

The unresolved question of the responsibilities in relation to the secession regions made them especially prone to criminal activities. Trafficking in drugs and weapons, and participation in smuggling through their territories became a profitable business for the criminal groups of all conflict sides: Russians, Georgians, Abkhazians and Ossetians (Kukhianidze et al. 2007). Organized crime in Georgia, which many associate with the Soviet past, has not disappeared, but has instead undergone a transformation during the tumultuous period of the 90s. Through their links to high level protectors in Parliament, the State Chancellery and law enforcement, the professional criminals have managed to become the beneficiaries of the property distribution processes (Shelley 2007) and to play the central role in the capture of the state, the mode of governance which was peculiar to the highly corrupted post communist states (Hellman et al. 2000).

After the Rose Revolution the new government under president Saakashvili declared the restoration of Georgia's territorial integrity as the first priority and expressed a desire to "consolidate" the country by seeking resolutions (German 2006). The first success was the re-integration of Adjara into the Georgian fold. As a consequence of the strategy of pressure imposed by Tbilisi, the Adjarian leader Abashidze escalated the conflict unacceptably, a process that eventually caused his flight to Moscow (Lynch 2006). In contrast to South Ossetia and Abkhazia the conflict with Adjara did not have any ethnic dimension as the inhabitants have never questioned their affiliation with Georgia.

In July 2006 the government of Saakashvili managed to oust the local authority and warlord Emzar Kvitsiani[1] and gain control in all villages of Kodori Gorge. The area was given the new name of Upper Abkhazia and turned into the temporary administrative centre of the Abkhazian government in exile.

In May 2007 an interim administrative unit was created in South Ossetia. In the alternative poll supported by the Georgian government, Dmitriy Sanakoev, who advocated an autonomous South Ossetia within Georgia, was elected president of South Ossetia (Nodia 2008).

In an effort to arm the forces according to NATO standards, military expenditures were substantially increased from the former level of 0.5 in 2003 to 3 percent of GDP in 2006 (Halbach 2006b). In 2007, Georgia's defence spending was even roughly 10 percent of GDP, putting Georgia on the second place in the world in terms of the military spending

[1] Emzar Kvitsiani rearmed his former paramilitary group and defied control by Georgia's central authorities.

share (UNDP 2008). With the financial assistance and advising and training programmes of the US, the new government could impel the modernisation of the Georgian military force. The reforms of the Ministry of Defence, reorganisation of the Ministry of Internal Affairs (MIA) and its merger with the former Ministry of State Security as well as the integration of the Border Guard Department into the MIA were important steps in the clarification and streamlining of functions, tasks and responsibilities of each service (Fluri/Lortkipanidze 2006).

Georgia's capacity to fight organized crime has improved too. Dozens of Georgian high-level criminals, so-called *thieves in-law*, have been arrested domestically and abroad (Shelley 2006). Information sharing with foreign governments, determined prosecutors and improved investigation capacities have facilitated concrete actions against organized crime. The reforms of state bureaucracy and increased salaries for public service officials have been important in overcoming flourishing corruption and economic crime (Machavariani 2006). The dismissal of sixteen thousand law enforcement personnel broke the institutionalised bonds between them and the criminals at the operational level (Shelley 2007). Georgian legislation and law enforcement against smuggling and shadow economy has enhanced. A visible example is the elimination of Ergneti illegal markets in the South Ossetian conflict zone[2] (Kukhianidze et al. 2007).

The described developments show the new government's clear efforts to improve the security situation. However, the attempts to repeat the success of Adjara in other breakaway regions resulted in conflict escalation and increasing tensions with Russia. In August 2008, the war in South Ossetia, which ended in a large-scale Russian military intervention across Georgia, has of course substantially wreaked Georgia's security status and reignited hostilities. The unilateral recognition of the separatist regions as independent states by Russia has further compounded the conflict solution perspectives.

2 The wholesale market in Ergneti, near Tskhinvali, was the intermediating point for the smuggling of goods from Russia. According to the estimation of experts, only 10 percent of the funds extracted from the market were transferred to the Georgian budget (Kukhianidze et al. 2007). In June 2004, the Georgian section of Ergneti market was closed.

3.2 Welfare

After 1991 the Georgian living standard declined dramatically as the consequence of civil war, inefficient governance and the dissolution of previous economic and trade ties. Measured by per capita consumption and real income, Georgia, as part of the USSR, was among the richest Soviet Republics along with Estonia and Latvia (Jawad 2006). The major impediment of economic development can be seen in the secessionist conflicts. Furthermore, the structural problems, such as backlog in reforms, influence of entrenched interests, poor infrastructure, pervasive corruption and shadow economy, made it enormously difficult to normalize the economy. The country suffered from the negative effects of brain drain as the most educated and productive members of Georgian society emigrated. Two wars created an influx of internally displaced people who needed state support. By 2003, the economy had deteriorated to only 41% of its real GDP level in 1989 (EBRD 2004). In this period the output fell by 70 percent and exports by 90 percent. It was the worst decline suffered by any transition economy (World Bank 2008). The hyperinflation was fuelled by the poor fiscal performance and undue monetary policy. In the middle of the 90s a group of reformers inside Shevardnadze's government started to successfully launch an IFM supported reform and stabilization program. In 1995, the Georgian Lari was introduced. Macroeconomic stabilization was achieved and the GDP began to grow. Nevertheless, recovery of the economy remained uncertain.

Stagnation contributed to the lack of state transfers and public services (Huber 2004). The tax collection level was extremely low, for instance in 2003 tax revenue was only 13.9 percent of GDP (UNDP 2008). Instead of public goods the state supplied so-called *network goods* to those who were part of a particular personalized network that linked them to an individual in a position of state power. Inequity and low levels of social participation were the results. Especially people without access to the personalized network, mainly in rural areas, became almost entirely disengaged from the public sphere (Wheatley 2005). Though, in 2001, the Gini Index[3] was estimated at 36.9 by UNDP and could be classified as medium rather than high, poverty was unequally distributed throughout the country and in many cases correlated with remote areas and low density of arable land (BTI 2006). According to the national statistics, the share of

[3] The Gini Index measures the distribution of income. A low Gini Index indicates more equality, while a high Gini Index indicates more unequal distribution. As a comparison, the Gini Index of Germany is 28.3.

the population living below the poverty level amounted to 54 percent in 2003, the percentage of extremely poor was 15-17 percent.

Due to low tax revenues the financial resources were limited, which led to underfunded infrastructure, education and health as well as a delay in the payments of social benefits and pensions. The basic infrastructure, roads or irrigation systems, were ruined. The supply of utilities like water and electricity was at best unreliable. Despite the fact that Georgia maintained an above average level of primary, secondary and tertiary education and literacy (UNDP 2008), the quality of the offered education began to deteriorate. Financing of education decreased from 7 percent of GDP in 1991 to 1 percent in 1994 (UNICEF 1999); in 2003, it still accounted for only 2.1 percent (UNDP 2008). Educational institutions suffered not only from the worsened physical infrastructure, but also from extremely low salaries of the teaching staff. Public spending on health was also relatively low: in 2003–2004 it amounted to only 1 percent of GDP. At the same time it was mostly limited to hospital care, a fact that rather benefited the relatively well off part of the population (ADB 2007).

After the Rose Revolution, Georgian economy experienced remarkable positive changes. One of the successes was a strong growth in GDP per capita, from 5.9 percent in 2004 to around 10 percent in 2006 and 12 percent in 2007 (UNDP 2008). A wide range of the implemented structural reforms, like the simplified tax code, amendments to laws dealing with the registration of real estate, ownership enforcement proceedings, lower import duties and lighter bureaucratic burdens for foreign investments, led to a significant improvement of the business climate. According to the World Bank's Ease of Doing Business survey, Georgia's rank moved from the 112th position in 2006 to the 15th in 2008 (World Bank 2008). Due to the more favourable environment, the share of the informal economy, which constituted 67% a few years ago, has been constantly shrinking (BTI 2008). By cutting down the bloated staffing and government bureaucracy, implementing several anti-corruption laws and initiating the process of confiscating unproved property, impressive results were achieved in terms of fighting corruption. In 2004, Georgia was given rank 130 on the corruption perception index, according to Transparency International. In 2008, the country moved forward to the 67th place. The tax receipts increased significantly from GEL 1.2 billion (USD 558 million) in 2003 to GEL 4.4 billion (USD 2.6 billion) in 2007 (UNDP 2008). Combined with considerable increases in revenue from privatisations, the central government was able to substantially raise its spending.

Infrastructure reforms have gathered pace in recent years. On the energy sector the problem of power shortages was alleviated and the financial situation of energy utilities was improved. The government financed the renovation of the hydroelectric power stations across the country. Among them, Inguri hydro dam is the most significant. Today it provides 40 percent of Georgia's electricity needs (UNDP 2008). Furthermore, the state intensified its efforts to optimise the management of drinkable water supply and to increase the funding of municipal infrastructure. The allocation of resources to the transport sector, particularly for the rehabilitation of road infrastructure, has also been stepped up significantly (ADB 2007).

Besides the productive and infrastructure sectors, the social targets of public policy, like education, healthcare and social protection, gained more attention. Measures taken in the higher education sector include the introduction of the merit-based selection and funding of the students through unified national examinations, re-examination of university teachers, substantial rise in salaries and significant reduction in the number of private universities, as a result of the demand to renew their licenses (BTI 2008). The reforms in healthcare have focused on the following elements: expansion of primary health care and of the health assistance targeted to socially vulnerable people, development of insurance and privatisation of public healthcare facilities. Despite the increased public expenditures in the last years, the spendings on health and education are still fairly low compared to international standards. There seems to be a similar assessment of several institutions that no progress can be reported in terms of poverty reduction (UNDP 2008). Despite the doubling of pensions and timely payment of social transfers, the poorly developed social net is not able to cope with the widespread poverty. The dramatic increase of consumer prices has further worsened the situation (BTI 2008). The overall enhancement of the economy and business climate could not reach the very poor in the short run, but has set the groundwork for future improvements.

3.3 Legitimacy ,Rule of Law

Due to its ambiguous character, the Georgian political system has been characterised as "Potemkin Democracy" (King 2001) or "Defective Democracy" (Croissant/Thiery 1999). The formal transformation from authoritarian to democratic system in post-communist Georgia was completed at the latest in 1995 with the election of Shevardnadze for head of state (and hence the decision for a presidential system of government). However, the consolidation of democratic regimes turned out to be a difficult task. Shevardnadze was mediating between different corrupt groups in order to maintain the general stability of the system, but he did not try to enforce the rules of law (Nodia 2005). To justify himself in the eyes of international donors he was engaged in convincingly simulating democratic reforms, but at the same time he was reluctant to implement them (Christophe 2004). The legitimacy of Shevardnadze's regime and the trust in his government diminished rapidly. This was mainly caused by the failure of the state to fulfil its core functions and not by the dictatorial character of the regime.

Although the Georgian political system was presidential with a rather strong executive power and not clearly separated responsibilities, the parliament had more legislative competences and was more transparent than its counterparts in most other member countries of the Commonwealth of Independent States (Halbach 2006a). To some extent, a pluralist party system evolved but remained weak and rather person than program-oriented. The dense network of Non-governmental Organizations compensated for this handicap. A vibrant free media played a valuable role in political life (Mitchell 2006).

The civil society and critical media were the backbones of the mass protests which followed after the parliamentary election fraud in 2003 became evident. President Shevardnadze resigned and the group of the "young reformers" led by Saakashvili came to power. The Rose Revolution was assessed by many observers as a "second wave of democratization jump" (McFaul/Tatic 2005 as cited in Kalandadze 2006). However, the new governments' records on issues such as building democratic institutions, strengthening government accountability and supporting civil society are not really successful (Mitchell 2006).

In February 2004, an amendment to the constitution was introduced that significantly enhanced the authority of the executive branch and weakened the parliament. The president acquired the right to dissolve parliament, for instance in the case that the legislators deliver three successive no-confidence votes to the cabinet. Secondly, the stable majority of the pro-presidential party "National Movement" in parliament limits legislature's ability to exercise its oversight functions (BTI 2008). Furthermore, many criticise the undue hastiness displayed by parliament, as a large amount of laws have been adopted without sufficient deliberation (Nodia 2008).

The hard-line approach of the new government, particularly in the context of the struggle against corruption, endangered civil liberties. The fact that only very few of the high profile arrests in 2004 have been brought to court has raised concerns with regard to the transparency of these measures and evoked the doubt that individuals were targeted for political reasons (BTI 2008). After the Rose Revolution the judiciary has been even less capable of withstanding the pressure on judges and the interference into court affairs by the executive power. The new elite has earned much criticism due to the setbacks in the implementation of legal reforms.

While prior to the Rose Revolution the civil society organisations used to play an important role in mobilizing the support for democratic causes, their visibility has declined over the last years as many of the most famous representatives have become members of the government. Most of the NGOs ceased to play the role of watchdogs with regard to the new political elite. The development in the media has been even more worrying. A number of formerly independent media outlets, the television station

Rustavi 2 as the most notable example, have been accused of becoming far less independent and critical (Schlindwein 2007). The private media owners systematically use the outlets in order to advance their political agenda instead of respecting journalist's editorial freedom (Nodia 2008).

The government has often been criticised for its ignorance of any oppositional opinion and refusal to enter into a dialogue with the members of the opposition (Lynch 2006). The arrest on corruption charges of Irakli Okruashvili, a former Minister of Defence who became an opposition politician, provoked a series of rallies. The demonstrators protested against the controversial decision to extend the term of the standing parliament from April to October–December 2008 and criticised Saakashvili's strong presidentialism as a trend towards an authoritarian regime (Nodia 2008). Lots of participants joined the rallies in order to articulate their dissatisfaction with socio-economic policies. Justified by the argument that the demonstrations posed a threat of government overthrow, the demonstrations were dispersed. Furthermore, the government suspended broadcasting of the two major opposition-oriented TV stations and introduced a state of emergency that lasted nine days. To deescalate the situation the new presidential elections were appointed for January 5, 2008. Saakashvili was re-elected to his second term with 53.47% of the votes. Levan Gachechiladze, the candidate nominated by the nine-party opposition coalition, had the assignment to transform Georgia into a parliamentary republic in the case of victory. As a rather inexperienced challenger and compromise candidate of the opposition, Gachechiladze was not capable to counterweigh the incumbent president. In the parliamentary elections in May 2008, the ruling party of Saakashvili won 119 of the 150 parliamentary seats and guaranteed itself the majority of two thirds, allowing the implementation of constitutional amendments. Such a result facilitates the fears of the limited political dialogue and can be considered as an impediment to the development of a pluralistic democracy.

The crisis at the end of 2007 had a significantly negative influence on the democratic development in Georgia. In the annual major findings of Freedom House, Arch Puddington stated "...disturbing was the substantial reversal suffered by Georgia, a country that had made important strides toward democracy since its 2003 Rose Revolution. Georgia saw both its political rights and its civil liberties ratings decline due to the imposition of a state of emergency following antigovernment demonstrations in November, restrictions on press freedom, and a systematic campaign to marginalize the political opposition" (2008).

4 Conclusion

The government of President Saakashvili inherited a "failing state" afflicted with separatism, corruption, economic decline and diminishing legitimacy. All key components of the *security* and *welfare* functions have been observed to be unsatisfactory. Despite major deficits in terms of *legitimacy/rule of law*, the democratisation in Georgia was seen as relatively advanced compared to the other countries in the proximal region.

As summarized in table 1, most functions are still only rudimentarily fulfilled after the change of government in November. There are a number of achievements within the *security* and *welfare* functions. The reforms in the security sector, modernization of public administration and combating widespread corruption have been assessed as positive. Despite the fact that the "trickle down" effect of economic growth was only very limited, the government could restore financial order, increase tax revenues, overcome the budgetary crisis and pay off the (...) debts to pensioners and public service staff. With regard to the achievements in the third function, *legitimacy/rule of law,* the results of Saakashvili's government are at best modest. The executive branch has been further strengthened at the expense of the parliament. The Civil society and media no longer play the role of government's watchdogs.

Table 1: The functions of stateness in Georgia

State functions	Before Rose Revolution (overall assessment)	After Rose Revolution (overall assessment)	Trend
Security	-/+	-/+	⇑
Degree of control over the entire national territory	-/+	-/+	⇑
Degree of control of external borders	-/+	-/+	⇑
Ongoing or repeated violent conflicts	-/+	-/+	⇔
Number and political relevance of non-governmental violent actors	-/+	-/+	⇑
Actual situation of the state security sector	-/+	+/-	⇑
Level of crime rates	-/+	-/+	⇑
Degree of threat executed by governmental bodies towards the physical safety of citizens	+/-	+/-	⇔

Table 1 : The functions of stateness in Georgia (continued)

Welfare	-/+	-/+	⇑
Degree of participation of certain population groups in economic resources	-/+	-/+	⇑
Continuing economic and / or monetary crises	-/+	+/-	⇑
Amount of tax and customs revenues	-/+	+/-	⇑
Amount and distribution of government expenditure	-/+	+/-	⇑
Inequality level	-/+	-/+	⇔
Amount of external debts	-/+	-/+	⇔
Unemployment or labour force participation rate	-/+	-/+	⇔
Level of human development (HDI)	+/-	+/-	⇔
State of social security systems	-/+	-/+	⇑
State of infrastructure	-/+	+/-	⇑
Education and health care	+/-	+/-	⇑
Legitimacy/*Rule* of Law	-/+	-/+	⇔
Extent of political freedoms (including expression, freedom of assembly)	+/-	-/+	⇓
Political participation rights	-/+	-/+	⇔
Dealing with the political opposition	-/+	-/+	⇓
Extent of election fraud	-/+	-/+	⇔
Degree of political participation of certain population groups (such as minorities)	-/+	-/+	⇔
Extent of judiciary independence	-/+	-/+	⇓
State of public administration	-/+	+/-	⇑
Extent of corruption and clientelism	-/+	+/-	⇑

Explanations: +/- functioning to some extent, -/+ functioning only rudimentary; ⇑ improvement

⇓ negative development; ⇔ no notable change

Source: Indicators are derived from Schneckener (2006); own assessment

The dispersion of the oppositional demonstrations clearly contradicts the proclaimed aim of the Rose Revolution leaders to reinforce democracy. These mixed results rather indicate the efforts to preserve the existing regime and not to promote democratisation. Whether such a strategy will be conducive to the successful long-term state-building with a high legitimacy is more than questionable.

5 References

Asian Development Bank ADB (2007): Country Economic Report: Georgia

Bertelsmann Transformation Index BTI (2008): Georgia Country Report, http://www.bertelsmann-transformation-index.de/150.0.html 19. 11. 2008

Christophe, Barbara (2005): Understanding politics in Georgia. DEMSTAR Research Report No.2, Department of Political Science, University of Aarhus

Coppieters, Bruno (2005): Locating Georgian Security in: Coppieters, Bruno/Legvold, Robert (eds): Statehood and Security: Georgia after the Rose Revolution, American Academy Studies in Global Security

Croissant, Aurel/Thiery, Peter (1999): Demokratien und Defekte Demokratien. Zur konzeptionellen Revision des Demokratiebegriffs in der Transformationsforschung. Arbeitspapier der Forschungsgruppe „Defekte Demokratien". Ruprecht-Karls-Universität Heidelberg.

Diamond, Larry (2002): Thinking about Hybrid Regimes; Journal of Democracy 13, 2 (2002)

European Bank for Reconstruction and Development EBRD (2004): Transition Report 2004: Infrustructure in Transition, London

Fairbanks, Charles H., Jr. (2004): Georgia's Rose Revolution, Journal of Democracy 15, 2, April

Fluri, Philipp H./Lortkipanidze, Shorena (2006): Introduction: Georgian Security Sector Governance after the Rose Revolution, in: Fluri, Philipp H./ Darchiashvili, David (eds.): After Shevardnadze: Georgian Security Sector Governance after the Rose Revolution, The Geneva Centre for the Democratic Control of Armed Forces

Fritz, Antje (2006): Security Sector Governance in Georgia: Status and Perspectives, in: Fluri, Philipp H/ Darchiashvili, David (eds.): After Shevardnadze: Georgian Security Sector Governance after the Rose Revolution, The Geneva Centre for the Democratic Control of Armed Forces

German, Tracey (2006): Abkhazia and South Ossetia: Collision of Georgian and Russian Interests. Ifri, Paris http://www.ifri.org/files/Russie/german_anglais.pdf 11.11.2008

Grimm, Dieter (ed.) (1996): Staatsaufgaben, Frankfurt am Main, Suhrkamp

Halbach, Uwe (2006a): Georgien: Staatsversagen als Folge von Korruption und territorialer Desintegration, in: Schneckener, Ulrich (Ed.): Fragile Staatlichkeit „States at Risk" zwischen Stabilität und Scheitern, Baden-Baden, 229-253.

Halbach, Uwe (2006b): Säbelrasseln und Friedenspolitik in Europas neuer Nachbarschaft, SWP-Aktuell 32, Berlin (SWP), July 2006

Hale, Henry (2005): Regime Cycles: Democracy, Autocracy, and Revolution in Post-Soviet Eurasia, World Politics 58:1

Hellmann, Joel/Jones, Geraint/Kaufmann, Daniel (2000):. Seize the State, Seize the Day:State Capture, Corruption, and Influence in Transition, Washington, D.C: World Bank. Policy Research Working Paper 2444

Huber, Martina (2004): State-building in Georgia. Unfinished and at Risk? Den Haag (Netherlands Institute of International Relations, "Clingendael" Conflict Research Unit)

Jawad, Pamela (2006): Diversity, Conflict, and State Failure: Chances and Challenges for Democratic Consolidation in Georgia after the "Rose Revolution", Cornell University Peace Studies Program

Jones, Stephen F (1993): Georgia: a failed democratic transition", in: Bremmer, Ian/Ray, Taras (eds.): Nations and Politics in the Soviet Successor States, Cambridge: Cambridge University

Kalandadze, Katya (2006): Populism vs. Democratization. The Georgian Rose Revolution, http://www.maxwell.syr.edu/moynihan/Programs/ces/Kalandadze.doc 03.11.2008

Karl, Terry Lynn (1995): The Hybrid Regimes of Central America, Journal of Democracy, vol. 6, no. 3, July

King, Charles (2001): Potemkin Democracy: Four Myths about Post-Soviet Georgia, in: The National Interest 64

Koyama, Shukuko (2002): Security sector reform in Georgia. Saferworld Briefing, London

Kukhianidze Alexander/Kupatadze, Alexander/Gotsiridze Roman (2007): Smuggling in Abkhazia and Tskhinvali region in 2003-2004, in: Shelley, Louise/Scott, Erik R/ Latta, Anthony (eds): Organized Crime and Corruption in Georgia, London et al, Routledge

Küchholz, Juliane (2005): Die Rosenrevolution in Georgien –Ausdruck der Demokratie oder ein von den USA erkaufter Putsch? Arbeitspapiere des Osteuropa-Instituts der Freien Universität Berlin, Heft 49.3 http://userpage.fu-berlin.de/~segbers/ downloads/ working_papers/AP49-3.pdf 19.11.2008

Linz, Juan J./Stepan, Alfred (1996): Problems of Democratic. Transition and Consolidation . Southern Europe, South. America, and Post-Communist Europe, Baltimore, MD Johns Hopkins University Press

Lynch, Dov (2006): Why Georgia matters, Chaillot Papers No. 86 Institute for Security Studies, Paris

Machavariani, Shalva (2007): Overcoming economic crime in georgia through public service reform, in Shelley, Louise/Scott, Erik R/ Latta, Anthony (Eds): Organized Crime and Corruption in Georgia, London et al., Routledge

McFaul, Michael/Tatic, Sanja (2005): Report on Russia in Sarah Repucci: Report on Russia in: Repucci, Sarah / Walker, Christopher (eds.):Countries at the Crossroads , A Survey of Democratic Governance, Freedom House

Merkel, Woflang (2004): Embedded and Defective Democracies, in: Democratization, vol.11, no.5, 33–58

Merkel, Wolfgang/Croissant, Aurel (2000): Formale und Informale Institutionen in defekten. Demokratien, in: Politische Vierteljahresschrift, vol. 41, no.1, 3-30

Milliken Jennifer/Krause, Keith (2003): State Failure, State Collapse and State Reconstruction: Concepts, Lessons and Strategies, in: Milliken, Jennifer (ed.): State Failure, Collapse and Reconstruction, Oxford et al, Blackwell

Mitchell, Lincoln A. (2006): Democracy in Georgia since the Rose Revolution, in: Orbis, Volume 50, Issue 4, 669-676

Nodia, Ghia (2005): Georgia: Dimensions of Insecurity, in Coppieters, Bruno/Legvold Robert (eds.): Statehood and Security: Georgia after the Rose Revolution, American Academy Studies in Global Security

Nodia, Ghia (2008): Nations in Transit 2008 Georgia, Freedom House, 2008 http://www.freedomhouse.hu/images/fdh_galleries/NIT2008/NT-Georgia-final.pdf, 11. 11. 2008

Ottaway, Marina (2003): Democracy Challenged: The Rise of Semi-Authoritarianism, Carnegie Endowment for International Peace

Pickel, Susanne/Müller, Thomas 2006: Systemvermessung – Schwächen der Konzepte und Verzerrungen der empirisch-quantitativen Bestimmung von Demokratie und Autokratie. In: Pickel, Gert/Pickel, Susanne (Ed.): Demokratisierung im internationalen Vergleich. Neue Erkenntnisse und Perspektiven. Wiesbaden

Puddington, Arch (2008): Findings of Freedom in the World 2008 -- Freedom in Retreat: Is the Tide Turning? Freedom House, http://www.freedomhouse.org/template.cfm?page=130&year=2008 03.11.2008

Risse, Thomas/Lehmkuhl, Ursula (2007): Governance in Räumen begrenzter Staatlichkeit, in: Aus Politik und Zeitgeschichte, 20-21: 2007, 3-9.

Schlindwein Simone (2007): Zwischen Propaganda und Kommerz - Medien(un)freiheit in Südost-, Mittelost- und Osteuropa, http://www.n-ost.de/cms/images//studie% 20 pressefreiheit.pdf 14.10.2007

Schneckener, Ulrich (2006): States at Risk. Zur Analyse fragiler Staatlichkeit, in: Schneckener, Ulrich (ed.): Fragile Staatlichkeit „States at Risk“ zwischen Stabilität und Scheitern, Baden-Baden, 9-40

Shelley, Louise (2007): Georgian organized crime, in: Shelley, Louise/Scott, Erik R/ Latta, Anthony (Eds): Organized Crime and Corruption in Georgia, London et al. Routledge

UNDP (2008): Georgia Human Development Report 2008. The Reforms and Beyond http://europeandcis.undp.org/home/show/02060DFC-F203-1EE9-B8EEBC823D 615E62 11.11.2008

UNICEF (1999), The analysis of the situation of children's school education in Georgia in 1998 – 1999, Tbilisi, Georgia

Weber, Max (1980): Wirtschaft und Gesellschaft, Tübingen#

Wheatley, Jonathan (2005): Georgia from national awakening to Rose Revolution: delayed transition in the former Soviet Union, Aldershot [et al.], Ashgate

World Bank (2008): Georgia: Country Brief 2008 http://web.worldbank.org/ WBSITE/ EXTERNAL/COUNTRIES/ECAEXT/GEORGIAEXTN/0,,menuPK:301755~pagePK: 141132~piPK:141107~theSitePK:301746,00.html 01.11.2008

World Bank (2008): The data query http://ddp-ext.worldbank.org/ext/DDPQQ/ report.do? method =showReport

CONFLICT RESOLUTION IN GEORGIA: A NEVER-ENDING STORY?

MARIETTA S. KÖNIG

1 Introduction

Georgia's process of development has been closely linked to its success in restoring its territorial integrity ever since the country gained its independence in April 1991. However, wars in Abkhazia and in South Ossetia have led to the factual, though internationally never recognized, independence of both entities. Unsteady cease-fires preserved these statuses for more than a decade, while negotiations were hardly of success. Instead, positions were drifting further and further apart. This situation, however, was no longer acceptable for Georgia, when Mikheil Saakasvhili took over power in January 2004. Restoring territorial integrity became the top priority for Georgia's new leaders. In seeking to achieve this, police and quasi-military operations have often been used, contributing to worrying escalations in the conflict zones, and also leading to harsh criticism within the international community.

When Georgian troops moved towards Tskhinvali, the capital of South Ossetia, on 7 August 2008, the conflict between Georgia and the seceding South Ossetia finally broke out again. Large parts of Tskhinvali were reported to be destroyed, while the degree of humanitarian damage left room for wild speculations. Meanwhile, Georgia's move led to a five-days war between Georgia and the Russian Federation, making the worst nightmare of Georgian society become reality: Georgia was invaded and attacked by Russian military forces. Although a cease-fire agreement was finally signed between Georgia and the Russian Federation, the present conflict has not yet reached its peak. The deployment of additional Russian troops in and around South Ossetia and Abkhazia, and particularly Russia's unexpected official recognition of both entities' independence on 26 August has led to worldwide irritation. Nonetheless, Russia's decision has accomplished facts, which will have a serious impact on the future Georgian-Abkhaz and Georgian-South Ossetian conflict resolution processes. Georgia is as far from its territorial integrity as ever before.

This article is concerned with the main factors that have hindered finding a solution to both conflicts up to today. By describing recent developments in the two conflict regions and comparing them with Saakashvili's policies so far I hope to give an overall view of Georgia's current political situation and its significance for the renewed deterioration in the ethno-political conflicts with Abkhazia and South Ossetia.

2 Background to the Georgian-South Ossetian Conflict

Both the conflicts of Georgia with Abkhazia and with South Ossetia have deep historical roots. Although the Abkhaz and Ossetes can scarcely be compared in terms of ethnogeny and language, there are certain similarities in the explanations of their contrasts with Georgia. Unlike the Georgians, a South-Caucasian ethnic group, and the Abkhaz, a North-Caucasian tribe, the Ossetes are not of Caucasian origin. They designate themselves as Iron and accordingly South Ossetia as *Xusar Iryston (Southern Iryston)* and the Republic of North Ossetia-Alania as *Cægat Irystony Alanijy Respublikæ (Northern Iryston-Alanian Republic)*. The Ossetes are considered to be an autochthon ethnic group directly descended from the north-eastern Scythes and Sarmatians. These, in turn, constituted the Alan, a nomadic tribe, which controlled what is today's North Caucasus in the first four centuries A.D. before merging with other tribes of the region. Between the 16th and 18th centuries the present Ossetian people evolved. The most widespread dialects are Digiron in the West and the even more widespread Iron in the East, which is also the Ossetians' literary language.[1] The Ossetian language is assigned to a northern subgroup of the eastern Iranian languages, and thus is part of the Indo-European language family. Ossetian is considered to be the only living version of the languages that used to be spoken by the Sarmatian tribes.[2]

While the Ossetes argue that their forefathers had been living in the North Caucasus for centuries, the Georgians on the other hand contend that the Ossetes settled down in the territories at Russia's southern borders no earlier than the seventeenth century. The Georgian argumentation is further concentrated on political developments in the 19th and 20th centuries, when the Ossetes' settlement area was constantly re-divided between Russia and Georgia and thus was never considered an independent territory. The Ossetes had already voluntarily submitted to Russian supremacy in 1774. As they now stress, when negotiating with Russia, there was no distinction made between

1 Ossetian has been written in Cyrillic since 1954. Before that time, Russia and Georgia fought for alphabetic supremacy in the newly created Ossetian literary language.

2 Thordarson, F.: Ossetic. In: Schmitt, R. (Hg.), Compendium Linguarum Iranicarum, Wiesbaden 1989, pp. 456-79.

North and South Ossetes.[3] This is a central point in the contemporary arguments of the Ossetes who see themselves as a single ethnic group, albeit divided into several branches. For Georgia, however, this argumentation is irrelevant, because to them the territory of today's South Ossetia has always been part of Georgia under the name Samachablo (named after the Georgian family Machabeli, which ruled this territory for a long time). Samachablo would thus be located within the Georgian region Shida Kartli (Central Kartli, region in Northern Georgia).

In May 1918, following the revolutionary events in Russia, the Democratic Republic of Georgia, which included Samachablo/South Ossetia, was founded. North Ossetia, however, was taken over by Russia. Georgia invaded Samachablo/South Ossetia to prevent it from taking further steps towards independence, which were regarded as a threat against territorial integrity. There followed in the years 1918 to 1920, a series of uprisings, which were put down bloodily by Georgian troops. About 5,000 people alone died during the clashes between April and June 1920, and about 20,000 South Ossetians fled to North Ossetia then. Even today, these events continue to burden Georgian-South Ossetian relations. In 1922 South Ossetia became an Autonomous Region (*Oblast'*) of Georgia, while North Ossetia was granted the status of an Autonomous Soviet Socialist Republic within Russia in 1936. Thus, the two Ossetias ultimately became integral parts of two different Soviet Republics.

Interethnic relations, however, harmonized during the years of the Soviet regime. According to the last Soviet census in 1989, there were 164,000 Ossetians in Georgia, 65,000 of them in South Ossetia, whose multiethnic total population was 98,000. Georgian-South Ossetian marriages were quite common. South Ossetia, however, continuously felt economically disadvantaged in comparison to the Georgian heartland. In the late 1980s, a national movement came together in South Ossetia, largely under the leadership of the *Adamon Nikhas* ("Voice of the People") movement. This national movement basically discussed possibilities for reunifying with North Ossetia. This was considered an affront on the Georgian side, which sought state independence and considered South Ossetia to be part of its territory. On 20 September 1990 South Ossetia declared its sovereignty and in December of the same year it carried out elections for its own parliament. As a result, the Georgian Supreme Soviet voted to abolish South Ossetia's status as an Autonomous Region.

3 For more details about Georgian history see, *Passim*, Lang, D.: A modern history of Georgia, London 1962; Suny, R. G.: The Making of the Georgian Nation, Indiana 1988/1994.

In January 1991, still prior to the break-up of the Soviet Union, the Georgian-South Ossetian conflict flared into violence. Both sides suffered significant military and civilian losses. Starting in February 1991, the South Ossetian capital, Tskhinvali, was blockaded by Georgian troops for several months. In April 1991, some 500 soldiers of the Soviet Interior Ministry were stationed in South Ossetia, where they more or less openly took the side of the South Ossetians before being withdrawn a year later during the break-up of the Soviet Union. Their withdrawal was a further cause for disagreement between the parties to the conflict as the departing Soviet army had left the South Ossetian side various weapons, including large-calibre arms. South Ossetia was also actively supported by volunteers from the Confederation of the Mountain People of the Caucasus and, with reference to a referendum conducted on 19 January 1992 South Ossetia finally declared its independence on 29 May 1992. As a result, fighting broke out once again, and this time lasted until July 1992. It cost around 1,000 people their lives and created some 60,000 displaced persons. Due to the war, South Ossetia's population shrank by about one quarter. Only few refugees have returned.

On 24 June 1992 an agreement on the principles of a settlement of the Georgian-Ossetian conflict was signed. Known as the Sochi Agreement, this document also provided for the creation of Joint Peacekeeping Forces (JPKF), consisting of Georgian, Russian (including North Ossetian), and South Ossetian units. Russia has *de facto* overall command and has assumed the role of chief mediator in the resulting conflict-resolution process. The key task of the tripartite peacekeeping force is to monitor the cease-fire, to keep the conflicting parties apart, and to guarantee security in the conflict zone. In addition, the cease-fire agreement provides for the creation of a security corridor along the main lines of confrontation, the pull-out of armed groups, and the disarmament of self-defence units.[4] The Georgian side accuses Russia of having already stationed 1,000 men in South Ossetia, which they consider a partial occupation of Georgian territory. Apart from that, many of them have been involved in the region's shadow economy. Thus, Georgia demands that the present framework for negotiation be changed: a stronger involvement of international actors and the replacement of the

[4] Cf. Agreement on Principles of Settlement of the Georgian-Ossetian Conflict (English Version), in: Tamaz Diasamidze (ed.), Regional Conflicts in Georgia – the Autonomous Oblast of South Ossetia, the Autonomous SS Republic of Abkhazia (1989-2002); The Collection of Political-Legal Acts in English Language, Tbilisi 2003, at: http:// www.rrc.ge/, pp. 94f.

JPKF by international troops. All these points, however, have been rejected by the Russian and the Ossetian sides.[5]

On 3 December 1992, an OSCE mission of long duration was finally established in Georgia. One aspect of its remit was to promote negotiations between the parties to the conflict, thereby contributing to the search for a speedy resolution.[6] The mission was also charged with monitoring the work of the JPKF in South Ossetia, and joint OSCE-JPKF projects were set up. In addition to high-level meetings, two further negotiation mechanisms have a key role in the negotiating process that aims to end the conflict: the Joint Control Commission (JCC), and the Experts' Group on the Georgian-South Ossetian conflict-resolution process. Both generally meet at the initiative and under the aegis of the OSCE Mission to Georgia. Since Eduard Kokoity took over the "presidency" in 2001, however, Tskhinvali's sole goal has been reunification with North Ossetia and integration into the Russian Federation. These attempts were increasingly supported by Russia which started to provide Russian passports, paying pensions, and establishing the Russian ruble as common currency in South Ossetia.

3 Background to the Georgian-Abkhazian Conflict

In the case of Abkhazia the conflict is also connected with the ethnic and language differences between the Abkhaz and Georgians as well as with Russian and Soviet influences. The territorial claims are grounded on a subjective vision of historical developments. The Abkhaz claim to be the titular nation of the Abkhaz territory. Unlike the Georgians, the Abkhaz belong to the North Caucasian ethnic and language group. They are considered to be southern relatives of the Adygheans, the Cherkess and the Karbadians. While linguistic features of the Abkhaz clearly indicate that the Abkhaz and the Georgian languages have been in close contact for a considerable period, place names in Abkhazia clearly mark it as having been the home for the Abkhaz for a very long time.[7] However, Abkhaz' still young literary language made it difficult for them to

[5] On the resolutions of the Georgian parliament concerning withdrawal of Russian and CIS-troops from the conflict regions see: König, M.: Not Frozen but Red Hot: Conflict Resolution in Georgia Following the Change of Government. In: Institute for Peace Research and Security Policy at the University of Hamburg (Ed.), OSCE Yearbook 2006, Nr. 12, Baden-Baden 2007, pp. 93f.

[6] Cf. 17th Committee of Senior Officials, Prague 1992, Journal No. 2, 6 November 1992, Annex 2.

[7] Cf. Colarusso, John, Abkhazia, Central Asian Survey No. 14(1), Abingdon 1995, pp. 76ff.

prove territory tenures, compared to the Georgians whose alphabet was created sometime in the 4-5th century AD. In addition, large parts of Abkhazia's present territory have been under Greek, and later Roman and Ottoman rule for long periods of time. The Abkhaz people themselves, with reference to their specific tribal rules and traditions, did not make significant attempts to create structures of statehood until the eighth century AD. In the Abkhaz perception, the Abkhaz Kingdom was allied with Georgia since the 10th century AD. It existed more or less independently until 1810, when it came under the protection of the Russian Empire, with, however, the status of a separate state until the year 1864. From 1864 to 1917, Abkhazia was subordinated to the Russian Tsarist administration for the Caucasus. After the Russian Revolution in 1917, Abkhazia first obtained the status of an Autonomous Republic within the Soviet Union, but later, in 1931, it was incorporated as an Autonomous Soviet Socialistic Republic (ASSR) into the Georgian SSR. Thus, according to Abkhaz understanding, Abkhazia did not come under Georgian rule until 1931.[8] Stalin's Sovietization policies in the following years led to continuous discrimination against Abkhaz culture and language. At the same time the policy of settlement of Georgians in Abkhazia was pursued, thus turning the Abkhaz into a minority in their own claimed area of settlement. According to the last Soviet population census in 1989, only 18% of the population of Abkhazia (about 93,000 people) were ethnic Abkhaz, while Georgians, with 45.7% (almost 240,000), represented the largest population group. This created a well-founded fear of assimilation and loss of cultural identity amongst the Abkhaz population.

At the end of the 1980s the wish for secession grew stronger within the Abkhazian SSR. While the Georgian SSR sought independence, the Abkhazians first voted for the Soviet Union Treaty. With the downfall of the Soviet Union they considered their national sovereignty to be legally legitimized. For Georgia, however, this was not acceptable. A "War of Rules" between Sukhumi and Tbilisi, which had already started in 1988, escalated in 1992, when the Georgian constitution from 1921 and the constitution of the Abkhaz SSR from 1925 were re-implemented. According to the 1921 constitution, Abkhazia was considered a sovereign republic, which however remained linked with Georgia via a union treaty.

In August 1992 Georgian military units without authority moved into Abkhazia, and the conflict turned into a months-long war. Several cease-fire agreements were violated by

8 For a profound Abkhaz point of view of Abkhazia's history see works by the well-known Abkhaz historian Stanislav Lak'oba, such as: Lak'oba, S.: History: 18th century-1917, in: Hewitt, George (ed.), The Abkhazians. A Handbook, Richmond 1999, pp. 67-101.

both parties to the conflict. With the help of Cossacks, former Russian army officers and volunteers from the Confederation of the Mountain People of the Caucasus, the Abkhazians finally pushed the Georgian National Guard out of Abkhazia. On 14 May 1994 a cease-fire agreement which turned out to be durable, was signed. It is monitored by a 18,000-man strong CIS-peacekeeping force which is mainly staffed by the Russian Federation. Presently, a further 133 military personnel from the United Nation Observer Mission in Georgia (UNOMIG) in Sukhumi also observe the situation on the ground. UNOMIG began operating in August 1993. Since December 2003, according to recent data, some 19 police officers have been supporting the local police on the Georgian side of the demarcation zone.[9]

During combat operations from 1992 until 1994, when both sides to the conflict committed human rights violations, about 8,000 people were killed, and more than 250,000 people, mostly Georgians, fled from Abkhazia. With this exodus, the Abkhazians succeeded in changing the ethnic composition of Abkhazia to their own advantage. Thus, during the Georgian-Abkhazian negotiations on the future status of Abkhazia, the Georgian request for repatriation of refugees became particularly problematic. Despite these tricky negotiations, many Georgians, ignoring Abkhazian resistance, began returning to the Abkhazian border district Gali. In the years that followed, violent clashes were common, especially in Gali. In 1998 alone 350 people died and some 35,000 again fled from Abkhazia. The Abkhaz government was at first willing to accept a confederal agreement with Georgia. However, this position changed when Georgia refused to sign a joint treaty in 1997, and after a referendum was conducted and on the basis of its outcome, Abkhazia declared its independence on 12 October 1999.[10] With that the UN-led peace negotiations finally came to a halt. Russia, meanwhile, initiated a parallel framework for negotiation, the so-called Sochi-process, and in 2002 Russia encouraged the Abkhaz side to ignore any negotiations based on the UN-document entitled "Basic Principles for the Distribution of Competences between Tbilisi and Sukhumi", since it would grant Abkhazia only some broad autonomy-based rights.

In August 2001 violence escalated again, when some 450-800-man strong guerilla troops moved into the partly Georgian, partly Abkhaz-controlled Kodori-valley. In the

9 UN, S/2008/38, Report of the Secretary-General on the situation in Abkhazia, Georgia, New York, 23 January 2008.

10 Kvarchelia, Liana, An Abkhaz Perspective, in: Cohen, Jonathan (Hrsg.): A Question of sovereignty. The Georgia-Abkhazia Peace Process, Accord issue 7, London 1999. Online at: http://www.c-r.org/our-work/accord/georgia-abkhazia/abkhaz-perspective.php.

following years the trouble spot focused on the Gori district. Here, every now and then people die in combat and villages are shelled by non-identified airplanes. Observers consider the shelling as Russian attempts to further destabilize the region and to legitimate its continuing military presence. In fact, Abkhazia still considers Russia to be its only effective protective power. Having supported Abkhazia during war, Russian troops are, despite international agreements, still deployed at the military base of Gudauta. The CIS imposed a political and economic embargo against Abkhazia in January 1996. Being Abkhazia's most important trade partner, however, Russia, only adheres to it in principle. More than seventy per cent of the Abkhaz population has already acquired Russian citizenship since 2002. And beginning in September 2004 the railroad between Sukhumi and Moscow was reopened, an event which was very much criticized not only by Georgia but also by UNOMIG.

4 Conflict Resolution Mechanisms under Saakashvili

4.1 The Situation in South Ossetia

When Mikheil Saakashvili came to power in November 2003, the recreation of Georgia's territorial integrity became the highest priority. Encouraged by its first and quick success in reintegrating Ajaria into Georgia's administration in May 2004, Georgia started to focus on South Ossetia. On 23 May 2004, parliamentary elections were held in South Ossetia, from which the Unity movement of *de facto* President Kokoity emerged the strongest party. The elections were not recognized by the international community. Once more, anti-Georgian resentment grew in volume and was ignored by Tbilisi. Instead, on 31 May 2004, President Saakashvili sent Georgian Interior Ministry troops to South Ossetia to establish checkpoints and roadblocks at key road junctions to, officially, combat widespread smuggling. This step was justified by the new Georgian government in terms of its aspiration to achieve regional stability and improve economic performance in the country. However, the conflict escalated again and led to the most serious combat since the end of war. Dozens of people were killed; at least 80 more were seriously injured. It was international pressure in particular that led to a still-valid cease-fire agreement in August 2004.

The security situation, however, seriously deteriorated after this crisis. While the OSCE border monitors had to be withdrawn in 2005 due to Russian veto, the OSCE-Mission and the delegation of the European Commission started to strongly support education programs for Georgian border soldiers. However, the background to the violence and who exactly its initiators were in the zone of conflict often remained unclear. Every now and then bursts of escalation threatened the conflict zone, due to actions carried out by one or another party to the conflict. In November 2004, Georgia established a reservists' camp in the zone of conflict. This had to be closed down again in April 2005 due to pressure from the OSCE. Since the beginning of 2007, however, Georgia has been building a military base close to Gori/Georgia proper, which again was considered as an aggressive act by Tskhinvali. Russia, meanwhile, stood accused of establishing a military base as well, close to Java on South Ossetian territory.

A further crisis emerged in autumn 2006 after Edward Kokoity was reelected "president" in South Ossetia on 12 November. At the same time, official figures showed that some 99 per cent of the South Ossetian population confirmed in a referendum their will for reunification with North Ossetia. Parallel to this, however, alternative presidential elections were carried out in the Georgian villages of South Ossetia. Here, the former South Ossetian "prime minister", Dimitri Sanakoev, was elected "president". Furthermore, the Georgian population agreed in a referendum to reunify with Georgia. Since then there have been, *de facto*, two governments in South Ossetia.

In March 2007 Georgia restored the administrative district of South Ossetia, which had been abolished in 1990. Thus, Sanakoev became the official representative of the Georgian government in South Ossetia, and a significant part of South Ossetia again came under Georgian control. Georgia started to build up the infrastructure there. Strategically important for Georgia, and at the same time fiercely opposed by Kokoity's regime, was the construction of a road to connect the Georgian villages south and west of Tskhinvali. Furthermore, in April 2007, the Georgian government decided to annul the agreement with Russia for the economic rehabilitation of the zone of conflict.

Two months later, the water supply system of South Ossetia turned into a political issue with potential for a violent outburst of the conflict. The water pipelines leading from Georgian controlled territory to Tskhinvali had been set off by residents to water their fields. However, this damaged the pipelines and water supply to Tskhinvali was cut off. As a result, many parts of South Ossetia remained without water for several weeks. Mutual escalations continued: while Georgia built up its police forces in the zone of conflict, both sides to the conflict repeatedly blocked roads and introduced passport controls. Combat with mortar and grenade shelling followed; villages on both sides repeatedly suffered attacks. Furthermore, Russia's role in the ongoing escalations became increasingly incomprehensible. In August 2007, for example a threatening scenario occurred, when a supposedly Russian military plane fired a rocket on Georgian territory close to the zone of conflict. Fortunately, however, the rocket did not explode. Russia repeatedly flew over the Georgian-South Ossetian area of conflict, thereby violating not only the Sochi-cease-fire agreement but also Georgia's sovereign rights.[11]

Given the continuous instability, negotiations between Tbilisi and Tskhinvali have remained blocked. Demonstrating the ineffectiveness of negotiating with Tbilisi, Tskhinvali repeatedly declared it was suspending negotiations within the framework of the JCC. Tbilisi, likewise, regularly refers to the fact that the existing patterns hardly promote a de-escalation, not to mention a solution of the conflict.

4.2 The Situation in Abkhazia

The 2004 summer crisis in South Ossetia also had an impact on the Georgian-Abkhazian conflict resolution process. Abkhazia instantly armed its borders. Georgia sought to calm tempers and disarmed and dissolved guerilla troops which, up until then, had been operating very close to the zone of conflict. Abkhazia meanwhile carried out an internationally unrecognized presidential election in autumn 2004. After a week-long debacle including violent protests, the opposition candidate, Sergej Bagapsh, was finally elected President, while his Russian-backed opponent, Raul Khadshimba, was elected Vice-President. With these elections Abkhazia for the first time demonstrated some political independence from Russia. However, Bagapsh adheres to traditional Abkhaz positions, and likewise claims international recognition for Abkhazia's state independence, while accepting a solely associate membership in the Russian Federation.

The status quo, however, is not simply a result of the opposing purposes of the parties to the conflict, but also a guarantee for the prospering shadow economy along the demarcation line. About 80 per cent of the 90,000 Georgian refugees from the Gali region have returned to their homes during recent years to work on their fields and to sell their products. Abkhaz likewise sell their products, however, primarily at the Russian border. Apart from agricultural products, petrol and narcotics are commonly smuggled via Abkhazia's borders. The flourishing black market is controlled by Georgian and Abkhaz militia. Members of the CIS-peacekeeping forces also participate

[11] Russia e.g. later confirmed that an aircraft of the Russian air force conducted a brief flight over South Ossetia on 8 July 2008, "to cool hot heads in Tbilisi", as was announced by the Russian Foreign Ministry, see: Russian MFA Information and Press Department Commentary Concerning the Situation in South Ossetia, 10 July 2008, at: http://www.un.int/russia/new/ MainRoot/ docs/off_news/ 100708/ newen3.htm.

financially in this business. The increased criminality and bomb attacks in the zone of conflict since 2006 probably hint to a competency conflict between individual bands.

A new critical turn in the conflict emerged in July 2006, when Georgian police troops moved into the Kodori valley to destroy a 300-400-man strong guerilla group, which was no longer willing to cooperate with Tbilisi. After only a few days of this operation, the Georgian government declared it to have been successful. About 60 people were arrested; about 60 casualties were reported, not, however, verified. The UN-Security Council issued a resolution in which the Georgian action in the Kodori valley was strongly criticized.

Abkhazia and Russia declared this action a military maneuver to forcibly return Abkhazia to Georgian territory, and thus a violation of the Moscow cease-fire agreement of 1994. According to this document the upper part of the valley is Georgian-administered, however only accessible via the demilitarized zone. The situation continued to be tense, because on 27 September, Georgia decided to move the Abkhaz government-in-exile from Tbilisi to the upper Kodori valley, which was then named "Upper-Abkhazia" and became a Georgian administrative district. Shortly afterwards "Upper-Abkhazia" participated in the Georgian local elections.

Since this officially declared police operation, the situation in the Kodori valley, as well as within the entire border region has remained tense. Shelling and combat occur regularly, most often clashes between Georgian police forces and Russian peacekeepers. This further undermines Russia's role in defusing the conflict in the region.

Especially troublesome was the opening on 26 May 2007 of a Georgian so-called patriotic youth camp in Ganmukhuri, not too far from the demarcation line on the Georgian side of the Inguri-river. Despite the UN Secretary General's pressure to close this provocative camp, Saakashvili announced an expansion of the camp scheduled for September. Explosive too were the rumors about Georgia's plans to establish an additional fifth brigade of 2,500 men close to the Gali region. With approximately 32,000 men the Georgian army would then be twice as big as originally recommended by Western military advisers.

The police action in the Kodori valley also led to a deadlock in Georgian-Abkhazian negotiations. Since October 2006 any direct talks had been frozen, and Abkhazia was not willing to resume talks unless Georgia withdrew from the Kodori valley. The Group of Friends of the UN Secretary General continued to hold its biannual meetings. Direct meetings, however, were not taking place. Thus, pressure on Georgia increased to present well-grounded concepts and incentives to regain Abkhazia as soon as possible and to resume direct negotiations with the leaderships of the two de-facto states.

5 Georgian Initiatives

In his speech to the Munich Conference on Security Policy on 2 March 2006, Mikheil Saakashvili clearly expressed Georgia's national interest, which is concentrated on achieving the following goals:

- The peaceful and complete restoration of Georgia's territorial integrity;
- The establishment of lasting security through energy diversification and the creation of a modern national defense force;
- Broad economic growth and investment in education, health care and good governance;
- And integration into NATO and other European institutions, including constructive relations with all the neighbours.[12]

The international community, including Georgia's allies in the Euro-Atlantic region, are to help it implement these goals. In order to involve the international community more closely in the conflict resolution process, President Saakashvili presented a three-step plan at the General Assembly of the United Nations, on 21 September 2004, aimed at settling the South Ossetian and Abkhaz conflicts. The plan stipulated measures to build confidence between the ethnic groups, demilitarization and the search for a format for an international peacekeeping mission, and, finally, the resolution of the status question. These steps were supposed to be dealt with one by one.[13] On 27 October 2005, the Georgian Prime Minister, Zurab Nogaideli, presented an updated version of the "South Ossetia Conflict Resolution Plan" to the Permanent Council of the OSCE.[14] Because Georgia called for changes to the current negotiating framework and a rapid solution to the status question – before the end of 2006 – this plan was rejected by Moscow and Tskhinvali. Thereafter, the Georgian government reverted to emphasizing Saakashvili's three-step plan, which omits the two controversial points. The "Statement on Georgia" adopted by the Ministerial Council on 6 December 2005 underlined the significance of this plan as a basis for the peaceful settlement of the conflicts.

[12] Mikheil Saakashvili, Georgia's role in International Security Policy, at: http://www.securitycon ference.de/ konferenzen/rede.php?sprache=en&id=163&, 2 March 2006

[13] Address by Mikhail Saakashvili, President of the Republic of Georgia, in: United Nations, General Assembly, Fifty-ninth session, 4th plenary meeting, 21 September 2004, A/59/PV.4, pp. 11-16, here: p. 14.

[14] South Ossetia Conflict Resolution Plan – Presented by Prime Minister of Georgia Mr. Zurab Nogaideli, Vienna, 27 October 2005, in: OSCE, *Conference* Services, PC.DEL/1060/05, 27 October 2005.

In spring 2007, Tbilisi came up with a new proposal for a conflict-resolution plan. On 10 April 2007, Zurab Nogaideli addressed the United Nations Security Council on Georgia's peace initiatives in Abkhazia. According to media reports of Nogaideli's presentation, "Georgia today is prepared to offer a lesser degree of autonomy than was the case in earlier peace plans". The new three-stage plan foresaw (1) permission for repatriation of Georgian displaced persons who had fled Abkhazia during the 1992-1993 war; (2) recognition by Abkhazian authorities of Georgia's territorial integrity; and (3) a referendum on Abkhazia's future status, to be held only after the return of all displaced persons.[15] It is clear that the acceptance of these three points by Abkhazian authorities would require a U-turn in their current policies aiming to further promote the potential recognition of Abkhazia's independence. On 12 April, in response to the proposed plan, Abkhazia's de facto Foreign Minister, Sergey Shamba, demanded that Tbilisi should abandon 'illusions' about Sukhumi's willingness to accept autonomous status within Georgia, claiming that the parties to the conflict could only talk about 'intergovernmental relations'.[16]

Shortly thereafter, a five-point conflict-resolution plan for South Ossetia was also proposed by Saakashvili's government. In particular, Tbilisi suggested reestablishment of an administrative district of South Ossetia within Georgia and offered Eduard Kokoity's secessionist authorities and the Sanakoev-led alternative (Tbilisi-loyal) government the possibility of creating a provisional administration, to work over the region's autonomous status and introduce a position of the head of the provisional administration. According to the plan, Georgian authorities would also allow the provisional administration to appoint deputy ministers of internal affairs, finance, economics, education and science, healthcare and social protection, culture, justice, agriculture, and environment, who would be responsible for issues relating to the region.[17] Since the initiative to create a transitional administration in South Ossetia aimed at a peaceful conflict resolution, as Saakashvili noted at a session of the National Security Council on 23 April 2007, the proposals were not time-limited and Tbilisi was ready to give everybody an opportunity to participate in the resolution and

[15] Liz Fuller, Georgia: Abkhazia Certain To Reject New 'Peace Plan', Radio Free Europe / Radio Liberty, 13 April 2007, at: http://www.rferl.org/featuresarticle/ 2007/04/84b2d194-36b0-4ae9-bb41-802bd3c2b981.html.

[16] Molly Corso, Georgia promotes its conflict-resolution plan, Civil Society, Eurasianet, 24 April 2007, at: http://www.eurasianet.org/departments/insight/ articles/eav042407.shtml.

[17] Civil Georgia, Saakashvili Offers Cooperation within New Administration of South Ossetia, 23 April 2007, at: http://www.civil.ge/eng/article.php?id=14997.

decision-making process. However, both *de facto* governments were expected to respond to the proposals before 1 May 2007, which actually left only one week for the *de facto* authorities to consider Saakashvili's plan. Eduard Kokoity's secessionist authorities saw in the proposed plan an attempt of the Georgian side to promote the Tbilisi-loyal *de facto* president of South Ossetia, Dimitry Sanakoev, which eventually led to an aggravation of tensions between the parties to the conflict instead of contributing to the peace-building process.

Given little progress in the conflict resolution process and Abkhazia's and South Ossetia's reluctance to negotiate any plans touching their claimed status of independence, Georgia's three-step plan was not further elaborated for almost one year. It proved to be a trick that moderate Georgian mediators were regularly replaced by others. In July 2006, Giorgi Khaindrava, former Minister for Conflict Resolution, was replaced by the Deputy Foreign Minister, Merab Antadze. In the same month, Saakashvili appointed Irakli Alasania, the former chief negotiator on Abkhazia, as Georgia's ambassador to the United Nations. Papers produced by those with no formal role in the conflicts and discussed by the parties to the conflict, the media, NGOs, and the opposition, are officially not registered by the Georgian government.[18]

However, the political crisis in Georgia in November 2007, which led to early presidential elections on 5 February this year, further increased the pressure on Georgia's government to present some initial results in the conflict resolution process. These circumstances not only led to further tensions between Georgia and the *de facto* authorities, but also between Georgia and its Western partners, especially the EU, and western EU member states. These countries point out the importance of confidence-building measures, and prefer a bottom-up approach to solve both conflicts in the long-term, while Georgia wants to have the status question solved first, before negotiating any further measures, and thus prefers a top-down approach. A bottom-up approach is not in Georgia's interests, because this would take time, a commodity they consider is running short. Thus, a quick resolution in both conflicts has top priority for Georgia, while this haste is considered to be counterproductive by the European side. A solution for this dilemma is not yet in sight. Instead, the positions of the parties to the conflicts

[18] One example is Paata Zakareishvili, et al., *A concept on the special status of Abkhazia in the Georgian state (Project)*, Tbilisi 2004 (unpublished).

are hardening still further, and rhetoric is becoming sharper.[19] On 28 March 2008, Saakashvili finally presented a modified version of the three-steps plan, with a stronger focus on economic issues. Specifically new is his proposition of the creation of a joint Georgian-Abkhaz free economic zone. Furthermore, he offers the Abkhaz the position of a vice-president within commonly agreed-on state structures. Sukhumi, however, repels any negotiations on the basis of Tbilisi's propositions.

6 The Kosovo Factor

The Kosovo status question played in increasing role in the Georgian-Abkhaz and Georgian-South Ossetian conflict resolution processes since late 2005. The possibility of Kosovan independence was explicitly raised in early 2006. The possibility of a precedent which might reinterpret the peoples' right for self-determination raised hopes, particularly in Abkhazia, that internationally recognized independence nevertheless might be realistic one day. In June 2007 the de facto presidents of Abkhazia and South Ossetia appealed to the Secretary General of the United Nations with a joint declaration according to which they would have the same right to independence as Kosovo. Tbilisi responded by setting up a state commission for the definition of South Ossetia's autonomous status within Georgia. However, Tskhinvali neglected any cooperation with this commission.

Meanwhile, the Kosovo case unnoticeably developed into a ticking bomb. Since early 2006, the Russian government let it be known that, first, Kosovar independence went against Serbia's interests and was therefore unacceptable, and, secondly, that if Kosovar independence were to receive international recognition, Russia would respond by recognizing the independence of other *de facto* states, namely Abkhazia, South Ossetia, and Transdniestria. The negotiations on Kosovo's status have completely stalled since June 2007. Until then, there had still been hope in Europe and the USA that Russia could be brought round to support Kosovar independence as envisaged in the Ahtisaari plan if the right offers and concessions were made in other areas. However, in this case, this tried-and-tested method proved fruitless.

[19] Interviews carried out by the author in Berlin, February 2008. See also Coppieters, according to whom Georgia's impatience vs. the EU's thumping for more time presently complicates EU-Georgian communication: Coppieters, B.: The EU and Georgia: Time Perspectives in Conflict Resolution, ISS Occasional Paper No. 70, December 2007.

On 16 February 2008, Kosovo unilaterally declared its independence. Since the European Union had, for months, not been able to find consensus on how to deal with Kosovo's aspiring independence, only a few EU member states immediately recognized Kosovo's independence (among them Germany, France and Great Britain), while several others (such as Spain, Greece or Cyprus) will, for the moment, refrain from doing so. Thus, the fact that major EU member states such as Germany and France were not able to convince other EU member states to find a joint solution to the Kosovo status question, finally demonstrated the EU's current internal conflict. The EU lacks a common foreign policy, and thus has no clear strategies, which would be essential if it is to be taken as a serious partner by other major international actors. The inconsistency on decisive issues weakens the EU's international position and endangers cooperation with strategically important actors.

Regardless of the questions of whether Kosovo is considered to be a precedent case or not, and what Russia's interests are in repeatedly referring to other *de facto* states in the former Soviet Union, the international community has to come to terms with the fact that the Kosovo status negotiations are already playing a key role in the Abkhazian, South Ossetian and Transdniestrian arguments. Despite increasing tension between the various conflict parties there is no movement in the negotiation processes. No one is willing to compromise, and no incentives are being created that make revising positions more attractive.[20]

In prosecuting, for example, its conflict with Georgia, Russia has always encouraged and exploited the separatist tendencies in Abkhazia and South Ossetia in order to increase Georgia's dependence on Russia and to emphasize its failure to reintegrate the two separatist entities, and thereby, among other things, to hamper Georgia's chances of NATO membership. The NATO-summit in Bukarest in April this year was therefore a setback for Georgia, even though the alliance promised to admit Georgia sometime in the future. After that, mutual provocations in both regions of conflict achieved a new level of intensity, because Russia's problematic role became more and more obvious here. After Russia had already announced on 6 March that it would withdraw from the 1996 CIS treaty imposing sanctions on Georgia's breakaway region of Abkhazia, it now, in April, declared to officially establish legal acts with Abkhazian

[20] On the question of applicability of the Kosovo case see also: Coppieters, B.: Kosovo and the Principles of Just Secession, CEPS Policy Brief No. 146, November 2007. Koenig, M.: The Effects of the Kosovo Status Negotiations on the Relationship Between Russia and the EU and on the De Facto States in the Post-Soviet Space, in: IFSH (ed.), OSCE Yearbook 2007, Baden-Baden 2008, pp.37-50.

and South Ossetian authorities. International attention, however, was focused on another threatening incident in Abkhazia: a Russian fighter jet had downed an unarmed Georgian drone over the Georgian-Abkhaz zone of conflict, an act of mutual aggression confirmed by a UNOMIG investigative report issued on 26 May.

Even though Russia's motives for recognizing Abkhazia's and South Ossetia's independence following a short but painful war in Georgia in last August are, despite an endless list of incidents and motivations, not yet fully understood the consequences of this move are expected to be problematic for all sides. While even close allies within the CIS refrained from supporting Russia's move, Georgia's territorial integrity is durably put into question, and Abkhazia's and South Ossetia's dependence from Russia's goodwill determined for now.

7 Prospects for Settlement

After the war between Georgia and Russia, followed by Russia's formal recognition of Abkhazia's and South Ossetia's independence on 26 August 2008, conflict resolution in Georgia bears the danger of becoming a never-ending story. Despite mutual and months-long escalations it eventually was Tbilisi's own responsibility to move into South Ossetia and start open war. Georgia's allies had repeatedly stressed that military actions would never be tolerated to accelerate the settlements of conflict in Abkhazia and South Ossetia. Despite existing moderate voices in Tbilisi, the radical circle of consultants around the president finally succeeded in convincing Saakashvili that military force would be the only solution. Saakashvili's pressure to solve these ethno-political conflicts had significantly increased in winter 2007/2008. Due to a domestic crisis in November 2007 leading to the declaration of a state of emergency, Saakashvili was finally forced to move the presidential elections forward. On 5 January 2008 he was barely re-elected, on the informal condition that he will solve the conflicts in his second term. Saakashvili urgently needed a success story. Lack of international support in resisting Russian pressure, lacking sensitivity in Georgia that good conflict resolution takes a long time and is only successful with permanent confidence building, overestimation of its own abilities, and, even more problematic to digest, Georgia's limited relevance in world politics, finally led to the catastrophic war with Russia. However, apart from Georgia's and Russia's problematic roles in this war, it has also demonstrated the West's failure to seek and maintain cooperative relations with Russia, and to take Georgia's worries seriously. In fact, the EU has always remained reluctant to engage politically in the South Caucasus with its still unresolved ethno-political conflicts. Since the 1990s the entire South Caucasus has been considered to be part of the Russian periphery, where the Euro-Atlantic Alliance was reluctant to interfere. Building a new cooperative relationship with Moscow had top priority. In the

EU perception only little has changed here, despite the fact that EU's confrontations with Russia were already a reality for quite some time, and the EU's search for harmony in Euro-Russian relations left out that a critical debate on mutual moves is a natural part of a progressive partnership.

Meanwhile, since Russia's recognition of Abkhazia and South Ossetia, it would be naïve to deny the impact of Kosovo's independence on the Georgian-South Ossetian and Georgian-Abkhazian conflict resolution processes. As soon as the conditions are set to restart negotiations, the Abkhaz and the South Ossetian sides, ignoring any international interpretations, however now officially supported by Russia, will refer to Kosovo, which is a precedent in their eyes. This, however, will make it difficult for Georgia, which is not willing to further accept the status quo, to continue insisting on the restoration of Georgia's territorial integrity. Thus, the international community, especially those, who have recognized Kosovo's independence, will now have to do their part in engaging more actively in the conflict resolution processes. Any hesitation here will lead to further regional instability, because Georgia's present political fragility constantly increases due to rising critical voices against Saakashvili's actions carried out in South Ossetia. In fact, the consequences of the war are disastrous for Georgia: Georgian villages in South Ossetia have been destroyed and the entire Georgian population was forced to leave South Ossetia, Georgia has lost any control over South Ossetia and Abkhazia. This is because Abkhazia established a second front in the upper Kodori valley, with Georgia consecutively losing control of its "Upper-Abkhazia". In fact, Georgia's military success has always been unlikely, because Russia repeatedly announced that it will have to support Abkhazians as well as South Ossetians which, due to their Russian passports, have to be protected like every other Russian citizen. Renewed human rights violations and losses on all sides to the conflict have further deepened traumatic memories of entire generations and made conflict resolution processes a never-ending story.

The international community is now in charge of breaking this infinite loop and of actively supporting the parties to the conflict in finding a solution. Here, the upcoming negotiations on the future mandates of UNOMIG and the OSCE Mission to Georgia will be very crucial. Abkhazia, South Ossetia, as well as Russia have already announced that the present mandates lack significant modifications. A special responsibility here is carried by the EU, which is already sharing common maritime borders with Georgia via the Black Sea. As already mentioned, the EU is presently the only external actor which is welcomed by all sides to the conflicts. Since conflict resolution in Abkhazia and South Ossetia is not possible without Russian support, the EU's first actions will have to focus on presenting itself as a cooperative partner and to actively include Russia in the elaboration of conflict resolution strategies. Up until now, however, the European Union has very little with which it can counter any of Russia's objections. Meanwhile, in

Georgia disappointment towards the EU has started to emerge, as well as impatience on when these years-long conflicts will finally be solved. Georgia itself is not willing to propose more than "broadest autonomy" to Abkhazia and South Ossetia. Nationalism and regionalism, however, is still a serious issue in Saakashvili's Georgia. Minority rights in Georgia are still in poor condition, and most of the regions where minorities create a majority belong to the least developed.[21] Though repeatedly averted, multi-ethnic Georgia will only gain a peaceful future by adopting a federal constitution which openly welcomes and includes present and future minorities. Local and international proposals on what this could look like already exist. The parties to the conflict as well as their allies are now in charge of mutually motivating each other in rejoining negotiations and finally opening the way for regional stability and prosperity; a long and difficult road, however manageable.

8 References

Birch, J.: The Georgian/South Ossetian Territorial and Boundary Dispute. In: Wright, J./ Goldenberg, S. and Schofield, R. (eds.), Transcaucasian Boundaries, London 1996, pp. 151-189.

Colarusso, J.: Abkhazia, Central Asian Survey No. 14(1), Abingdon 1995, pp. 75-96.

Coopieters, B.: Kosovo and the Principles of Just Secession, CEPS Policy Brief No. 146, November 2007.

Coppieters, B.: The EU and Georgia: Time Perspectives in Conflict Resolution, ISS Occasional Paper No. 70, December 2007.

Coppieters, B.: The South Ossetia Case. In: Cohen, Jonathan (ed.), A Question of Sovereignty. The Georgian-Abkhazia Peace Process, Accord Issue 7, London 1999.

CSCE Mission to Georgia: Draft Proposal on a Future Political Status for South-Ossetia, Tbilisi 1994.

Diasamidze, T.: Regional Conflicts in Georgia – The Autonomous Oblast of South Ossetia, The Autonomous Republic of Abkhazia (1989-2002). The Legal Collection of Political-Legal Acts, Regionalism Research Centre, Tbilisi 2003.

Dzugayev, K.: South Ossetia's President Clamps Down, IWPR Caucasus Reporting Service (CRS) No. 185, 4. July 2003.

[21] See Azeri in Kvemo Kartli and Armeniens in Djavakheti. 30 per cent of Georgia's population are of non-Georgian origin.

German, T.: Abkhazia and South Ossetia: Collision of Georgian and Russian Interests. Ifri, Paris 2006. Online at: http://www.ifri.org/files/ Russie/ german_anglais.pdf, 25. May 2007

Gremy, F.: Eduard Kokoity. President of the Unrecognized Republic of South Ossetia. In: Caucaz, Issue 4. January 2005. Online at: http://www.caucaz.com/ home_eng/breve_contenu.php?id=100, 20. May 2007.

Halbach, U.: Ethno-territoriale Konflikte in der GUS. In: Berichte des Bundesinstituts für ostwissenschaftliche und internationale Studien, Nr. 31, 1992.

International Crisis Group: Georgia: Avoiding War in South Ossetia, Europe Report Nr. 159, Tbilissi/ Brüssel 26. November 2004.

International Crisis Group: Georgia's South Ossetia Conflict: Make Haste Slowly, Europe Report Nr. 183, Tbilissi/ Brüssel 7. Juni 2007.

International Crisis Group: Georgia-South Ossetia: Refugee Return the Path to Peace, Europe Briefing No. 38, 19. April 2005, Tbilissi/ Brüssel.

König, M.: Not Frozen but Red Hot: Conflict Resolution in Georgia Following the Change of Government. In: Institute for Peace Research and Security Policy at the University of Hamburg (Ed.), OSCE Yearbook 2006, Nr. 12, Baden-Baden 2007, pp. 85-96.

König, M.: The Effects of the Kosovo Status Negotiations on the Relationship Between Russia and the EU and on the De Facto States in the Post-Soviet Space, in: IFSH (ed.), OSCE Yearbook 2007, Baden-Baden 2008, pp.37-50.

König, M.S.: Researcher and doctoral candidate at the Centre for OSCE-Research (CORE)/Institute for Peace Research and Security Policy at the University of Hamburg (IFSH); Hamburg

Lak'oba, S.: History: 18th century-1917, in: Hewitt, George (ed.), The Abkhazians. A Handbook, Richmond 1999, pp. 67-101.

Lang, D.: A modern history of Georgia, London 1962.

Ria Novosti: South Ossetia de Facto Part of Russia - Region's President, 6. February 2006. Online at: http://en.rian.ru/russia/20060602 /48956927.html, 14. April 2007.

Russetski, A./ Woeber, S.: Abkhazia and South Ossetia: Basic Principles for an effective transformation of the peace process ('7D'). In: Helsinki Monitor (Den Haag), 16 (2005), pp. 150-153.

Suny, R. G.: The Making of the Georgian Nation, Indiana 1988/1994.

Thordarson, F.: Ossetic. In: Schmitt, R. (Hg.), Compendium Linguarum Iranicarum, Wiesbaden 1989, pp. 456-79.

ETHNIC MINORITIES DURING THE TRANSITIONAL PERIOD

MAMUKA KOMAKHIA

1 Introduction

Georgia is a multiethnic state which, following the dissolution of the Soviet Union went through a difficult transitional period. The ethnic policy of the Soviet authorities and the grave socio-economic environment and political instability reigning in the 1990s had a serious impact on ethnic minorities living in Georgia, which, during the transitional period, turned into secondary citizens.

The most acute problems emerged in Kvemo Kartli and Javakheti, where ethnic minorities are settled compactly. In the 1990s, Azerbaijanis and Armenians living in these regions were disconnected from the ongoing public-political or economic processes in the country, which deepened the isolation of the Kvemo Kartli and Javakheti populations.

The present article deals with the causes of isolation of ethnic minorities and the main challenges of their integration.[1] First of all, the drawing-up of an ethnic map in Georgia and the contemporary migration processes will be discussed. Afterwards, main challenges hindering the integration of ethnic minorities will be analyzed: bad or lack of knowledge of the state language, information isolation and the media, participation in public-political and economic life. Finally, legal and institutional aspects of the integration of ethnic minorities will be examined.

1 This article does not review the problems existing between ethnic groups living in Abkhazia and the Tskhinvali Region, as military confrontations, which emerged at the beginning of the 1990s, practically detached these regions from the jurisdiction of Georgia's central authorities and therefore ethnic groups residing there faced different challenges during the transition period.

2 Drawing up an Ethnic Map and Contemporary Migration Processes

Due to political or socio-economic factors, the ethnic composition of the Georgian population varied in different historical epochs. Under the impact of these factors certain trends of migratory processes formed in relation to altering numbers of this or that ethnic group. Changes in the ethnic composition of the Georgian population may be divided into four main stages:

The *first* period lasted from medieval times until the end of the 18th century, when, as a result of the invitation of Georgian kings/nobles or historical cataclysms, a non-Georgian population settled in Georgia. During this period, the non-Georgian population occupied territories not settled by Georgians. The *second* period lasted from the beginning of the 19th century until the Soviet occupation of Georgia, when changes in numbers of various ethnic groups were mainly caused by a forced type of migrations. During this period, group settlements of the non-Georgian population in the territory of Georgia took place mainly due to military-political and economic considerations of the Russian empire. The *third* period lasted from 1921 until the dissolution of the Soviet Union, when, during the industrialization and the industrial boom in Soviet Georgia, hundreds of thousands of migrants from the Soviet Republics settled in Georgia.[2] The *fourth* period comprises contemporary migratory processes and has lasted from the independence of Georgia in 1991 until the present day, when, due to political and socio-economic cataclysms emerging from the dissolution of the Soviet Union, representatives of ethnic minorities emigrated from Georgia.

The ethnic composition of the Georgian population witnessed serious changes following the dissolution of the Soviet Union. If, according to the data of the population census of 1989, ethnic minorities constituted 30% of the country's population, the 2002 census demonstrated that this number (excluding territories of Abkhazia and the Tskhinvali Region) decreased to 16% in Georgia.

Emigration is a main reason for the decreasing number of ethnic minorities in Georgia. Following the dissolution of the Soviet Union, the harsh deterioration of the socio-economic state and of the political environment in the 1990s triggered a first wave of emigration. Emigration developed a mass character with regard to the country's entire

[2] Vakhtang Jaoshvili, *Population of Georgia in the XVIII-XX Centuries*, (Tbilisi, publishing house Metsniereba, 1984), 209. (In Georgian)

population, i.e. with regard to Georgians as well as to representatives of ethnic minorities.

The discriminatory policy of the political groups which had come into power in Georgia after the dissolution of the Soviet Union was also a significant factor that led to the emigration of ethnic minorities at the outset of 1990s. On several occasions, nationalist rhetoric of the authorities under President Zviad Gamsakhurdia was reflected in the oppression of representatives of ethnic minorities. A slogan "Georgia for the Georgians" became popular among the public, and the non-Georgian population gained a "guest" status. Oppression had not developed a mass character; however, nationalist political organizations in various districts (Dmanisi, Borjomi, Tetritskaro, Gori, Akhalgori, Lagodekhi, Mtskheta) forced mainly Azerbaijanis and Ossetians to leave/sell their houses and emigrate.[3]

Regarding the ethnic minorities, emigration of the Slavs (mainly Russians and Ukrainians) occurred in especially intense forms. A lot of them were military and civil servants of the Soviet army and their family members, who left Georgia massively from the beginning of 1990s in connection with changing postings of several military units. The Slavs exceeded 60% of the population emigrating from Georgia. The majority of them resided in the cities and almost all of them left to live in the cities of the Russian Federation.

On the other hand, the majority (60%) of the Azerbaijanis, the second largest group of emigrants, were village residents and they nevertheless also preferred to leave, mainly for the cities of other countries. More than a fourth of the Azerbaijanis who left Georgia in 1992 went to live in the Russian Federation, while the rest – went to Azerbaijan. The Armenians were the third largest emigrating group and their 71% were city residents. The majority of Armenians registered out of Georgia in order to resettle in the Russian Federation, while only a fourth of the emigrants planned to settle in Armenia.[4]

Following the military coup in 1992, i.e. after the overthrow of Zviad Gamsakhurdia, a former leader of the Soviet Georgia and a former Minister of Foreign Affairs of the Soviet Union, Eduard Shevardnadze, came into power. He refused to continue with nationalist rhetoric, but this has not reduced the migratory activity rate among the ethnic minorities. Unemployment, deterioration of the level of life, and the impossibility

[3] Toma Gugushvili, External Migratory-Demographic Problems of Georgia (1990-1998), (Tbilisi, 1998), 70-78. (In Georgian)

[4] Revaz Gachechiladze, *Population Migration in Georgia and its Socio-Economic Results,* (United Nations Development Program – Georgia, Tbilisi, 1997), 37-38. (In Georgian)

to realize one's self became the basic factors for emigration after Shevardnadze's coming into power.

Throughout the 1990s, a most attractive country for the ethnic minorities to migrate to was the Russian Federation, which due to linguistic (Russian was a main communication language among the ethnic groups during the Soviet period) and psychological considerations was closest to them. Notably, several ethnic minorities preferred to migrate to their historical homeland. This was further instigated by the feeling of living a non-desirable an life in Georgia and the attractive force of one's historical homeland: the Russians left for Russia, the Greeks went to Greece, the Jews to Israel, the Ukrainians to Ukraine, part of the Armenians to Armenia, and a part of the Azerbaijanis to Azerbaijan. More importantly, out of the two largest ethnic minorities of Georgia – the Azerbaijanis and the Armenians – ethnically oriented emigration was more characteristic to the Azerbaijanis.[5]

Socio-economic types of factors appear to be the main reasons for emigration in recent years. Unemployment, deterioration of the level of life, a low level of realization of one's self, loss of trust in the nearest future, and a feeling of political instability are the most significant among these factors. So-called discriminatory factors have less impact on emigrational attitudes: negative attitude of the authorities and population towards their ethnic origin, inability to develop their native tongue and culture, low level of satisfaction of religious feelings among the ethnic minority groups.[6]

A vigorous wave of emigration from Georgia has already gone by and currently only a small portion of ethnic minorities leaves Georgia. Apart from this, there are factors which prevent emigration of ethnic minorities, such as a high level of integration into Georgian society, a lack of necessary financial funds for migration, or living in mixed families. Bad health conditions and problems encountered at the consulate services are comparatively less impeding factors. It has to be noted that a need not to leave farming, established over the years, patriotic feelings, and satisfaction with the existing situation also represent preventive factors for the Azerbaijanis.[7]

5 Guram Svanidze, Constantine Kokoev, "Emigration from Georgia and its Reasons". *Central Asia and the Caucasus*, 4 (2002): 133. (In Russian)

6 Guram Svanidze, David Svanidze, "Emigration from Georgia and its Reasons", in: *Migration in the Caucasus,* (Conference Materials, Caucasus Media Institute, Yerevan, 2003), 129. (In Russian)

7 G. Svanidze, D. Svanidze ... 130.

There are two types of settlement of ethnic minorities in Georgia – dispersed and compact. Compact settlements mainly exist in Kvemo Kartli and Samtskhe-Javakheti, where both Azerbaijanis and Armenians mainly live. Compact settlements are to be found in other regions of Georgia as well, but these two regions are especially distinct from the perspective of a number of ethnic minorities and the specific development of the regions during the transitional period. Ethnic groups living dispersedly in Georgia are mainly Russians, Kurds, Greeks, Ukrainians, Jews, Assyrians and others.

3 Main Challenges During the Transitional Period

3.1 Proficiency in the State Language

During the Soviet years, Russian was a language of communication among ethnic groups living in Georgia. Soviet authorities facilitated the inculcation of the Russian language as the *lingua franca*. Proficiency in the Russian language was a decisive factor for the career path, thus motivation to learn Georgian was low. Because of this factor, the Russians along with other ethnic groups including Georgians represented a significant number among students of the Russian language schools active in Georgia.

After Georgia gained independence, Russian lost its function as the everyday language of general communication. In view of the fact that the Georgian youth seriously lacks motivation to learn Russian, and ethnic minorities are still not proficient in Georgian, a problem of communication among the ethnic groups has emerged. Exceptions in this regard are the urban areas only, where ethnic minorities have to deal with a Georgian-speaking population on a daily basis, be it on a common or a professional level, and where a level of their proficiency in Georgian is more or less satisfactory.

Unlike in urban areas, the population practically does not speak Georgian in regions compactly settled with ethnic minorities. In Kvemo Kartli and Javakheti, the lack of or bad knowledge of the Georgian language became a serious impeding factor for the local population with regard to a participation in political and public life, thus diminishing chances for ethnic minorities to achieve professional and economic success in Georgia during the transitional period. Besides, lack of or bad knowledge of the state language raises doubts among the Georgian population whether ethnic minorities are loyal towards the Georgian state, thus widening a gap between the majority of ethnic Georgians and ethnic minorities, practically excluding ethnic minorities' involvement in the state building process.

Lack of contact with the Georgian-speaking population and several gaps which existed in the education system in the 1990s have led to a low level of proficiency in Georgian in the Kvemo Kartli and Javakheti. Finding qualified teachers of Georgian for non-

Georgian schools was a problem of special concern. Often ethnic Georgian teachers sent to Azerbaijani/Armenian schools did not know the students' mother tongue and therefore communication between teacher and the students was simply impossible. Apart from this, the local population complained about a low qualification of teachers.[8] Hence, teaching Georgian carried a formal character only and did not raise the level of proficiency in the state language among the ethnic minorities.

In line with the opinions found among the representatives of ethnic minorities, it would, from the perspective of integration, have had a larger effect to send children to Georgian language schools. However, in this regard they also fear that the existing teaching system would lead to their assimilation and thus diminish ethnic self-esteem. For instance, there are doubts among religious Muslim Azerbaijanis that in case their children are sent to the Georgian language schools, lessons of religion and practical teachings about Christian religion will have an impact on childrens' religious stance.

Since the 1990s, representatives of ethnic groups have mainly sent children to schools teaching in their native tongues. The Georgians go to Georgian language schools, Armenians to Armenian, and Azerbaijanis to Azerbaijani schools. As a result of this trend the number of Russian language schools decreased in Georgia and the level of teaching the Russian language dropped as well.

Starting in 2004, raising the level of proficiency in the state language among ethnic minorities has been one of the main objectives of a reform initiated in the field of education. This should facilitate the minorities' integration. Ongoing reforms raise hopes that in the years to come the level of proficiency in the Georgian language among ethnic minorities will increase. Nevertheless, to eliminate similar problems among the elderly population will be practically impossible. For a short while, the young population will still continue to emigrate in order to receive an education. According to the trend established during the transitional period, young Russians go to the Russian, Armenians to the Armenian, and Azerbaijanis to the Azerbaijani higher education institutions to continue their studies. As a rule, the young population, which has emigrated for educational purposes, prefers to stay put, as according to their perception a return to Georgia is not associated with any prospects. As a result, the

[8] Baazov, Zaza, "Georgian Azeris locked out by language," *Institute for War and Peace Reporting,* 145, 5 September 2002, http://www.iwpr.net/?s=f&o=159818&p=crs&l=EN& apc_state=hena-Georgian%20Azeris%20locked%20out%20by%20language_2 ______publish_date_1_10_compact (accessed January 10, 2008).

most promising of the young population of ethnic minorities become potential emigrants due to a lack of knowledge of the state language.

3.2 Informative Isolation and the Media

Lack of knowledge of the state language has also fostered the informative isolation of ethnic minorities during the transitional period. In addition, non-professional coverage of events in the Kvemo Kartli and Javakheti by the Georgian media had a serious impact on forming a negative attitude towards ethnic minorities among the Georgian public and the ruling elite. All this widened a gap between ethnic minorities and the rest of the society and created a skeptic attitude among ethnic minorities towards the Georgian-speaking media.

Faced with the conditions of media pluralism, which emerged after the dissolution of the Soviet Union, ethnic minorities mainly chose non-Georgian media. Such a choice was preconditioned by lack of or bad knowledge of the state language by ethnic minorities. When ethnic minorities, especially the ones compactly settled in Kvemo Kartli and Javakheti, which do not speak Georgian, were confronted with Georgian television channels broadcasting in Georgian, they - from the perspective of receiving information - became more dependent on the media means of neighboring countries (Russia, Azerbaijan, Armenia, Turkey). As a result, ethnic minorities are far better informed about developing events in neighboring countries and practically do not possess information about processes developing in the Georgian capital and other regions.

Moreover, biased coverage of developing processes in Georgia by foreign media periodically created problems in the relations between ethnic minorities and the central authorities. For example, with respect to the withdrawal of the Russian military base #62 located in Javakheti in Akhalkalaki, the Russian and Armenian media often spoke about the Georgian authorities' intention to replace Russian trops with the NATO troops in Javakheti, and particularly with the Turkish military contingent. This was perceived by local Armenians as an attempt to force them to migrate from Javakheti.

Printing media of ethnic minorities are weakly developed in Georgia, and central or local television channels in Kvemo Kartli and Javakheti have a low rating. Central television channels, despite a high quality of coverage of areas settled with ethnic minorities, only enjoy a small audience due to Georgian language broadcasting. Furthermore, the majority of televisions is privately owned and a topic related to ethnic minorities only becomes of concern to them when rallies, confrontations between various ethnic groups, or notorious criminal events take place in Kvemo Kartli and Javakheti. The only exception is a "public broadcaster" which, since 2005, produces news programs in several languages of ethnic minorities (in Abkhazian, Ossetian, Armenian, Azerbaijani, and Russian). However, due to the non-efficiency of these news programs' broadcastings, they have a low rating among ethnic minorities.

Local television channels operating in Kvemo Kartli and Javakheti, which broadcast in the language of ethnic minorities, also have a low rating due to their limited broadcasting area and extremely low quality of broadcasting. Due to financial or political problems in the region, a majority of these channels does not have the possibility of independent operation and development. The operation of mass media mainly depends on the support of the authority of influential individuals. Due to the non-professionalism of journalists, there is a high degree of danger that an aired report will cause certain dissatisfaction, and therefore owners of media means abstain from covering acute issues.[9] The local population jokes that the main function of local television operating in Kvemo Kartli and Javakheti is to only announce obituaries and congratulations.

3.3 Political and Civil Participation

The degree of participation of ethnic minorities in public-political life during a transitional period is quite low. From the perspective of forming policies in various fields, those individuals or organizations which are exceptions in this regard, exert a certain influence at a regional level only. Their role in the process of important decision making by central authorities is insignificant. Notably, organizations operating in Javakheti are far more agile and politically active in the public-political sphere than organizations based in Kvemo Kartli. As for a non-regional level, the positions of representatives of Georgia's two largest ethnic minorities – Azerbaijanis and Armenians – are weak.

In general, several factors predetermine a low degree of participation of ethnic minorities in public-political life: due to the lack or bad knowledge of the state language it is more difficult to find a job in a public office, where cases are processed in Georgian; ethnic minorities are distinct for their political passiveness and weak political mobilization; during the post-Soviet period, a vicious tradition has been established in the areas compactly settled with ethnic minorities to support political groups in the

9 Vartanian, Olesey, "Mass Media Means of Javakheti, Bulletins of the Center for Extreme Journalism, http://www.cjes.ru/bulletin/?bulletin_id=2314 (accessed May 17, 2007).

elections, which again support the authorities; after the 1990s, professional cadres from ethnic minorities have emigrated en masse.

The public organizations of Kvemo Kartli and Javakheti were exclusively active in Georgia's public-political life during the first half of the 1990s. Against the background of nationalism which was rampant in the country at the outset of the 1990s and ethnic conflicts that developed in the South Caucasus, organizations with an ethnic profile became popular in Kvemo Kartli and Javakheti. Demands to grant autonomous status to Kvemo Kartli and Javakheti were invigorated as well.

The "Javakhk" movement, established on ethnic grounds in Javakheti, demanded a protection of the Armenians' cultural heritage and the teaching of Armenia's history and other subjects in Javakheti-based schools. During the post-Gamsakhurdia period, this movement became further organized and oriented itself towards the protection of rights of the Javakheti-based Armenians. In 1994 these organizations and the Javakheti population were especially dissatisfied with the ordinance of the president, by virtue of which Samtskhe and Javakheti were transformed into one administrative unit, with the president appointing his representative (an ethnic Georgian) to head the unit. Local Armenians stated that by such a merger the authorities attempted to misbalance the demographic picture of the region. Following this merger, the Armenians did no longer represent an absolute majority in the new administrative unit of Samtskhe-Javakheti, Akhaltsikhe being its center.

Nowadays the local organizations "Javakhk" and "Virk" periodically raise their voices to demand autonomy within Georgia, but their influence in recent years has significantly diminished. To a certain extent, this was predetermined by the Russian Federation's closing of military base #62, located in Akhalkalaki, in 2007. During the tense Georgian-Russian relations, this military base was one of the main supporters of anti-government protests in Javakheti. Owing to this and other factors, the influence of Georgia's central authorities in Javakheti was very weak. For instance, in August of 1998 the local population did not let the troops of the Georgian army into the region to attend planned military trainings. Georgia's Minister of Defense was obliged to call Georgian military troops back in order to avoid clashes with the local population.

Demands of these organizations have a weak back-up from Armenia, which during the period of the Karabagh conflict at the outset of the 1990s facilitated a mobilization of volunteers and material assistance from Javakheti. In the following years, during Armenia's isolation, Georgia became a vital transit country for Armenia and support of organizations like "Javakhk" and "Virk" decreased to a minimum.

Demands of the autonomy status were less vigorous in Kvemo Kartli. At the beginning of the 1990s, several confrontations occurred between the Georgian and Azerbaijani populations in Kvemo Kartli, and particularly in the Marneuli and Bolnisi districts. No bloodshed was caused by those confrontations, but the Azerbaijanis living in Bolnisi had to leave town. Against this background, nationalist groups operating in Kvemo Kartli voiced demands to set up the "Borchalo[10] Autonomy" in the territory of the Kvemo Kartli districts settled by Azerbaijanis. Authors of the autonomy idea were nationalistically-minded groups which did not enjoy support among the population.

Owing to the ethnic conflicts initiated in "South Ossetia" and Abkhazia, autonomous units of Georgia at the outset of the 1990s, the authorities were wary of creating a tense point in another region of Georgia and managed to quieten down the situation in Kvemo Kartli. After this agiotage there are no more open talks about the "Borchalo Autonomy" and practically this issue is not on the agenda any more. Any statement demanding autonomy provokes a negative attitude among the local population, and the authors of such ideas are viewed as instigators.[11]

Throughout the transitional period, the degree of representation of ethnic minorities in the government agencies, political parties, or non-governmental organizations has been low. The presence of representatives of ethnic minorities in the country's supreme legislative body is perceived as a tool of the authorities to demonstrate the country's multi-ethnicity. It is mainly the ruling parties which offer the representatives of ethnic minorities guaranteed places on the party lists or support them less vigorously in the majoritarian districts.

After 1992, the smallest number of ethnic minorities was represented in the calling of the parliament of 1992-95. Afterwards this number increased significantly in the callings of the parliaments of 1995-99 and 1999-2004. An absolute majority of representatives of ethnic minorities ended up in the parliament as a result of the authorities' support in either proportional or majoritarian elections. In the parliament elected after the Rose Revolution, their number decreased to nine. This can be explained by a political competition when the allocation of places to representatives of ethnic minorities on the party lists turned into a luxurious undertaking.

In various periods, representatives of different ethnic minorities were elected to the Parliament of Georgia, but the ruling political force primarily tries to keep the balance between the two largest ethnic groups – Azerbaijanis and Armenians. Therefore, their

[10] Borchalo is a historic region in Kvemo Kartli. The name comes from the Turkmen tribe Borchalu, settled by Shah-Abas in the Debedi Gorge at the beginning of XVII century.

[11] Gia Nodia (edit.), Ethnic-Confession Groups and Problems of Civil Integration in Georgia: Azerbaijanis, Javakheti Armenians, and Muslim Meskhetians (Tbilisi, Caucasian Institute for Peace, Democracy, and Development, 2002), 8-9, 17-18. (In Georgian)

representatives have guaranteed places in the Parliament in more or less equal numbers. In the current Parliament, out of the nine ethnic minority MPs there are five Armenians and three Azerbaijanis. When the Azerbaijanis became the largest ethnic minority during the transitional period, this balance periodically caused dissatisfaction. Owing to this, the government that came to power as a result of the Rose Revolution in November 2003 is frequently accused of pursuing the "pro-Armenian" and "anti-Azerbaijani" policy.[12]

Due to a lack of proficiency in the state language, the degree of parliamentary activity of MPs from Kvemo Kartli and Javakheti is very low. Their activities are more a formal attempt to demonstrate that ethnic minorities are involved in political life. Society is not informed about their activities, as they either do not attend the sessions or do not deliver speeches at the parliamentary sessions, due to their lack or bad knowledge of the language. In this regard it should be noted that their activities in recent years are mainly related to a commemoration of the date of mass killings of Armenians in the Ottoman Empire in 1915. Demands to recognize this fact as genocide were actively voiced in 2002. After that, ethnic Armenian MPs make a special statement on the Armenian genocide every year, to which ethnic Azerbaijani MPs have a harsh reaction.[13] Activities of ethnic minority MPs are mainly confined to greeting the country's leadership when visiting Kvemo Kartli and Javakheti and visiting Armenia and Azerbaijan together with the parliamentary delegations.

Viewed from the perspective of elections, a vicious tradition was established in Kvemo Kartli and Javakheti during the post-Soviet period to support pro-authorities political groups in the elections. This factor is actively abused by every government, hence fostering political indifference among the ethnic minorities. As a rule, the pro-government political groups in Kvemo Kartli and Javakheti receive a guaranteed number of votes and the opposition political groups do not, in practice, conduct election campaigns in these regions. A distinguished picture was created at the time of the November 2003 parliamentary elections, when the opposition party "National Movement" conducted an active campaign in Kvemo Kartli. At the same time, an extremely negative evaluation was given of an unprecedented fact in Georgian political reality, when a person of Azerbaijani nationality became one of the faces (Kamal

[12] E. Alekperov, "We Support Eduard Shevardnadze", *Echo*, 10 March 2003, http://207.218.206.33/~admin213/facts.shtml#2 (accessed March 10, 2003). (In Russian)

[13] "Armenian-Azerbaijani "War" in the Parliament", *Alia*, 25-26 April 2002. (In Georgian)

Muradkhanov) of the opposition party. As one part of the Azerbaijanis put it, such a decision simply disgraced the Azerbaijanis.[14]

However, the existing attitude against the "National Movement" has changed after their coming to power, when the leader of the "National Movement" Mikhail Saakashvili collected the absolute majority of votes on 4 January 2004 in the snap presidential elections. In the snap parliamentary elections of 28 March 2004, Saakashvili's political organization "National Movement – Democrats" collected 76% of the votes in Kvemo Kartli.[15]

Similar results were achieved by Shevardnadze during his rule. The changing loyalty of the Kvemo Kartli Azerbaijanis clearly demonstrates the delicate relations that exist between the largest ethnic minority in Georgia and the central authorities. According to a perception which exists among Azerbaijanis, they are afraid the opposition may change its opinion and forget about them. Therefore, they support the pro-government political groups in all the elections.[16]

Since the 1990s, the participation of ethnic minorities in the elections is viewed as an expression of loyalty towards the state. Therefore, they have always supported the pro-government political groups. Apart from this, the lack or bad knowledge of the state language and the existence of ballots only in Georgian[17] makes manipulations of their votes a simple process. Owing to the mentioned factors the most serious violations were noted in Kvemo Kartli and Javakheti.[18]

[14] T. Mchedlishvili, *"The Most Dangerous Election Scenario"*, 24 Hours, 27 September 2003. (In Georgian)

[15] M. Miri, "Saakashvili Counts on Support of the Azerbaijanis", Zerkalo, 15 December 2003, www.zerkalo.az (accessed December 15, 2003). (In Russian)

[16] Lobzhanidze, Tea, "Georgia: Azerbaijanis Back Saakashvili", Institute for War and Peace Reporting, 225, April 1 2004, http://www.iwpr.net/?p=crs&s=f&o=161063&apc_state=henicrs2004, (accessed January 10, 2008).

[17] Ballots and election materials were first translated into the languages of ethnic minorities for the spring parliamentary elections of 2004. For the next elections, the Central Election Commission was more active in this direction.

[18] It should be noted that violations of the election process took place all over Georgia, but Samtskhe-Javakheti and Kvemo Kartli should still be distinguished seperately. For example, please refer to: Office for Democratic Institutions and Human Rights, Georgia, Parliamentary Elections 31 October – 14 November, 1999 Final Report, Warsaw, 7 February 2000; Office for Democratic Institutions and Human Rights, Republic of Georgia, Presidential Election, 9 April 2000, Final Report, Warsaw, 9 June 2000.

In the areas of dispersed settlements, ethnic minorities are almost not represented in the self-governance agencies. Compact settlement areas are an exception. For instance in Javakheti, offices are mostly occupied by Armenians, whereas in Kvemo Kartli it is rare that to find Azerbaijanis in a high-level office. Unlike the Azerbaijanis, Armenians are more active and organized in the public-political life and respectively hold both elective and appointing offices at the regional level.

Specificity of elections and political passiveness of the local population in Kvemo Kartli might be factors which predetermine a small number of the Azerbaijanis in the elective agencies. Since the 1990s, central and local authorities have made decisions on a practical level for the local population of Kvemo Kartli. Thus, President Shevardnadze and his supporting political organizations almost always enjoyed a hundred percent support.[19] Opposition parties simply could not collect any votes. Azerbaijani authorities, who reacted harshly to any protests against the Shevardnadze government, also urged local Azerbaijanis to support Shevardnadze.[20]

The degree of participation of ethnic minorities in the political parties and especially in the opposition parties is low. Representatives of ethnic minorities are practically never nominated for high party positions. Following the Rose Revolution, a Kabardian and an Ossetian were appointed to the government as ministers. The latter was a State Minister for Civil Integration. Traditionally, the authorities try to observe a balance, mainly between the Azerbaijanis and the Armenians.

In the recent period, regardless of the more active role of non-governmental organizations, their influence within the civil sector is still weak. Their pro-activeness is more noticeable at the regional level, in Javakheti and Kvemo Kartli. Activities of organizations founded by representatives of ethnic minorities serve mainly as preservers of language, culture, and traditions of their own ethnic groups.

[19] Despite the fact that during the Shevardnadze rule socio-economic conditions of the Azerbaijanis did not witness any substantial changes, there was an undisputed trust in him among the population. For one thing, this was also preconditioned by the experience of the recent past, when during the Gamsakhurdia rule nationalistically-minded organizations were active, as well as by Shevardnadze's friendship with Azerbaijan's leader Heidar Aliev, who enjoyed a vast authority among the Azerbaijani population of Georgia.

[20] Azerbaijani authorities used to send their close relatives to support President Shevardnadze. For instance, prior to the parliamentary elections of 2 November 2003, Heidar Aliev's brother arrived in Kvemo Kartli to propagate Shevardnadze's political movement: *Z. Kechakmadze, "Aliev's Brother Assists Georgian Authorities Prior to the elections"*, Rezonansi, 7 October 2003. (In Georgian)

4 Involvement in Economic Life

Ethnic conflicts, political instability, and the breakdown of traditional economic ties between the Soviet Republics emerging after the dissolution of the Soviet Union led to a collapse of economic life all over the country, including Kvemo Kartli and Javakheti. In the 1990s the economic infrastructure was practically completely destroyed, enterprises were shut down, communications went out of order, and serious problems emerged with regard to power supplies. Owing to these reasons a big part of Georgia's able-to-work population, regardless of ethnic belonging, left Georgia on a massive scale in labor emigration. It is noteworthy that money remittances sent by labor migrants have become a significant source of income for the population during the transitional period.

Regardless of the country's socio-economic problems which emerged in the 1990s, the population of Kvemo Kartli and of Javakheti managed to create a more self-sustainable economy as compared to other regions of Georgia. On account of exporting agricultural products to internal Georgian markets, especially in Tbilisi, as well as to neighboring Armenia, Azerbaijan, and Russia, it became more or less possible to maintain traditional fields of agriculture in these regions.

Small and medium business was a main source of income for a majority of Kvemo Kartli Azerbaijanis during the transitional period. Since the 1990s, Azerbaijanis have earned their income mainly by importing home appliances and other goods from Azerbaijan and selling these on wholesale markets. The import of goods to Georgia was chiefly connected with corrupt bargains with the state structures, which turned into a "traditional" economic activity during the transitional period.

Besides illegal trade on the Georgian-Azerbaijani border, agriculture still plays a significant role in the economic life of Azerbaijanis as one of the most developed fields on a country scale. Agricultural products of the region are of the best quality, and owing to the short distance from the capital, Tbilisi's population is also supplied with these goods.

These factors' integrity has ensured the economic self-sustainability of the Azerbaijanis over the years. However, as a result of developments after the Rose Revolution, the population was faced with new economic challenges. The shut down of the Sadakhlo wholesale market on the territory adjacent to Georgia-Azerbaijan-Armenia, which was accused of trading with smuggled goods and of tax evasion, caused big harm to small and medium businesses in the region. Not only Azerbaijanis, but also people from Azerbaijan, Armenia, and other regions of Georgia used to trade on this market. In addition, giving a legislative framework to the import of goods at the Georgian-Azerbaijani border caused a serious blow to the import of smuggled goods, thus making an import of goods from Azerbaijan less attractive income-wise.

As a result of impeded "traditional" economic activities, the population drew its attention to the agricultural lands, a fact which turned into an economic and political problem in Kvemo Kartli after the 1990s. Due to the non-awareness of legislation adopted in the 1990s, a big portion of the land was taken by large lessees, and a majority of the population was left without lands. Over the years the population has therefore had to sub-lease land parcels from the large lessors for comparatively high prices. Moreover, existing legislation pronounced a 21km border line area as an exclusive zone, occupied by non-profitable military agricultural enterprises. According to the statements of local Azerbaijanis, these lands were the ones which were best suited for agricultural purposes, but the legislation prohibited privatization of these land parcels.[21] Local Azerbaijanis viewed the existing system of land distribution as a discriminatory policy against them, aimed at oppressing them and forcing them to emigrate. After the Rose Revolution a process of land distribution commenced among the Azerbaijani population of Kvemo Kartli, but nevertheless the income generated from the sales of agricultural products still does not fully compensate for incomes lost due to limitations imposed on the trade with smuggled goods.

Unlike Kvemo Kartli which is located near the capital of Georgia, Javakheti, which is located in a high mountainous zone and is called the "Siberia of Georgia" due to harsh climate conditions, was completely detached from other regions of Georgia. Citizens of neighboring countries and employees of the Russian military base located in Akhalkalaki were the main customers of agricultural goods produced in Javakheti. Because of the conditions of corruption that exist on Georgia's highways, trading on internal Georgian markets was not profitable for the population of Javakheti, due to which the economic ties between Javakheti and the rest of Georgia weakened significantly.

First signs of Javakheti's return to the economic life of Georgia have been noticeable over the past years. The launch of a large-scale road and infrastructure program[22], which was effected in 2004 after the Rose Revolution, and a real extermination of

21 Jonathan Wheatley, Obstacles Impeding the Regional Integration of the Kvemo Kartli Region of Georgia, (ECMI Working Paper #23, February 2005), 7-8.

22 Within the framework of the Samtskhe-Javakheti Road Rehabilitation Project of the Millennium Challenge Georgia program, restoration of the road-transport network in Samtskhe-Javakheti is underway.

corruption on the highways of Georgia[23] will more or less resolve a problem of Javakheti's isolation and accelerate its integration with other regions of Georgia in the years to come.

Nowadays, ethnic minorities are less represented in the business elite due to a lack of the necessary political connections required for conducting large business operations. Representatives of ethnic minorities are mainly involved in small and medium businesses. For instance, they hold leading positions on the wholesale markets of home appliances and agricultural goods.

The Azerbaijanis and Armenians residing in Kvemo Kartli and Javakheti are involved in relatively large businesses. Their businesses are backed by clan relations established in the 1990s and existing ties with their historical homelands (Azerbaijan, Armenia).

5 Legal and Institutional Aspects of Integration

The protection of ethnic minorities became an issue of utmost concern after the dissolution of the Soviet Union, when in Georgia as a newly independent state Georgians turned out to be in majority and other ethnic groups in minority. Since the 1990s, more practical rather than legal problems have arisen in light of the protection of ethnic minorities. From a legal perspective, the Georgian Constitution and other laws directly or indirectly provide for the protection of ethnic minorities.

An important step for the protection of ethnic minorities was made in 1999, when parallel to the affiliation with the Council of Europe, Georgia signed the European Framework Convention on the Protection of Ethnic Minorities. Regardless of a signature, ratification of the Convention was protracted in time, as the political elite doubted that obligations foreseen under the Convention would instigate separatist aspirations among the ethnic minorities. After the Rose Revolution, on 13 October 2005, the Parliament of Georgia finally ratified the Convention, which foresees enactment of relevant legislative acts in order to enforce principles conveyed in the Convention.

At the same time, the execution of international obligations creates an institutional type of problems for the state. According to the ratification of the Convention the state is

[23] Within a framework of Samtskhe-Javakheti Road Rehabilitation Project of the Millennium Challenge Georgia program, restoration of the road-transporting network in Samtskhe-Javakheti is underway.

obliged to implement specific programs in all fields of ethnic minorities' economic, social, political, and cultural life in order to carry out their integration.

Since 2004, the authorities have begun to take active steps towards their integration and have established the position of State Minister for Civil Integration in order to coordinate the policy of civil integration of ethnic minorities. The State Minister's staff is responsible for the enforcement of the civil integration policy. Apart from this, a permanent committee working on ethnic minorities is active in the parliament of Georgia since 1995. One of the objectives of the committee is a full fledged integration of the representatives of ethnic minorities into society.

In recent years the state appears to be ready to take further practical steps from the civil integration perspective and go beyond the frames of rhetoric statements. Efficiency of the civil integration policy will largely depend on state agencies operating in the field of integration and on the coordinated work of local non-governmental and international organizations. At this stage, steps made towards civil integration are devoid of systemic approach, which might lead to an ineffective spending of resources in the integration process and to dragging out the process itself.

6 Conclusion

The problems which emerged in Georgia in the 1990s as a result of the dissolution of the Soviet Union led to the detachment of ethnic minorities from public-political and economic life and their isolation. The established unstable environment gave birth to the feeling that life in Georgia was unrealized, thus fostering massive emigration of ethnic minorities. A part of the ethnic minorities remaining in Georgia, particularly the Azerbaijanis and Armenians, compactly settled in Kvemo Kartli and Javakheti and encountered the most difficult challenges of civil integration. Steps taken in recent years towards their integration provide a perspective of full fledged involvement of ethnic minorities in the country's life. Nevertheless, acute problems amassed during the transitional period turn this perspective of shaking off isolation into a long-term one.

II. ECONOMICS AND SOCIETY

CHALLENGES OF DECENTRALISATION IN THE REPUBLIC OF GEORGIA

TANJA KIRN, ELGUJA KHOKRISHVILI

1 Introduction

Over the past decades, a global trend towards fiscal decentralisation has emerged and a number of countries have undertaken comprehensive reforms of their intergovernmental fiscal systems. This trend is evident in developed, transition and developing countries The global trend is based on the wide acceptance of the principle of subsidiarity and on the view that the realization of this principle results in improved efficiency and delivery of public services and, hence, a more efficient allocation of resources in the economy. Besides being "economically efficient", fiscal federalism can promote democratic and participatory forms of government, improving the accountability and responsiveness of politicians and bureaucrats. Although the positive effects of decentralisation are well acknowledged, the success of decentralisation depends on its design. The right mix of policies, the question of sequencing and pace, and the balance of countervailing powers affect the outcome of the decentralisation process.

After the "Rose Revolution" an ambitious reform agenda was launched to improve the fiscal system and to ensure Georgia's territorial integrity. It is far from clear whether decentralisation in response to regional tensions can increase instability or political stability. We will outline the process of decentralisation and try to identify obstacles and challenges for the future reform process.

2 Dimensions of Decentralisation

Decentralisation is a broad topic that includes aspects of the institutional framework, the assignment of expenditure responsibilities and revenue sources to subnational governments. The World Bank, among many others, makes the distinction between fiscal, administrative and political decentralisation.

According to Schneider (2003), we identify basic elements that capture some key aspects needed for successful decentralisation reforms in transition countries.

- Fiscal decentralisation refers to the extent to which central governments cede fiscal power to non-central government entities. It requires a clear assignment of expenditure responsibilities and a transparent revenue assignment system. In most of the transition countries, the expenditure autonomy is still limited at the sub national level, and the revenue assignment leads to a weak fiscal capacity at the local level.
- Administrative decentralisation refers to how much administrative autonomy non-central government entities have. Administrative decentralisation is generally categorized by the terms of deconcentration, delegation and devolution (Rondinelli, 1990).
- Political decentralisation refers to the degree to which central governments allow lower level government entities to undertake the political functions of governance. It requires local elections in which local parties/political actors compete. Due to the bequeathed vertical power structure of the Soviet era the administrative and political decentralisation requires far reaching restructuring processes.

These dimensions are interrelated and can interact. One possible pattern is that increasing decentralisation in one dimension leads to an increase in decentralisation in another dimension. Nevertheless, the attempts to find causal relations between these dimensions have produced mixed results. For example, changes in fiscal decentralisation can affect administrative or political decentralisation in a positive or negative way. Therefore, the issue of matching fiscal, political and administrative arrangements to achieve the optimal mix is difficult. Another related question of designing decentralisation is that of proper sequencing. Case studies show various results concerning the question whether fiscal decentralisation should precede or follow political decentralisation

3 Challenges of Decentralisation

Due to the complex fiscal, administrative, and political interactions the process of decentralisation becomes a complicated matter. This is particularly true in former socialist countries. Tax bases have eroded with the economic decline related to the transformation process while corruption and the shadow economy have spread. Most of the transition countries have inherited centralized structures and intra-state tensions occurred after the collapse of the Soviet Union.

The rise of Georgian nationalism during the early years of the transformation process evoked fears of ethnic minorities that escalated in bloody conflicts and secessionist movements. To freeze the conflict, wider political, administrative and fiscal powers were provided to the Autonomous Republic of Adjara, which was one of the seceding regions. Since the Autonomous Republic of Adjara enjoys a distinct form of autonomy, the State of Georgia can be characterized as a partly decentralized unitary state with an asymmetrical federalism. Although the case of the Autonomous Republic of Adjara, however, does not serve as an example of a successful decentralisation strategy, since the autonomy is very weak. The status of Abkhazia and South Ossetia is still unclear, and the intra-state tensions remain unsolved.

Decentralisation reforms can contribute to the resolution of regional conflicts by assuring local governments to perform their targets free from central interference, with the exception that the central government should keep the role of coordinator and assessor of the local government's achievements. The first steps toward decentralisation are made, but further advances in fiscal, administrative and political decentralisation have to be made.

4 Intra-state Tensions Escalated

The modern Republic of Georgia[1], as well as the ancient Kingdom of Georgia and the historically independent state of Georgia, encompass different ethnic groups. The major ethno-linguistic groups within Georgia are Georgians, Abhazians and Ossetians. Ethnic tensions always existed in Georgia but were suppressed by imperial rule. During the Soviet era Georgia was a union republic, the regions of Abkhazia, Adjara, and South Ossetia had the status of an "Autonomous Soviet Socialist Republic" within the Georgian SSR. The delineation of the autonomous republics corresponded to the settlements of ethno-linguistic groups in Abkhazia and South Ossetia. Despite the autonomous status, the ethnical tensions were not resolved. The autonomy of the autonomous republics was limited due to the centralized power structure. The splitting of ethnic groups over several territories, forcible settlements and the artificial merging of different people into single republics has enforced ethnic conflicts.

Shortly before the collapse of the USSR, the Republic of Georgia declared its independence on April 9th, 1991. First steps towards decentralisation were taken

[1] The roots of statehood reach back to the 4th century B.C. when a Unified Kingdom of Georgia was established. After a changeful history, the Georgian Kingdom was incorporated into the Russian Empire in 1801. Following a brief independence period (1918-1921), Georgia became a republic of the Soviet Union.

through the introduction of a four-tiered government hierarchy (Bolashvili, 2002). Although the Soviet institutional structure largely remained, independence led to a rapid destruction of political structures. Tbilisi failed to assert its central power at the regional and local level, which caused a power vacuum. Basic municipal services were no longer provided in a sufficient manner and the socio-economic decline worsened the situation (Jackson, 2004).

As Georgian nationalism flourished in the late 80s, the Abkhaz ethnic minority feared their possible cultural and ethnic disappearance within an independent Georgia. Several thousand Abkhaz signed the so called "Lykhny Declaration" which called for the independence of the region (Lynch, 2006). A bloody conflict, which lasted for more than a year (1992-1993), escalated between Abkhaz forces and Georgian National Troops; more than 8,000 people lost their lives. The conflict led to ethnic cleansing and the displacement of 300,000 people (most of them belonged to the Georgian population), a significant part of the pre-war population of 525,000. Although in 1994 the Abkhaz and the Georgian sides agreed to a cease-fire, the situation is tense (UNDP, 2005).

The conflict about the status of South Ossetia escalated in 1991, when President Gamsakhurdia unilaterally abolished the autonomous status of the region. After a bloody conflict between Georgian troops and Ossetian militia, South Ossetia declared its independence in 1992, which is not internationally recognized. Until today, the status of South Ossetia is still unclear (Reisner/Kvatchadze, 2005).

In the Autonomous Republic of Adjara the political situation remained stable during these years of conflict due to the authoritarian rule by the regional leader Abashidze. During the presidency of Shevardnadze the central government in Tbilisi ignored the situation in Adjara, even though Abashidze was involved in organized crime and human rights violations. This changed during the "Rose Revolution" (January 2004), when Saakashvili came to power and claimed to reunite the country. The eruption of an armed conflict was prevented by Saakashvili's ultimatums and mass protests against the autocratic rule; Abashidze resigned and went into exile in Russia. After the elections of the Supreme Council, a constitutional law on the Status of the Autonomous Republic was enacted (ICG, 2004).

Adjara is the only autonomous region in Georgia that succeeded in establishing a power sharing arrangement with Tbilisi. Unlike the regions of South Ossetia and Abkhazia, Adjara did not claim full independence and did not argue for its status on the principle of national self-determination, since the people of Adjara consider themselves ethnically as Georgians.

The status of South Ossetia and Abkhazia is quite different. Although these *'quasi-states'* are weak with regard to their economic situation and deficient state-building capacity, they are remarkably robust. The quasi-states have survived since the collapse of the Soviet Union, and it seems they will continue to survive. The viability of these quasi-states is based on a symbolic nation-building and the militarization of the society. This is further aggravated since Ossetia and Abkhazia enjoy the support of Russia (Kolstø, 2006). Although these factors back the "statehood' of the seceding regions, they are regarded as temporary phenomena. In the future they may be included into the Russian Federation or be reabsorbed into the Republic of Georgia or unite with the Republic of Georgia in a (con)-federal arrangement. After the declaration of their interdependence, an independent state became their maximal negotiable aim, the minimal non-negotiable aim is some form of autonomy in a confederal system (Lynch, 2006). The peace proposals between Tbilisi and the breakaway regions, which were based on a model of federal power sharing, produced no results. Since the separatist authorities have little faith in the guaranteed autonomy the solution of the conflict becomes a complicated matter.

The case of the Autonomous Republic of Adjara could serve as a model for the integration process of the seceding regions. It offers the central government a chance to demonstrate their willingness for decentralisation.

5 Fiscal Decentralisation is Hesitant

When Saakashvili came to power in 2004, he launched an ambitious reform agenda to insure the integrity of the country. The institutional framework was changed into a three-level structure. The first level is formed of 66 Local Government Units (LGUs); the second level consists of nine regions and the Autonomous Republics of Adjara and Abkhazia and the third level is the central state level. Within the State of Georgia the Autonomous Republic of Adjara maintains a special status, while the status of the Autonomous Republic of Abkhazia and South-Ossetia is still undefined (see Petersen, 2008).

Even after this "second round of reforms", Georgia still follows a top-down fiscal policy approach. The financial transfers from the central level to the regional level are negotiated. Since the assignment of expenditure responsibilities is linked to the negotiation based assignment of revenues, there is a lack of basic and encoded criteria for the allocation of responsibilities between the regions, their respective LGUs, and the central state.

5.1 Expenditure Assignment

The degree of fiscal decentralisation can be roughly expressed by the ratio of subnational expenditures to national expenditures, since the assignment of the expenditures to the institutional levels reflects the relative importance of subnational governments. Nevertheless, it has to be taken into account that the assigned expenditures to the local governments can be delegated functions (contract management) or exclusive local functions. In the case of delegated functions the decision about the provision of the public good is made at the central level. The local government and, therefore, the local constituency do not have the competence to decide on the provision of the delegated functions. In the case of exclusive local functions the principle of fiscal equivalence would be achieved if the beneficiaries of public goods and the taxpayers were identical. In this case the local government should not only have the competence to decide on the provision of local public goods. Also the tax base should be sufficient to cover the respective expenditure needs.

Government spending comprises the expenditures at the local and central level and the extrabudgetary funds. Extrabudgetary funds and state budget are under the budget authority of the central government, the LGUs are responsible for the local budgets. The share of subnational governments in the consolidated budget has declined continuously over the last years. The volume of extrabudgetary funds has been increased after the Rose Revolution in order to finance the deferred payment of pensions. To increase the transparency of the budgeting process, the extrabudgetary funds were abolished in 2006. Pensions and the other functions that were paid by this fund are now financed by the state budget (table 1).

Table 1: Central, extrabudgetary and local budgets (2000-2006)

(at current prices, m.GEL)	2000	2001	2002	2003	2004	2005**	2006
state budget	681.4	756.1	1,049.4	1,207.1	1,930.2	2,618.6	3,821.4
extrabudgetary funds	179.9	206.8	238.1	234.5	452.3	428.8	0
local budget	324.7	397.5	405.2	457.6	637.5	826.0	873.9
total	1,186.0	1,360.4	1,692.7	1,899.2	3,020.0	3,873.4	4,695.3
share of local budget* (in %)	27.4	29.2	23.9	24.1	21.1	21.3	18.6
* local government expenditure / total government expenditures							
** extrabudgetary funds excluding infrastructure funds							

Source: Ministry of Finance (MoF), Republic of Georgia.

The revenues from the personal income tax are directed towards the LGUs to finance - as a delegated function – social spending at the local level. The regional level, which was previously involved in the planning and the implementation of regional development programmes, has no expenditure functions. The power of the regional level is limited to a representative and control function, like in Romania, Poland and Hungary. This has led to a de facto two-tier government structure according to the expenditure assignment.

In 2006, the assignment of local expenditure responsibility was revised. The revision of the Organic Law on Local Self-Government and Government[2] led to a stronger decentralisation. The LGUs now have exclusive functions in the education and cultural sphere. Delegated functions at the local level comprise health and social affairs at a minimum service level. Special investment projects, such as regional infrastructure programmes, which are part of a central government framework program, are also delegated to the local government (Narmania, 2008). The expenditure assignment is different in the Autonomous Republic of Adjara, which has a de facto three-level government structure.[3] The competencies of the central state level are confined to the field of public security; the government of Adjara has competencies in the fields of higher education, the cultural sphere and regional development.

Table 2: Regional and local expenditures in Adjara

(thsd.GEL)	2001	2002	2003	2004	2005	2006
regional budget	55,738.1	74,551.7	77,256.7	47,621.8	50,518.8	80,398.1
local budget	19,183.0	27,867.2	30,024.3	36,335.1	52,119.1	47,918.6
share of local budget (in %)*	25.6	27.2	28.0	43.3	50.8	37.3
*local government expenditures / total government expenditures						

Source: MoF, Autonomous Republic of Adjara.

The volume of the regional budget decreased after the Rose Revolution in 2004, and recovered with the revision of the expenditure assignment in 2006, which strengthened

2 The Organic Law on Local Self-Government and Government also applies in the Autonomous Republic of Adjara. Additional competencies may be assigned to the local level in accordance with the Law of the Autonomous Republic of Adjara.

3 According to the Law of the Status of the Autonomous Republic of Adjara (adopted in 2004).

the regional competencies (table 2). The share of local government spending could not be quantified exactly, since the spendings of the central state level in Adjara are not available. If we consider the total of the regional and local budget as a proxy, this rough measurement of the share of local level expenditures yield 37%, which is a fairly higher degree of expenditure decentralisation than the average share of local budgets across the nation (see table 1).

Table 3 shows the share of exclusive and delegated functions as a percentage of the total expenditures at the LGU-level in Adjara. Considering the share of order management, which amounts to 35% of the local budget, the "true" degree of expenditure decentralisation, measured as the share of exclusive functions at the local level, was approximately 24% (65% of 37,3%) in 2006. The share of exclusive and delegated functions varies significantly within the considered period. This illustrates the unsteady assignment of expenditure responsibilities.

Table 3: Share of exclusive and delegated functions in Adjara (2006)

(thsd. GEL)	total	Delegated functions	share of delegated functions (in %)
total budget	128,316.74	43,685.50	34.0
regional budget	80,398.10	27,000.00	33.6
local budgets	47,918.64	16,685.50	34.8

Source: Ministry of Finance in Adjara.

The reform of the Organic Law on Local Self-Government and Government led to a more decentralized provision of public goods and services. Nevertheless, the effective expenditure autonomy at the subnational level is also limited in Adjara. The regional level has no explicit competencies and the burden of delegated functions restrains the autonomy at the local level.

Depending upon the type of spending responsibilities assigned to local governments, there are some potential problems with having many small local governments and a vanishing intermediate tier. First, some services may not be delivered efficiently at the local level for reasons of economies of scale, resulting in both low service levels and high average cost. An intermediate level could provide these services efficiently or could coordinate spending on some certain services among localities. Secondly, without an intermediate tier it is practically impossible for the central government to maintain direct bilateral relations with so many units and the fragmentation may reduce the voice of local government as a whole (Wallich, 1996).

5.2 Revenue Assignment

A concrete and stable expenditure assignment is the basis for an effective decentralisation strategy. It should be supplemented by an adequate tax revenue assignment and fiscal equalization mechanism to guarantee the continuous provision of services. This protects local government from ad hoc transfers of responsibilities and increases the reliability of budget planning at all levels of government. Subnational governments should have the autonomy to set tax rates on significant tax sources that are assigned to the subnational level to ensure accountability.

5.2.1 Assignment of Tax Revenues

In Georgia, the tax bases and tax rates are defined by the central state government; concurrent tax powers do not exist. The revenues of the state budget consist of the tax revenues from VAT, excise taxes, customs duty, and profit tax. The revenues arising from income tax are transferred to the LGUs according to the Budget Law of the State of Georgia, which is valid for only one household period and can be easily changed by the Parliament of Georgia. Property taxes and taxes on gambling business are assigned to the local level.

The state taxes (VAT, excise taxes, customs duty, and profit tax) are the most important sources of tax revenues (about two thirds of total tax collection), whereas the local taxes generate fairly low revenues for the majority of the LGUs. Revenues from income tax barely reach 12% of total tax revenues (table 4). More than 90% of LGUs do not have revenues from the property tax on natural persons and, with a few exceptions no LGU receives revenues from taxes on gambling business. The contribution of the property tax is low due to a variety of tax exemptions. In 2005 the average household income from property was 525 GEL (Jastrzembski, 2008); a family, whose annual income does not exceed GEL 40,000, is exempt from the property tax of a natural person. A farmland plot (up to five hectares) is exempt from a farmland tax. LGUs with many land plots of more than five hectares are rare, they probably account for less than 20% of all self-government units.

Unlike other regions, the Autonomous Republic of Adjara receives revenues from profit taxes collected in its jurisdiction. According to the principle of fiscal equivalence the profit tax is generally considered one of the best taxes to be shared with or assigned to the local government. In the case of the Georgian tax system the principle of fiscal equivalence is not fully assured, since most of the larger taxpayers are registered in Tbilisi and therefore the tax collection takes place in the capital. The Autonomous Republic does not receive an equivalent share of the tax revenues corresponding to the economic activities in its jurisdiction. This could be eliminated, if the tax revenues were shared according to the economic activities across the different jurisdictions.

Table 4: Tax revenues by source (2000-2006)

	2000	2001	2002	2003	2004	2005	2006
income tax (in %)	12.7	14.2	13.5	12.8	14.7	12.1	12.3
profit tax (in %)	9.3	6.8	7.8	8.5	8.8	8.7	10.8
value added tax (in %)	32.7	36.4	38.4	34.1	34.4	40.9	42.3
excise tax (in %)	10.5	9.2	8.2	8.4	9.0	11.9	10.7
customs duty (in %)	6.2	5.8	5.6	5.9	5.5	5.1	4.2
other taxes (in %)	10.2	9.5	10.8	10.7	7.3	3.5	3.8
social tax (in %)	18.3	18.1	15.7	19.5	20.3	17.8	16.0
total (m. GEL)	854.3	954.7	1,054.7	1,192.5	1,827.7	2,411.4	3,149.1

Source: MoF, Republic of Georgia.

The local fiscal autonomy remains limited since the tax rates are defined by the central state level, and the tax revenues are very low. The local tax bases are weak; their contribution to the local budgets is small. This leads to a high dependency of local jurisdictions on transfers from the central level government. Vertical imbalance could be reduced by revenue sharing. For instance, the profit tax could be shared between the central level and local level government. Revenue sharing decreases the vertical imbalance and increases local autonomy and therefore accountability and efficiency. Due to the low local tax revenues and the uneven distribution of the tax bases across the jurisdictions, the tax assignment has to be supplemented by grants.

5.2.2 Assignment of Transfers

The local level receives additional conditional transfers, special transfers and unconditional transfers from the central government. The composition of the local budgets varies from LGU to LGU and among the regions. In 2007[4], transfers financed about 1.9% of local budgets on average. Four regions finance more than 10 percent of their budgets with revenues from transfers (table 5).

[4] According to the projections of the Ministry of Finance (MoF) and own calculations.

Table 5: Categories of transfers per region (2007)

(m. GEL)	con-ditional transfers	special transfers	uncon-ditional transfers	tax and non-tax revenues	total**	transfers as share of total of revenues (%)
Adjara	0.0	0.0	4,555.0	53,888.9	58,443.9	7.8
Tbilisi	0.0	0.0	0.0	771,000.0	771,000.0	0.0
Kaxeti	180.0	0.0	268.0	21,039.0	21,487.0	2.1
Imereti	1,215.0	4,263.0	1,205.0	38,885.2	45,568.2	14.7
Samegrelo-Svaneti	309.0	664.0	793.0	34,035.4	35,801.4	4.9
Shida-Kartli	171.0	482.0	1,208.0	14,541.7	16,402.7	11.3
Kvemo-Kartli	139.0	83.0	1,377.0	44,807.5	46,406.5	3.4
Guria	126.0	0.0	223.0	5,893.6	6,242.6	5.6
Samcxe-Javaxeti	117.0	219.0	126.0	13,707.2	14,169.2	3.3
Mcxeta-Mtianeti	291.0	416.0	757.0	10,168.3	11,632.3	12.6
Racha-Lechxumi	0.0	0.0	891.0	2,645.3	3,536.3	25.2
Total	2,548.0	6,127.0	11,403.0	1,010,612.1	1,030,690.1	1.9
** Includes revenues from transfers, tax and non-tax revenues						

Source: Own calculations, MoF, Republic of Georgia.

Conditional transfers are designed to finance delegated functions. The allocation of the conditional transfers is based on negotiations between the representative of the LGU (sakrebulo) and the Minister of Finance. Unlike other regions the LGUs of Adjara receive conditional transfers from the government of the Autonomous Republic of Adjara. The transfers are negotiated between the representatives of the LGU and the Adjara Minister of Finance. Due to the assignment of transfers based on negotiations the autonomy at the local level is restricted and the reliability of budget planning is limited (Narmania, 2008). This could lead to an insufficient provision of public goods at the local level and undermine the autonomy at the subordinated jurisdictions.

According to the Organic Law on Local Self-Government and Government the special transfers are designed for cases of emergency, for example, natural disasters, special infrastructure programs, and current expenditures. The purpose of the special transfers is quite vaguely defined as it comprises exceptional cases like the case of emergency as well as current expenditures. Furthermore, it makes no distinction between delegated and exclusive functions.

The allocation of unconditional transfers is formula based. It is encoded in the Law on Budget Authorities of Local Self-Government Units, which was adopted in 2007. The transfer is designed as an equalization transfer that reduces differences in fiscal capacity whereby the extent of equalization depends on geographical factors. The unconditional transfers have more the character of an earmarked transfer for mountainous regions, since the transfer volume is low for all other regions. The partial coverage of the fiscal gap does not accomplish the target of fiscal equalisation. To achieve fiscal capacity equalization a higher transfer volume would be required, which could be difficult to finance in the transition process (see Kirn/Khokrishvili 2008).

Altogether, the reliability of the transfers toward the LGUs is limited. Conditional and special transfers depend on negotiations with the superior government, and the assignment of unconditional transfers is still unclear and not reliable. The reliability of the unconditional transfers could be increased if the allocation were to rely on a clearly defined and codified allocation rule, since the transfer volume depends upon the estimation of the MoF about the "potential" fiscal capacity; it seems to be quite subjective (Kirn/Khokrishvili 2008).

A fair allocation of revenues with a low transfer volume could be accomplished by an expenditure-based approach of fiscal decentralisation. However, an expenditure-based approach is difficult to implement, since it requires a broad knowledge of the fiscal need. Historical data for indicators are related to the supply, not the demand of public services. If past decisions do not reflect the current demand of public services, due to political influences in the past or a shift in the demand of public services, regions with excess capacities, other things equal, would continue to receive larger transfers.

6 Lack of Administrative Decentralisation

During the presidency of Shevardnadze corruption had spread to all sectors, and the executive power was known as dysfunctional and corrupt. It was the first priority of the new government to fight corruption and tax evasion. A considerable number of officials were fired, and staff cuts were carried out. The government declared its intention to carry out a comprehensive public sector reform to improve public governance.

However, the administrative decentralisation made little progress, due to the dual character of budget planning with its role in financial planning and strategic planning for the activities at the local level. Most of the decisions and plans were centrally planned and carried out at the local level. The executive authority on the local level is also limited by its subordination to the corresponding ministries. For instance, the local tax administration in Adjara is subordinated to the Ministry of Finance of Georgia, which aggravates the level of dependency of the Ministry of Finance of Adjara on the central level government (Petersen, 2008). Some progress has been made in the modernisation of the tax administration. However, significant tax administration problems remain, encompassing the inability to handle large numbers of new taxpayers, the lack of specialised personnel to detect tax evasion techniques and insufficient resources to build adequate information and monitoring systems. These deficiencies have caused considerable tax evasion and a growth in tax arrears (Martinez-Vasquez/McNab 2000).

7 Insufficient Political Decentralisation

Besides the administrative competence, democratic representation plays an important role in a sound budget process. Equal partners in the sense of competence and legitimacy should bear the coordination and cooperation between different levels of government.

But the constitutional amendments after the Rose Revolution strengthened the executive branch and weakened parliament. The central role of the President was broadened and the post of a Prime Minister was created. The Venice Commission concluded that "this role of the President of Georgia seems democratically questionable and scarcely compatible with a status of autonomy" (Council of Europe, 2004). In addition, the process of political decentralisation gets little support through the weak civil society and self-censored media (PACE, 2004).

The presidential power also dominates the policy at the regional level, especially because the President of Georgia appoints the governors of the regions. The governor has a twofold function: he represents the central government at the regional level and vice versa. Due to the dependency of the governor on presidential approval and appointment, the political scope might be limited and therefore the political autonomy of the regions could be restricted.

Due to the autonomous status of Adjara, the institutions within that region differ to some extend from those of Georgia. The head of the Adjara government is appointed by the President of Georgia, but with regard to other regions, his competences are backed by the Law of the Status of the Autonomous Republic of Adjara. The power of the Government of Adjara is greater compared to those of the regional governors, but still limited[5]. At the local level (Local Government Units, LGU), the respective residents elect the so-called „sakrebulo"(representative body of local government), which in turn appoints the mayors of the villages. Every resident can apply for the mandate of the mayor; the sakrebulo appoints the mayor from the circle of the applicants. If the mayors were to be elected directly (by the residents of the municipality) or indirectly (by a municipal council), the relationship between the local government and its respective constituency would be strengthened.

8 Will an Asymmetrical System Resolve the Conflicts?

It is far from being clear whether decentralisation in response to regional tensions would increase instability or political stability. The issue of asymmetrical federalism arose in nationally diverse states where national minority communities seek self-determination (Hannum, 1996, Laipoth 1997, Watts 1999). Asymmetrical federalism might resolve conflicts within a state, but state nationalists or integrationists fear that it facilitates secession and that it could promote irredentism when minority regions unite with their 'kin-state' across national borders. In addition, asymmetrical federalism persists because of the 'domino' effect, i.e. other regions demand the same degree of autonomy. On the other hand, lessons from the states with asymmetrical federalism do not support the hypothesis that asymmetrical federalism promotes instability. The outcome depends rather on the details of the autonomy arrangements and the motivations of the parties involved. Voluntary agreements are likely to be more stable than those that are the result of dictates. Where autonomy gives rise to serious instability it might require changes in the autonomy agreement rather then its destruction. In order to prevent some sort of polarisation making autonomy insufficient for a minority, it is better to offer autonomy earlier than later and to be generous from the very beginning (McGarry, 2005). Therefore, the degree of local autonomy could serve as an indicator for the outcome of the decentralisation strategy.

Unquestionably, the limited autonomy at the local and regional level in the Autonomous Republic of Adjara is unsatisfactory. Further approaches towards decentralisation are required to enhance transparency and accountability at all levels of government. An

5 For example, the Parliament of Adjara could be suspended by the President of Georgia and acts of the Adjarian Parliament could be set off. Political decentralisation might improve if the Head of Government were elected either directly by the residents of Adjara or indirectly via the Parliament of Adjara

asymmetric system of decentralisation might be an appropriate framework for further progress.

Nevertheless, it has to be taken into account that one important reason for the break-up of Georgia was the fact that the regions were extremely diverse. The break up was not a result of the asymmetrical autonomy, but rather because of the thoughts of unilateral re-centralisation. The experiences with armed hostilities and displacements have aggravated the potential for conflict. The willingness of both conflicting parties to overcome the past is necessary in order to negotiate a common future. A clear commitment of the central state government towards decentralisation could pave the way for a reintegration of the renegade states. Up to now, the case of the Autonomous Republic of Adjara does not serve as an example of a successful decentralisation strategy, since the autonomy is very weak.

An asymmetric system could be the first step towards the integration of the regions Abchazia and South Ossetia, but would probably not lead to a stable solution. An asymmetric system could provoke the "domino effect" where other regions demand the same degree of autonomy. Therefore, it would perhaps be better in the long run to introduce a federal state rather than an asymmetric system. An asymmetric system could serve temporarily as a model for the decentralisation process. In a long-term view the structure of a federal state with strong regions would provide a more stable framework than a central state with regions.

9 References

Bolashvili (2002); Fiscal Autonomy Problems of Local Government in Georgia. In: K. Davey (ed.). Fiscal Autonomy and Efficiency. Reforms in the Former Soviet Union. Budapest: OSI/LGI. p. 59–87.

Council of Europe (2004), Venice Commission, Draft Opinion on the Draft Constitutional Law of Georgia on the Status of the Autonomous Republic of Adjara, Opinion No. 291/2004, Strasbourg.

Ehrke, J. (2008); Zentralisieren durch Dezentralisieren – Die Reform der Kommunalfinanzen in Georgien, Special Series No. G-3, Universität Potsdam.

Hannum, H. (1996); Autonomy, Sovereignty, and Self-Determination: The Accommodation of Conflicting Rights, University of Pennsylvania Press, Philadelphia.

ICG [International Crisis Group] (2004); Saakashvili's Ajara success: Repeatable elsewhere in Georgia?, ICG Europe Briefing, International Crisis Group (ICG), Tbilisi.

Jackson, P. (2004); Ethnicity, decentralisation and the fissile state in Georgia, in: Public Administration and Development, 24, (1), p. 75 - 86

Jastrzembski, A. (2008) Einkommensteuerschätzung in Georgien, Special Series No. G-5, Universität Potsdam.

Kirn, T. and E. Khokrishvili (2008) Will an Asymmetrical System of Fiscal Decentralisation Resolve the Conflicts in the Republic of Georgia?, Special Series No. G-9, Universität Potsdam.

Kolstø, P. (2006); The sustainability and future of unrecognized quasi-states, Journal of Peace Research, 43; 723.

Laipoth, R. 1997, Autonomy: Flexible Solutions to Ethnic Conflicts. United States Institute of Peace Press, Washington D.C.

Lynch, D. (2006); Why Georgia Matters, Chaillot Paper No. 86, EU Institute for Security Studies, Paris.

Martinez-Vazquez, J. and J. McNaab (2000); The Tax Reform Experiment in Transitional Countries. International Studies Program Working Paper. Atlanta: Andrew Young School of Policy Studies, Georgia State University, Georgia/USA.

Narmania, D. (2008); Rights of Local Jurisdictions and Tax Revenue Distribution in Georgia, Special Series G-6, Universität Potsdam.

PACE (2004); Committee on the Honouring of Obligations and Commitments by Member States of the Council of Europe, Honouring of obligations and commitments by Georgia, Doc. 10383, 21 December 2004.

Petersen, H.-G. (2008); Nachhaltigkeit in Finanz- und Sozialpolitik: Probleme und Lösungsansätze für den Transformationsprozess in Georgien, Special Series No. G-1, Universität Potsdam.

Reisner, O. and L. Kvatchadze (2005); Georgien, Studien zur länderbezogenen Konfliktanalyse, Friedrich-Ebert Stiftung.

Rondinelli, D. (1990); Decentralization, Territorial Power, and the State: A Critical Response;, Development and Change, 21 (3), p. 491-500.

Schneider, A. (2003); Decentralization: Conceptualization and measurement, Studies in Comparative International Development, 38 (3), 3256.

UNDP (2005); Fiscal Decentralisation in Transition Economies: Case Studies from the Balkans and Caucasus; Bratislava, 2005. www.undp.org/europeandcis.

Wallich, C. I.; Intergovernmental Finance In The Transition Economies, MOCT-MOST (6), p. 63-86.

Watts, R. L. (1999); Comparing Federal Systems, McGill-Queen's University Press, Montreal.

INVESTMENT CLIMATE: CHALLENGES OF TRANSITION

GEORGE TOMARADZE

1 Introduction

After the collapse of the planned economy in the former Eastern Bloc the transition process became one of the main challenges for former socialist countries. Depending on their initial situation and the consistent implementation (accomplishment) of the chosen transition strategy, these countries achieved different levels of success. One of the main and most interesting challenges was the creation of an appropriate investment climate, which would make it possible to catch the GDP-slumps and encourage a sustainable economic growth. Foreign direct investments have played a major role in its improvement, because after the collapse of the planned economy high inflation rates annihilated almost the whole savings of the affected economy. To reconstruct and restructure mostly not competitive companies, a high demand for investments emerged, which was mostly financed by foreign direct investment (FDI).[1]

Georgia was affected the most by the transition crises. Because of the collapse of the USSR and the Eastern-Bloc, followed by civil war from 1991 to 1994 and subsequent isolation, the GDP declined to quarter part of its value in 1989[2]. Hyperinflation has virtually annihilated the savings of a whole nation, making FDI the single source of investment in the country. The contribution of FDI was most important to the process of capital accumulation, employment, contribution to human capital accumulation and technology transfer. The appropriate investment climate for the Georgian economy was essential to achieve sustainable growth, reduce high rates of unemployment and finally to close supply gaps left by the Soviet system. Difficulties in this context are that, unlike other reforms, the investment climate has no direct reference field, but represents a

[1] About the role of FDI and investment climate in transition countries see: Bevan A. and Estrin S. (2000); Penev S., Rojec M. (2004); Dunning, J. H. (1993); Rojec M. /ET Al. (1995); Hunya, G. and Kalotay, K. (2000); Hewko, J. (2002);

[2] According to several indications, in 1989-1995 the GDP decreased to 25% of its initial value in 1989, not least because of civil war. However, according to first reliable data of 1997, the GDP reached 29.7% of its value in 1989 cp: Clement, H./et al. (2003)

coordinated mixture of various procedures, which mostly have no direct but an indirect impact on direct investments. Generally possible is the classification of factors affecting the investment climate in economic[3], legal[4] and political[5] aspects, respectively risks. However, the separation is not always exact[6]. In spite of the sustainable growth of FDI in Georgia (see Figure 7), per capita calculations were significantly lower in comparison with Central and Eastern Europe and Baltic countries (see Figure 1).

Figure 1: Commonwealth of Independent States (CIS) and Central and Eastern Europe and the Baltic States (CEB): Net FDI Inflows, 1995–2007

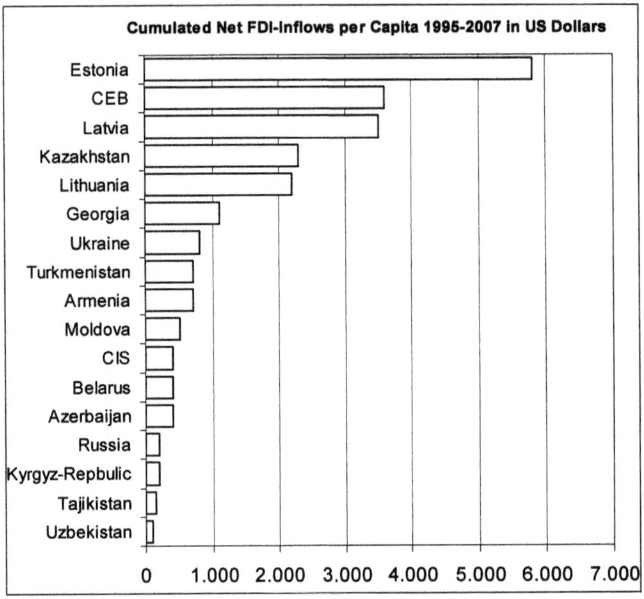

Source: EBRD, Transition Report 2007

3 Biswas, R. (2002); Le, Q. V. (2004); Khan, M. H. (2005); Oman, C. (2000); Kikeri, S./ ET all (2006); Smith, W. & Hallward-Driemeier, M. (2005)

4 Bates, R. H. (2004); Oman, C. (2000); Kikeri, S.,/ ET all (2006)

5 Le, Q. V. (2004); Chermak, J. M. (1992); Baum, M. A., Lake, D. A. (2003); Henisz, W. J.; Manfield, E. D. (2004); Freedom House (2001)

6 Economic and legal factors are especially affected by this: For example, ensuring property rights is the issue of economic frameworks, but on the other hand it requires legal implementation, which (as the Georgian example shows) does not always comply with the general course of economic reforms.

2 Transition progress in Georgia and economic freedom index

The most important precondition for a positive investment climate is a free market economy[7]. The more advanced the transition process is, the more attractive is the economy for foreign investors. Though Georgia's declared that its target after independence was the creation of a free market economy, the transition process was conducted sluggishly. Only after the "Rose Revolution" in 2003 has the new government become ready for radical transformations, with the economic reforms in Estonia chosen as the prototype.

Figure 2: Transition in Georgia, EBRD 2006

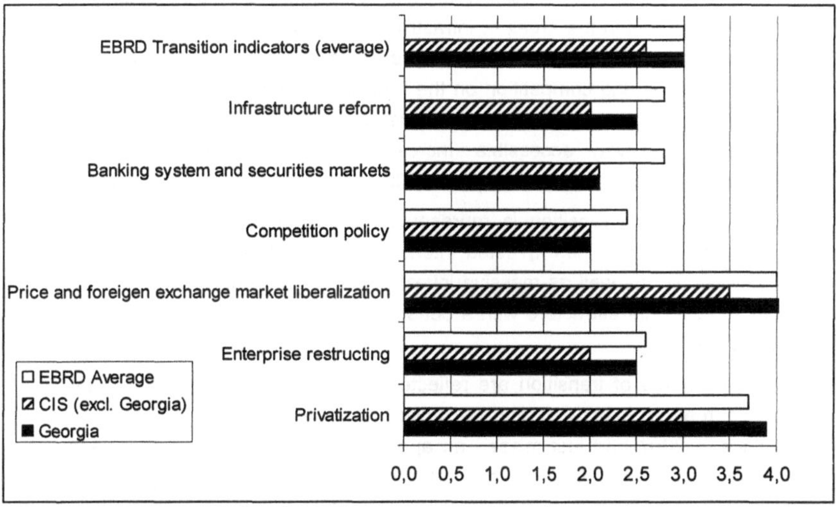

The measurement scale for the indicators ranges from 1 to 4+, where 1 reprsents a rigid centrally-planned economy and 4 + represents the standards of an industrialized market economy.

Source: International Monetary Fund (2007), p. 5

[7] The investment location of China was only successful with regard to trade-off after adopting the following reform: one country, two systems.

Figure 2 shows clearly that Georgia lay significantly above GUS-average with all key transition indices, but just near EBRD-average. It is clear to see that difficulties arise especially with regard to the processes of company restructuring, as well as regarding competition and the banking system. On the other hand, the government cannot directly affect the process of company restructuring. Competition and foreign investment, which brings the appropriate know-how into the country, can stimulate structural adjustments of companies. The World Bank report delivers another result[8].

Rushed and ambitious privatization[9], increasing expenditure for infrastructure purposes and tax-cuts are the trademark of this policy. On its way, Georgia has carried out important reforms, especially in the areas of price and currency stabilization[10], liberalization of external trade and the privatization process.

According to this, the main difficulties of Georgian companies did not mainly lie in the financial sphere, but in the areas of infrastructure, regulations and tax (Figure 3). This fact could partially be explained by the successful development of the banking system in recent years: Strong competition on the financial market, as well as investments of multinational bank groups like "Societe Generale", "Commerzbank" und "EBRD" have made credits relatively accessible. Increased government expenditures and the comparatively liberal monetary policy after 2003 have contributed to this process.

Another hot-spot of transition is represented by infrastructure problems. In recent years, the government has invested intensively in this field. Diversified energy and electricity supply, as well as massive investments in road networks makes one hope that an improved infrastructure with a better investment climate and economic growth will justify itself.

The achievements of transition are reflected in the economic freedom index. In the economic freedom index Georgia has moved from the 84th place in 2002-2003 to the 32nd place in 2007 and just missed an approval –"mostly free" – with a valuation of 69.2. In recent years, Georgia did not only beat the world-average but also the

8 Source: World Bank (2005)

9 Though the privatization of many corporations was discussed controversially, to the author the fast and not error-free privatization seems to be the smaller evil in comparison with the sluggish denationalization, which leads to the creation of lobbies and an antagonizing privatization process, not to mention budget-burdens and indirect macroeconomic costs related with inefficient governmental companies

10 The reforms were concluded extensively before the "Rose Revolution", so the Georgian National Bank became an example in this field.

European-average. However, the overview of 10 basic economic freedoms shows clearly where the weaknesses of transition processes in Georgia lie. Whereas conditions in labor, business, investment and financial freedom have continuously improved over the 4 recent years and as a rule lay above the average, no advances were reported in the areas of corruption and property rights (see table 1 and figure 4).

Figure 3: Constraints reported by firms – comparing Bulgaria, Georgia and Ukraine

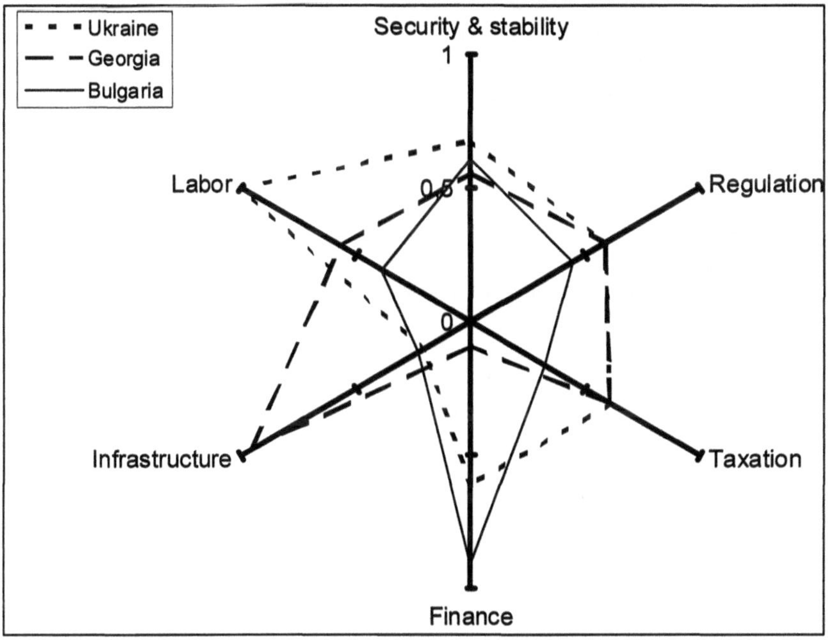

Source: World Bank (2005). p. 59

Note: Resulting indicators range from 0 (best) to 1 (worst). Indices are based on surveys of formal firms. Values are normalized by regional maxima and minima for each indicator. Countries selected to highlight differences.

Figure 4: Economic Freedom, Georgia 2008

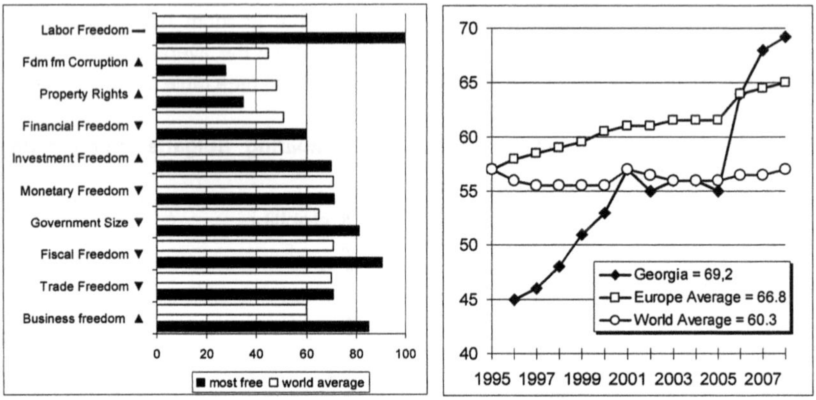

Source: The Heritage Foundation (2008), pg. 185-186

Table 1: Economic Freedom Index, Georgia 2002-2008

Ranking		Score	Business Freedom	Trade Freedom	Fiscal Freedom	Gov't Size	Monetary Freedom	Investment	Financial Freedom	Property Rights	from Corruptio	Labor Freedom
32	2008	69.2	85.0	71.0	90.7	81.3	71.4	70	60	35	28	99.9
31	2007	69.3	80.8	71.8	91.2	88.4	77.8	60	70	30	23	99.9
54	2006	63.5	73.9	67.6	94.1	90.8	75.2	50	70	30	20	63.3
93	2005	56.4	55.0	65.2	89.9	90.5	75.9	30	50	30	18.0	60.0
78	2004	58.9	55.0	65.2	90.6	90.1	75.5	50	50	30	24.0	
84	2003	58.6	55.0	64.8	90.6	88.9	74.8	50	50	30	23.0	
84	2002	58.6	55.0	64.8	90.6	88.9	74.8	50	50	30	23.0	

Source: The Heritage Foundation (2008)

2.1 PROPERTY RIGHTS AND CORRUPTION PROBLEMS

These fields can barley be regarded as a brake, especially for foreign investments. One could say that in this case the government has chosen the easy way. While it has abolished legal obstacles to business freedom, more complicated reforms concerning property rights and corruption containment were treated lightly. Measures for a simplification of doing business are namely a precondition for the abetment of corruption and the strengthening of property rights, however they have to be supported in legal fields and on the political level. The government cannot enforce or protect property rights itself[11]. This should be accomplished by an independent court, the decisions of which cannot be put into question or related to the government. This does not only imply legal frameworks but also reliance on the main actors within the economy. With regards to the political crisis of 2007 and the polarization of society, the best solution would be the involvement of foreign judges (US and EU) in a court of arbitration which would solely deal with property and economic questions. This would ward off the government's impact on cases concerning property rights and further improve the investment climate.

The other issue which is discussed controversially is anti-corruption measures of the government. Successes were achieved in some fields, for example in reforming the traffic police, which was corrupted before the reform. Furthermore, the simplification and optimization of different authorities, as registry and customs authorities, has to be mentioned. But these measures were mostly dedicated to small-scale corruption and in spite of spectacular scandals they did not develop into an overall fight against corruption, but were intended to be single actions, mostly without any political background. And as the freedom index of corruption shows, successes in this direction remained moderate. Thus corruption is to be considered as one of the main obstacles to Georgian transition.

In conclusion to the transition process in Georgia it has to be mentioned that the country has taken important steps towards a free market economy and that the double-digit economic growth lets us hope that sustainable economic growth could be ensured in spite of the first appearances of an overheated economy (see Figure 5).

3 Costs of doing business and tax policy

Besides extensive market frameworks, the transaction costs for doing business play an important role in the investment climate. Especially for transition countries it is of big importance. According to an EBRD survey, new and foreign companies are more

[11] The executive and – partially - the legislative branch of government are often regarded as the most important threats to property rights, in particular in post-Soviet countries, among them Georgia, not least because of the fact that in spite of many attempts it was not possible to create an independent judicial branch.

efficient than governmental companies made private.[12] For this reason, the simplification of doing business should be a priority of a transition country's government. It has to be appreciated that the simplification of doing business has indeed become one of the most important priorities of the government.

Figure 5: GBP, Real GBP growth and Inflation rate in Georgia 1996-2007

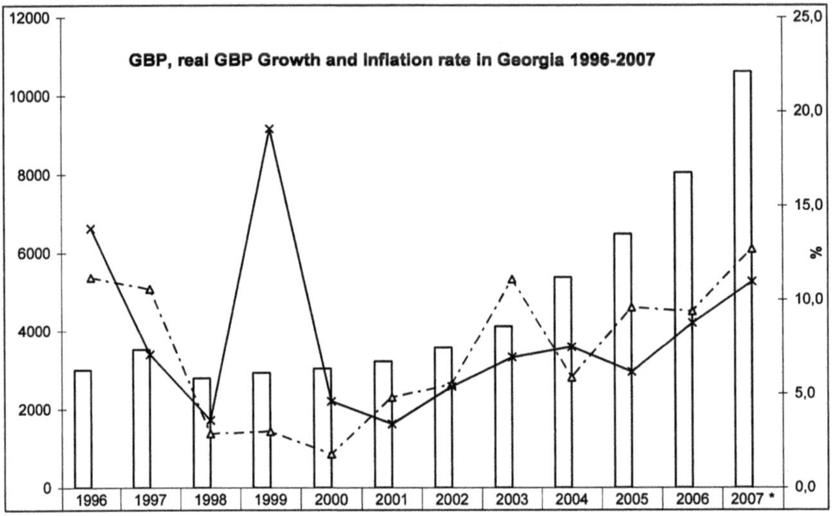

Source: NBG (2007)

After 2003, a wide range of reforms was carried out in order to simplify doing business, which led to an obvious improvement of the investment climate, bringing Georgia into the top 20 countries in international comparison. After 2004, successes were achieved in almost all fields: the number of taxes decreased from 21 to 9 and the duration of different procedures was reduced significantly.[13] Special successes were achieved in the fields of business foundation, license trade, registration of property rights and investment protection. On the other hand, the simplicity of tax-paying and liquidation is

[12] EBRD (2005) p. 17

[13] The number of licenses was reduced from 909 to 159

clearly below average. To some degree, the advancement of reforms achieved a few improvements, a fact for which taxes are a clear example. Indeed, tax-cuts were significant, but the procedures of preparing, filing and paying taxes were only partly simplified. For example, one needs twice the time than before to prepare, file and pay the three main groups of taxes.[14] Besides, the index of the "Tax Payment Number"[15] is almost twice as high as that of the OECD-member countries. Therefore, it would be more efficient for the investment climate to simplify the preparation of taxes, as well as the filing and paying process in comparison with further tax-cuts. Tax rates are already among the lowest in the region and significantly lower than in the EU, which is supposed to attract new investments. Consequently, it is not the tax-rates which are a problem, but the complicated procedure of tax preparation, filing and paying.

In regard to tax-cuts, the government accomplished a very successful policy in the years 2003-2007, both in view of budget growth and of significant tax-cuts, which had to encourage business activity. The most dramatic were income-tax-cuts and non-labor-cost-cuts: Until 2003 an employee had to pay a progressive income tax of up to 20%[16] and 2%[17] of non-labor-costs. Furthermore, the employer had to pay non-labor costs of up to 30% (social insurance, pension insurance, unemployment benefits). All together, costs amounted to about 50% of the overall labor costs of a company. After the reform, a flat income tax of 12% and an overall social insurance rate of 20% were introduced. This accounts for about 26.6% of labor costs. After 1st January 2008 the new overall "income and social tax" of 25% was adopted. VAT was also reduced to 18%[18].

[14] Corporation tax, VAT, income tax incl. non-wage labor costs

[15] When calculating this index, the International Finance Corporation does not only take into account the number of taxes, but also the number of actual tax payments, irrespective of how many tax payments took place. For example: Where full electronic filing is allowed - and it is used by the majority of medium-size businesses - the tax is counted as paid once a year, even if the payment is more frequent. For taxes paid through third parties, such as tax on interest paid by a financial institution, only one payment is included, even if payments are more frequent. Where 2 or more taxes or contributions are paid jointly, using the same form, each of these joint payments is counted once. Cp.: EBRD (2007a)

[16] Gradual increment from 12% to 20%

[17] 1% for social insurance and 1% for pension insurance respectively

[18] Even if the reduction of VAT were not as significant as was planned, this reduction had a positive impact on the investment climate.

As a whole, the adopted reforms are to be rated as positive with regard to the investment climate, even though there are a lot of optimization possibilities with regard to tax preparation, filing and paying procedures.

It is also promising that tax-cuts, accompanied by a drastic increase in government spending[19], did not result in a budget deficit. This, in connection with the huge balance of the payments deficit[20], would endanger the monetary stability and burden the investment climate in the country. Simultaneously with tax-cuts the government managed to quintuple public revenues in this period: In 2003, public revenues amounted to a total of 934.4 m. GEL (435 m. USD[21]), and only from January-November 2007 they reached 4523.3 m. GEL (2697.7 m. USD). It also has to be mentioned that tax-revenues increased to 26.7% of the GDP after the former amount of 10.9%.

4 Inflation and financial stability

The Georgian economy had a relatively stable price trend until 2003-2004, following the stabilization in 1995-1997.[22] As can be seen in Figure 6, the inflation developed largely synchronous with the exchange rates. However, the government introduced an expansionary policy after 2003 and inflation was fixed at the rate of about 7%. Even though increased government spendings were financed by tax-payments and budget deficit remained under control, the quick economic growth of the recent three years showed first indications of economic "overheating". For the first time since 1999, inflation reached the double-digit area. However, high defense spendings[23] extenuate the expansionary effect of increased government spendings. To contain inflation is important, as a successful restructuring of companies is only possible under the conditions of an increase in the purchasing power of households. However, inflation targeting may not be ignored, as only a moderate inflation of below 5-6% has a growth

[19] Especially in the fields of national security and the development of infrastructure

[20] Georgia's current account deficit grew continuously and reached 2,019 bn. in 2006, about 25% of GDP, with an index of import coverage by exports of 45% . Source: EBRD (2007)

[21] Average exchange rate in 2003

[22] Exception is an inflation jump in 1999 as a result of the crisis in Asia and Russia. Only the decisive action of the National Bank has softened the economic crisis and prevented further currency devaluation.

[23] In order to approach NATO-standards

supporting impact. In contrast, higher inflation rates represent a threat to macroeconomic stability. In this regard, the National Bank of Georgia has a great responsibility. The independence of the central bank's policy from fiscal authorities is the main supporting pillar of the national economy's monetary stability, which does not only prevent possible opportunistic monetary policies but is also a factor of confidence for foreign investors. Frequent changes in the National Bank's policy would strengthen uncertainty in investors and have a negative impact on the investment climate. For this reason, the newly suggested law modification concerning a strengthening of the "accountability and responsibility" of the National Bank's president to parliament or government cannot be acceptable, as they weaken the independence of the monetary board from fiscal authorities.[24]

The devaluation of the USD represents a further negative development concerning price increases. Since the exposure of Georgian's financial system to the USD is huge, it creates an awkward pressure on Georgian exports and intensifies the inflationary effect upon households. In this regard it is interesting that apparently there is larger coherence between the Euro's exchange rate and price increases in Georgia. This could be the reason to adjust the Georgian economic financial system to the Euro. On the one hand it would attract direct investments from the EU and on the other hand it would correspond to the development of Georgia's foreign trade structure. However, not enough statistical data are available to argue that the Georgian economy has to transfer from the USD to the EU, especially taking into account the world economy crisis and the dollar crisis. Bearing in mind Georgia's aims with regard to the EU, the government should strive more for cooperation with European companies. The EU-neighborhood-policy in particular could contribute positively to the investment climate. Opening EU-markets for Georgian products would be a great incentive for investing in Georgia. First of all, this would encourage economic growth and furthermore induce Georgian firms to implement faster reorganizations and adaptations to EU-standards.

[24] According to new law advancement, the National Bank's president will become more dependent on the Minister of Finance, which is actually unconstitutional. On the other hand, supervisory functions will be loosened. The introduction of a new law met with significant opposition in the budget committee of parliament and will eventually be adapted. Source: http://www.rustavi2.com/news/ news_text.php?id_news=24745&ct=4&im=main&ddd=&ddd2=&month=3&year=2008&srch_w=&srch=&wth=&rec_start=&rec_start_nav=&ddd2=11-02-08&month=3&year=08)

Figure 6: Cumulated Price Increase and Currency Devaluation (GEL/USD and GEL/EURO) 1995-2007 (1995=100)

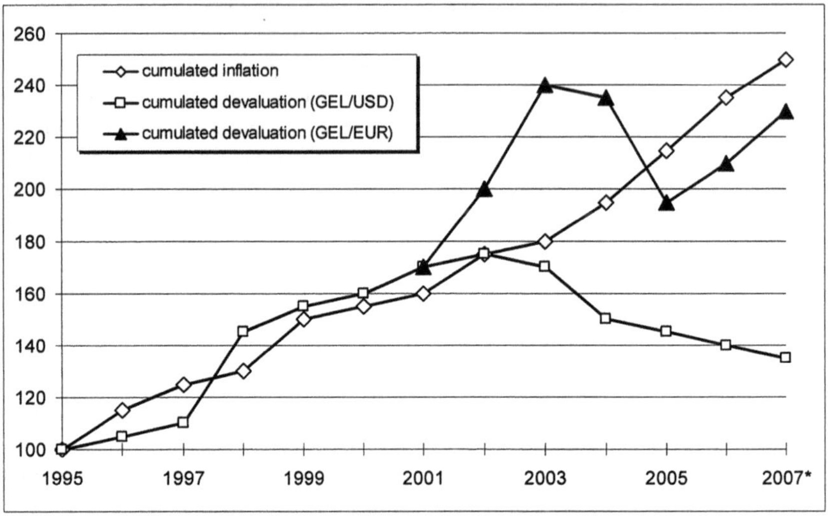

Source: compiled by the author from editions of NBG (2007) 2001-2007

5 Global economic risks – Russian sanctions

In consideration of Georgia's global economic risks it is important to mention the Russian sanctions. Russia cut down all transport connections to Georgia and sanctioned it economically after the arrest of two Russian spies. Even after the release and deportation of the two FSB-members the sanctioning still remains up to today. Also, the price for natural gas almost quadruplicated since then. The wine industry and the cultivation of fruits and vegetables were affected the most. It is probable that these sanctions exacerbate the foreign trade balance deficit and lead the country into an economic crisis, but since the year 2000 the rate of exports to Russia dropped from 30% to 16% in 2006[25]. Subsequently, according to the author's own presumptions, the

[25] It may be supposed that the lesson taught by the Russian crisisof 1999 was well learnt. The high percentage of foreign trade with Russia caused the crisis in Georgia by transferring the Russian crisis to the Georgian economy.

Georgian economy lost 2.5%-2.8%[26] of is gross domestic product. It is especially the raise of the price for natural gas which contributed to the situation, because 16% of Georgian imports consist of energy carriers[27]. The government was able to diversify the country's natural gas supply, so the exogenous shock could be absorbed and thus the economic growth of 9.4% in 2005 will presumably rise up to 12.4% in 2007. Taking into consideration that Russia is gaining more political power and that its actions are becoming more aggressive, a purposeful disconnection of the Georgian economy from the Russian should stabilize the Georgian market and minimize exogenous risks. This does not mean giving up the Russian market, but exploiting other markets to depress the foreign trade proportion of Russia to under 5%, with is now at 10.2%. In this case, the risks of the Russian economy would not reflect on Georgia as much as they do now, which could distinctly improve the investment climate.

In conclusion, the liberalization and transition of Georgia has been relatively successful: Tax reductions, simplification of regulations, actions taken for infrastructural development and an exemplary monetary policy of the central bank have a positive effect on the investment climate. But the government has reached the limits of such a policy, which is not efficiently realizable any further. The government needs to prioritize on other neglected components, such as the protection of property rights, to support the positive tendencies of the investment climate. The simplification of procedures regarding the preparation, filling and paying of taxes would increase the attractiveness of investments in Georgia as well.

6 Political factors – civil society

Besides the economic factors, political risks play an important role in the investment climate. A civil society is one of the most important conditions for stability on the domestic front. This is exactly the point where the problems and challenges for an improvement of the investment climate in Georgia are to be found. After the Rose Revolution and the events of November 2007, obvious proceedings received a severe setback. Georgia is still considered a partly free country[28]. Its main problems are a functioning government and independence of the courts. The elections on January 5th,

26 Data according to the Ministry of Finance cp.: http://www.bfai.de/fdb-SE, MKT200 61205 103912, Google.html

27 Cp.: NBG(2007) pg. 109f

28 Cp.: Freedom House (2007).

2008 have evidently shown the opposition's influence on the people and its popularity. Despite problems within the execution of the election, it was considered to be democratic and free, even though the government needs to eliminate the detriments until the election of parliament in April. According to experts, this political crisis leads to an, at least temporary, insecurity for the investors. Some have already withdrawn their projects until the power struggle between the president and the opposition is solved[29]. The German *Bundesagentur für Außenwirtschaft* states that the basically positive prospect for Georgia has not changed, but the future economic policy is unclear because of the political gridlock. Although the opposition is on a pro-Western course and does not rescind the liberal market economy reforms, the insecurity stresses the investment climate. The longer such a situation remains, the more it is likely that it will have a negative impact on the investment climate. It would be a milestone of the development of Georgian democracy and a civil society, if the government managed to perform the election of the parliament in an unequivocally democratic and free manner. Only reasonable discussions between the government and the opposition, as performed in the EU, could minimize the risk on the domestic front and help to improve the investment climate.

7 External risks and NATO-membership

External risks also contribute to the problems of domestic policy. Especially the Russian support of the separatist regime and the permanent threat to use the Kosovo as a precedent case, increase the risk of destabilization and stress Georgia's inflation climate correspondingly. In this case the government needs to aim for an explicitly Western course in cooperation with the NATO and the EU. The membership in NATO would unequivocally minimize foreign policy risks and lead to an improvement of investment possibilities. As the development of FDI in most Eastern European countries shows, the investment rate increased after their joining of NATO. In nearly every case this process was implemented simultaneously with the accession to the EU, which was probably the main cause of increasing investments. Nevertheless, in Poland and the Czech Republic, who joined NATO before joining the EU, foreign investments already increased before the EU membership (see Figure 7).

[29] Bundesagentur für Außenwirtschaft (2007). http://www.bfai.de/fdb-SE,MKT2007 1203140 425, Google.html

Figure 7: FDI in Poland, the Czech Republic, Romania, Latvia and Georgia

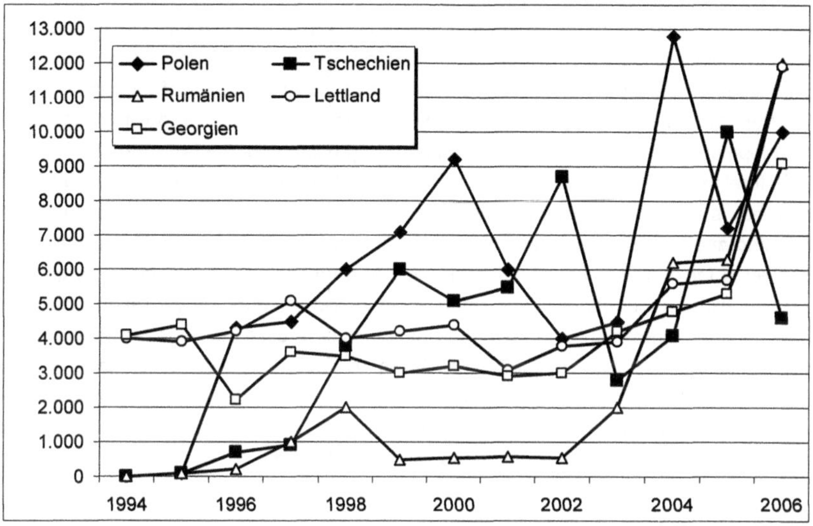

Source: EBRD (2007)

Even though there is no immediate connection between investment climate and the transatlantic alliance, it is obvious that besides the security of foreign policy and militarily security terms of admission such as a democratic and civil society and free market economy contribute to an evident improvement of the investment climate. On the one hand the security of foreign policy is guaranteed (as in case of the Baltic countries) and on the other hand the reforms that need to be implemented for a membership serve to strengthen the market economy and democracy. In reference to Georgia, which is not likely to join the EU in the near future, NATO membership could minimize the insecurities of foreign policy.

8 Conclusions

In conclusion it has to be stated that radical transition and the liberal market economy policy of the government have created a formidable positive investment climate in Georgia. Especially tax reductions, a simplification of doing business, as well as an exemplary monetary policy and regulatory restrictions are to be mentioned at this point. Admittedly, property rights and corruption remain the Achilles' heel of Georgian economy. An independent arbitrage-court consisting of experienced and respectable foreign judges could help to strengthen property rights[30]. The complicated and long processes of tax preparation, filing and paying reduce the effects of the tax reduction on the investment climate. Despite the fact that there are a positive economic development and an increase of FDI, the growing inflation of the last years and current and foreign trade account deficits must be held down, because they threaten the currency stability and thus the investment climate. In this context, the conservative-oriented and independent policy of the central bank has to remain untouched. Another challenge for domestic policy is the creation of a functioning civic society, as given in Europe, with an independent and respectable judicial system. Only a rational dialogue between the government and the opposition could neutralize the negative effects left by the governmental crisis of 2007. The Georgian government does not have a lot of influence on the challenges of foreign policy, but nevertheless an open cooperation with all adjoining countries, the integration into NATO and into European structures could contribute to a positive investment climate in Georgia.

9 References

Baum, M. A. & Lake, D. A. (2003): The Political Economy of Growth: Democracy and Human Capital. American Journal of Political Science, 47, 333-347

Bates, R. H. (2004): On The politics of Property Rights by Haber, Razo and Maurer. Journal of Economic Literature, 42, 494-500.

Bevan A. and Estrin S. (2000): The determinants of Foreign Direct Investment in transition economies, Discussion Paper No. 2638, Centre for Economic Policy Research, London.

Biswas, R. (2002): Determinants of Foreign Direct Investment. Review of Development Economics, 6, 492-504

Bundesagentur für Außenwirtschaft (2007): Wirtschaftsentwicklung Georgien 2007. http://www.bfai.de/fdb-SE,MKT20071203140425,Google.html

[30] In the long term this problem can only be solved by cardinal reforms of the court of law and strengthening of the legislative sovereignty.

Chermak, J. M. (1992): "Political Risk Analysis: Past and Present," Resources Policy, Vol. 18 (September), pp. 167–178.

Clement, H. / et al. (2003): Wachstum in schwierigem Umfeld – Wirtchaftslage und Reformprozesse in Ostmittel- und Südosteuropa sowie der Ukraine 2001/2002. (Working Papers des Osteuropa-Instituts München, Nr. 242), S. 129-137.

Clinton R. Shields (2003): FDI and the Investment Climate in the CIS Countries, 2003

Department of Statistics (2006): Statistical Yearbook of Georgia 2006, Tbilissi, 2006

Dunning, J. H. (1993): Multinational Enterprises and the Global Economy. Wokingham: Addison-Wesley. Merlevede, Bruno and Koen Schoors. 2004.

EBRD (2002): Transition Report 2002

EBRD (2005): Transition Report 2005, Business in Transition, Nov 2005

EBRD (2007): Selected economic Indicators – Macroeconomic, 2007 http://www.ebrd.com/country/sector/econo/stats/sei.xls

EBRD (2007a): Doing Business 2008. Georgia, Washington D.C. 2007 http://www.doingbusiness.org/MethodologySurveys/PayingTaxes.aspx

EBRD (2007b): Transition Report 2007, May 2007

Freedom House (2001): Nations in Transit 2001: Civil Society, Democracy, and Markets in East Central Europe and the Newly Independent States (Washington: Freedom House, Inc.)

Freedom House (2007): Freedom in The World 2007: Country Subscores http://www.freedomhouse.org/uploads/fiw/SubScoresFIW2007.xls

Henisz, W. J., Mansfield, E. D. (2004): Votes and Vetoes: The Political Determinants of Commercial Openness. The Wharton School and Department of Political Science, University of Pennsylvania.

Hewko, J. (2002): Foreign Direct Investment in Transitional Economies: Does the Rule of Law Matter? Democracy and Rule of Law Project. Washington, D.C., Carnegie Endowment for International Peace.

Hunya, G. and Kalotay, K. (2000): "Privatization and FDI in Central and Eastern Europe", Transnational Corporations, (9:11), pp. 39-66.

IFC - International Finance Corporation (1997): Foreign Direct Investment, Washington, D.C.: World Bank

International Monetary Fund (2007): Sixth Review Under the Three-Year Arrangement Under the Poverty Reduction and Growth Facility and Request for Waiver of Performance Criterion, Washington D.C. 2007 (Country Report No 07/299)

International Monetary Fund (2007a): Fifth Review Under the Three-Year Arrangement Under the Poverty Reduction and Growth Facility and Request for Waiver of Performance Criterion, Washington D.C. 2007 (Country Report No 07/107)

Kikeri S.,/ ET all (2006): Reforming the Investment Climate: Lessons for Practitioners, International Finance Corporation, Washington, D.C.

Khan, M. H. (2005): What is a "Good Investment Climate"? In: Kochendörferlucius, G. & Pleskovic, B. (Eds.) Investment Climate, Growth, and Poverty. Washington D.C, World Bank.

Le, Q. V. (2004): Political and Economic Determinants of Private Investment. Journal of International Development, 16, 589–604

Meyer, K. E., Jensen K. (2004): "Foreign direct investment and government policy in Central and Eastern Europe", Working Paper, Center for East European Studies, Copenhagen Business School

NBG (2007). National Bank of Georgia Monetary statistical Bulletin 2007, Tbilisi 2007

Oman, C. (2000): Policy Competition for Foreign Direct Investment, Paris: OECD

Penev S., Rojec M. (2004): Foreign Direct Investment and the Investment Climate in South-East Europe, Ekonomski anali br 163. http://www.doiserbia.nbs.bg.ac.yu/A(Wivo_dfiyAEkAAAAYWI4NjZIMGYtYzIzMS00YzM3LWJjMjQtYTI2ZmU3ZWJmYjRhzA-Vsv42eNeuLMbjfnC0Z AwtRW81))/img/doi/0013-3264/2004/0013-32640463071P. pdf)

Rojec M. /ET Al. (1995): Foreign Direct Investment and Privatisation in Central and Eastern Europe: A Comparative Analysis of Concepts, Experiences and Policy Issues in the Czech Republic, Hungary, Poland and Slovenia. Phare– ACE Project No. ACE– 92– 0108– R. Ljubljana: Centre for International Cooperation and Development

SMITH, W. & HALLWARD-DRIEMEIER, M. (2005): Understanding the investment climate. Finance And Development, A Quarterly Publication Of The International Monetary Fund, 42, 40-43

The Heritage Foundation (2008): 2008 Index of economic Freedom, Hong Kong, 2008.

Weltbank (2005): World Development Report 2005. A better investment climate for Everyone, Oxford, 2005.

Weltbank (2007): World Development Report 2007. Development and the next Generation, Washington D.C., 2007

THE EUROPEAN VECTOR OF ECONOMIC REFORMS IN POST-REVOLUTION GEORGIA

VLADIMER PAPAVA

1 Introduction

Several countries in the post-Soviet world have completed the transition to a European-type market economy and have been admitted to the European Union. In others—either partly or totally unsuccessful in transitioning—the question of whether or not this kind of market economy could be built is not even discussed. As to the potential of EU membership, such countries have either never set such a goal for themselves or, at best, are considering it in a long-term perspective.

The Rose Revolution of November 2003 was an outcome of the Georgian people's striving for the development of a democratic society and the improvement of a human rights record, the reduction of corruption, the enhancement of the national economy and the qualitative improvement of the social conditions of life (e.g. Ascherson, 2004, Coppieters, and Legvold, eds., 2005, Fairbanks, 2004, Karumidze, and Wertsch, eds., 2005, King, 2004, Miller, 2004, Welt, 2004). It is no surprise that the "Revolution" and its heroes, led by the present president of Georgia, Mr. Mikhail Saakashvili, have been supported by the US administration, the leadership of the EU and its member states, and the whole civilized world.

With the passage of two years after the "Rose Revolution", it is quite natural that the euphoria which had followed it both inside and outside the country has gradually diminished and been replaced by the state of "sobering down" which leads to a rather realistic evaluation of the results of post-revolution changes. It is important to note that such a sobering down from "revolutionary drunkenness" has taken place in Georgia rather quicker than oversees, which is quite understandable: Georgian citizens have been experiencing all "pleasures" of the revolution all on their own, whereas Georgia's international friends very often mistake what they would like to happen for what is really happening (Papava, 2006c). Nevertheless, even those analysts who have been a priori supportive of the revolutionary leaders cannot turn a blind eye to some recent antidemocratic developments in Georgia (such as the executive dominating the judiciary) (Jones, 2006). There were even more sobering accounts too (Kupchan, 2006), provoking discussions about the key values in post-revolution Georgia (Melikishvili, 2006).

From the very outset, the post-revolutionary political regime in Georgia was a kind of mixture of democratic and authoritarian elements (Nodia, 2005, p. 44-45). Such a hybrid regime is not a substantial novelty but has always been a characteristic of many post-Communist countries (e.g. Bunce, 1999) that have avoided "color" revolutions.

At the same time it is no secret that Georgia is not ready to join the EU in the near future. In view of continual official statements regarding Georgia's striving towards Euro-Atlantic organisations (see Papava and Tokmazishvili, 2006), however, we should know where we are going. One of the most important aspects is the vector of Georgia's economic development. If we see Georgia as a part of Europe—even if not fully accepted until the distant future—Georgia must transform into a European-type market economy.

2 EU's Economic Model and Georgia

It is not easy to describe the EU's economic model, which is still in formation (Fioretos, 2003). According to Albert (1991), the EU has been a battlefield of the two key models of capitalism; that is, the Anglo-American and the "Rhenish" (German-Japanese) ones.

In the Anglo-American model, the transfer of shareholding takes place quickly and without any obstacles, stock exchanges play a key role in companies' funding and some 40-60 percent of company shares are owned by institutional investors (such as insurance and pension providers). At the same time, the public sector is relatively small and social policy with respect to poverty and inequality is rather liberal; namely, inequality is understood as one of the incentives of competition and the fight against poverty is believed to not only be the government's function but, rather, also within the sphere of private charities as a part of morals and philanthropy.

In the Rhenish model, shareholding is rather stable and the banking sector plays a key role in funding shareholder companies. Stock markets are relatively constrained and, therefore, less active. The spheres of regulation are rather broad and the government's role in the distribution of the gross domestic product, ensuring equal competition and addressing social needs, is substantial.

Remarkably, the two models have merged in a number of EU member states under the pressure of the Anglo-American model over the Rhenish.

The EU countries do not share a *unified* economic model. Instead, there is a *common* economic model, which is not yet strong enough to ensure a full unification. At the same time, the integration underway in the EU is dominated by the transition from the common market to the unified market, which, in turn, has paved the way for economic and currency unions.

One of the most fundamental principles of the EU is the "preservation of what has been accomplished", which is important for any membership candidate to bear in mind. Whatever has been unified and achieved on the way to the integrated model has to be unconditionally copied by all candidates.

Georgia understands the tension between these two models. In particular, the Rhenish model prevails in shareholding and company funding which has yielded small and passive financial markets. This is a direct consequence of the official policy. On the other hand, the spheres of regulation are quite limited, which misleadingly suggest that Georgia fits the Anglo-American model. This is most evident in the government's policy towards poverty and inequality, which could be described as indifference (although sometimes inaccurately called "liberalism"). In Georgia, inequality is not the government's concern at all, whereas the alleviation of poverty has forcibly (under the government's pressure) become the headache of private charities and private entrepreneurs.

Under such circumstances it is natural to ask: Are we really headed for the EU? If we consider the government's economic agenda, we see movement in three different directions: North, Nowhere and East.

3 The Way to the North

In post-revolution Georgia, the so-called "Northern Way," or Georgia's integration into Russia's economic domain, has become prominent (Papava, 2006c). In 2003, Anatoliy Chubais, the President of the Management Board of RAO EES (Unified Energy Systems) Russia and prominent Russian statesman and political figure, wrote that Russia should establish a "Liberal Empire" in the post-Soviet world (Chubais, 2003).[1] Calling it "liberal" was meant in the sense that the new empire should be based on

[1] The origin of the Russian model of the Liberal Empire is based upon the Russian doctrine of Eurasianism (Dugin, 2005) which has deep historical roots in Russia (Gloveli, 2000). It should be emphasised that the idea of a Liberal Empire itself, following Krupnov's (2005) justified comment, is not Russia's but was first put forward as early as the second half of the 19th century in Great Britain (Matthew, 1973), was elaborated at the end of the 20th century (Reiff, 1999) and increasingly acquired a clearly American hue (Farrell, 2005). What is more, in all likelihood we should agree with the opinion that the acceleration of the formation of the Russian version of a Liberal Empire was given a particular boost (Torbakov, 2003) by the US's combat action in Afghanistan and Iraq as the possibility of forming an American Democratic Empire (Kurtz, 2003).

economics rather than coercion, wherein Russian companies (public and private) should take over the ownership of strategic companies in the former Soviet republics, which, in the long run, must lead to the re-establishment of Moscow's political influence over those countries.[2]

Russia started to fulfil its master plan in Armenia, which is its strategic partner in the region. In late 2002, on the eve of the presidential election in Armenia, a Russian-Armenian treaty called the "debt-for-equity" swap was executed.[3] In early 2006, Russia obtained new assets in exchange for not raising Armenia's price for natural gas.

As Russia and Armenia do not share a border, and in view of the frozen conflict between Armenia and Azerbaijan, the lynchpin in an economic space of Russia and Armenia is Georgia. If Georgia is dragged into the Liberal Empire, the fate of Azerbaijan will be determined as well, given the fact that all of its strategic economic projects are linked with Georgia (Papava, 2006a).

The first step towards snaring Georgia in the Liberal Empire's net was the summer 2003 takeover of the shares of the US-based company AES Silk Road by RAO EES (Gularidze, 2003).

The new Georgian government fully supported the entry of Russian capital into the Georgian economy during the broad-scale privatisation of government-owned enterprises after the Rose Revolution. The best example was the sale of Georgian gold

2 According to its architects, a Liberal Empire should be created not by forced armed occupation of the former Soviet Republics but by gaining control over the main economic facilities (by means of acquiring and developing assets) located on their territory. An essentially universal analysis of Russian investments in the CIS countries is presented in an article by Crane, Peterson and Oliker (2005). It is also important to note that in the Russian idea of a Liberal Empire, its developers and executors imply non-military methods of creation in the word "liberal" and not of the functioning (which they do not mention at all) of this "empire", which is essentially not surprising if we keep in mind the far from democratic and liberal nature of the regime in Putin's Russia (Åslund, 2005; Trenin, 2005). Herein we should note the interrelation in Russian policy in the post-Soviet space of energy dependence and political independence (Balzer, 2005), whereby when the former increases the latter weakens (Smith, 2004, pp. 5-8). It is no accident that for Russia, along with the formation of a Liberal Empire, a targeted advance towards creating an Energy Empire is of particular importance (Hill, 2004).

3 See *Armenia: Economic Hai-Lights*, July, 2002, online at http://www.bisnis.doc.gov/bisnis/bisdoc/ 020809ARHaiLights.htm.

and copper mining and processing companies to Stanton Equities, a subsidiary of the Russian holding group Industrial Investors.[4]

Encouraged by the government's affinity for Russian capital, the owners of the United Georgian Bank (privatised in 1995) sold it to Russia's Vneshtorgbank of which 99 percent are owned by the Russian Government.[5] This sale was the nationalisation of the United Georgian Bank by the *Russian* Government. This takeover came on the tails of Vneshtorgbank's acquisition of Armenia's Armsberbank.[6]

Gazprom is especially aggressive. It has made several attempts to take over the gas pipeline connecting Armenia with Russia. In late 2005 and early 2006, the Georgian Government negotiated with Gazprom, ostensibly to sell the gas pipeline (Voropaeva, 2005). It is not clear, however, why selling the pipeline to the Russian Government-owned gas company should be considered "privatisation." After America's intervention, the Georgian Government's talks with Gazprom came to an end,[7] which dealt a heavy blow to Russia's plans.

In late 2006, in the context of doubling the price for Russian gas, Gazprom once again tried to take over some of Georgia's energy assets (Socor, 2006).

It must be kept in mind that RAO EES has also not said its last word and it is very likely that sooner or later it will, at least, try to use the same scheme as it once successfully implemented with respect to AES Silk Road.

Regrettably, it is evident that Georgia is being culled into Russia's Liberal Empire (Papava, and Starr, 2006).

4 See: "Novosti: 97% aktsii GOK "Madneuli" prodano za $51 mln" (News: 97% of Shares JSC Madneuli are Sold at US $51.1 Million), *CITOH*, 7 November 2005, online at http://www.citon.com.ua/ news. php? id=10.

5 See: "Vneshtogbank priobriol kontrolnyi paket aktsii Ob'edinionnogo banka Gruzii" (Vneshtorgbank Purchases the Control Packet of Shares of the United Georgian Bank), *News.ru*, 18 January 2005, online at http://www.newsru.com/finance/ 18jan2005/vnesh.html.

6 See: "Russian Vneshtorgbank Continues Transcaucasian Expansion," *Mosnews.com*, 31 May 2004, online at http://www.mosnews.com/money/2004/05/31/vtbgeorgia.shtml.

7 See: "Millennium Challenge Corporation Board Approves $295.3 Million Compact with Georgia," *Georgian Business Week*, 16 August 2005, online at http://mca.gov/press/ eleases/ 005/release-081605-boardapprovesgeorgiacompact.php.

4 The Way to Nowhere

Some of the Georgian government's actions defy both economic theory and common sense. Chief amongst these was the president's generous summer 2006 initiative to add 50,000 people to a national employment programme.[8] The idea was to require private entrepreneurs to give three-month jobs to unemployed persons. For this, the latter were to be paid USD$85 a month out of the national budget. (In the summer of 2006, after the enactment of the new Labour Code, unemployment allowances [USD$12.40 per month] were abolished.) The programme cost the national budget USD$12.7 million. In principle, no business with the demand to expand needs any direct assistance from the national budget. All the government needs to do is to provide for the development of professional training or retraining programmes and the Georgian government has rightly assumed this duty.

This programme, however, employed only a few persons. In most cases, a simple deal was made. Businessmen agreed to subscribe to any contract under which they could pretend that they employed some people who did some job. Such a deal made an unemployed person happy, too, because he made USD$255 in three months for doing nothing. There were, however, rather perverted situations as well. Some businessmen agreed to subscribe to such contracts on the condition that new "employees" shared half of the sum with them.

Typically, the average share of salaries in the costs of production is twenty percent and so USD$12.7 million spent on salaries should produce goods and services worth about USD$60 million. Although there are no official employment statistics concerning this programme, we should expect that about ten percent of the targeted number of beneficiaries were actually employed. This means, then, that USD$12.7 million was spent out of the national budget for the production of goods and services with an aggregate value of perhaps USD$6 million.

In essence, it can be said that USD$12.7 million allocated from the national budget was nothing but an allowance for the unemployed. This amount went to market ostensibly with the purpose of producing consumer goods, but because no actual goods and services were produced the only contribution was to the growth of inflation. Consequently, any measures of this kind, however generous their objectives might be,

8 See: "Saakashvili Unveils New Economic Initiatives," Civil Georgia, 5 August 2006, online at http://www.civil.ge/eng/article.php?id=13256.

cannot, in fact, produce any desired results. On the contrary, all they can do, along with other similar measures, is to contribute to macroeconomic destabilisation.[9]

Unfortunately, the same type of national employment programme covering 100,000 people was started in December 2007.

The revolutionary wave also emboldened the government to reorganize some ministries and departments. In particular the State Department of Statistics, which before the revolution had been an independent agency accountable to the president, was folded into the Ministry of Economic Development—a blatant conflict of interests. In Georgia, statistics now play the same role as in the Soviet Union—a political function to proclaim annual improvement of the country's economy. In an "accidental leak" in August 2006, the department declared that in July 2006 the annual inflation rate in Georgia had reached 14.5 percent, which prompted criticism from the IMF. The government then effectively cut the inflation rate by dismissing the head of the department and his successor engineered a gradual reduction of the inflation rate until it dropped to 9.2% in December 2006; as it often did, the IMF turned a blind eye. In 2005, at the behest of the Prime Minister of Georgia, the IMF had named a new Mission Chief, who understood that any major disagreements with the Georgian government would be unwelcome.

In the context of the fight against corruption,[10] the government abolished some important institutions (for example, the Antimonopoly Service and the Food Quality Control Service). This left the public unprotected and the state weakened. Such measures could be compared to *fighting a headache by cutting off your head.*

The government's mistakes in the reorganization of the ministries and departments and the loss of "institutional memory" were supplemented with the non-professionalism of many post-revolution ministers. The most blatant example of non-professionalism was to entrust, in the spring of 2006, the Minister of Defense with the task to deal with the marketing of Georgian wines abroad. Another example of this kind was the abolition of the State's anti-monopoly service in late 2004 and Mr. Saakashvili's initiative in the fall of 2007 to empower the Ministry of the Interior (which in Georgia combines the police and the national security forces) to deal with the anti-monopoly regulation of the domestic market.

9 See: "IMF Warns Georgia on Inflation," Civil Georgia, 19 August 2006, online at http://www.civil.ge/eng/article.php?id=13350.

10 See: "WB Reports on 'Largest Reduction' of Corruption in Georgia," Civil Georgia, 27 July 2006, online at http://www.civil.ge/eng/article.php?id=13187.

One has to emphasize the ways of the transformation of corruption in post-revolution Georgia. The extra-budgetary accounts, which used to accumulate incomes of accused persons, are paying the "price of liberty." Because such accounts were outside the budgetary area, it is natural that there was no transparency in terms of spending too. However, the problem became even more serious as the government started to replenish those accounts by means of so-called "voluntary contributions" of businesses (e.g. Anjaparidze, 2006). As a result, if pre-revolutionary functionaries had pockets open for bribes, their post-revolutionary successors, having closed pockets, opened bank accounts. In the beginning, the IMF turned a blind eye to the existence of such accounts due to an erroneous assumption that all means are good to fight corruption. Obviously, it had forgotten a simple truth that it is a futile endeavor to beat corruption by means of corruption. Later, in the spring of 2006, the "extra-budgetary" accounts were abolished under the IMF's pressure.

Against the background of the successful fight with mass corruption, there still remains unsolved the problem of elite corruption involving high-level government officers who, on the one hand, in defiance of public procurement regulations, with total impunity, have been using budgetary resources for their own benefits, and, on the other hand, have been exploiting their official status to protect their favorite companies under their umbrella. The most palpable example of this kind of corruption is the former defense minister, Irakli Okruashvili, who was charged with corruption after his switching from the government team to the vigorous opposition of President Saakashvili. The Okruashvili lesson is exemplary for the other cabinet and parliamentary majority members to the extent that now everyone will know that nobody may be "insured" against the discovery of his or her corrupt behaviors should anybody dare to move into opposition.

The post-revolution government's disrespect for the constitution and laws became evident in the process of the privatization of public property when by means of bypassing the law, or much rather in complete defiance thereof, some new owners started to "emerge" from nowhere.

The process of "de-privatization" launched after the Rose Revolution is another reason for being worried. De-privatization in the post-revolution Georgia's context means that certain objects that had been privatized before the revolution were forcibly taken back by the government, which then offered them for privatization. In the process the law-enforcement organs (the General Prosecutor's Office and the Ministry of the Interior) got involved, which pressured the owners to "voluntarily" give up their property for the government's benefit. At the same time, there is no guarantee that sometime in the future a need of another de-privatization will not be brought onto the agenda (Papava, 2006b). The government's real goal was the redistribution of property for the benefit of the newly formed elite.

One should also be worried about some other examples of the post-revolution government's defiance of property rights, such as extrajudicial decisions to demolish privately owned residential houses built before the revolution, the owners of which had all relevant documents certifying both ownership and legitimacy of construction. The only argument presented by the government in justification of such demolitions was a desire to improve the city's "image". That the property rights were grossly infringed (Christiansen, 2006) the government seemed not to be concerned with at all. Paradoxically, the Georgian government's obvious disrespect for one's property rights never became subject to the West's criticism.

Multiple violations of human rights (including murders of young people by policemen for which – as is broadly believed by the public – not all responsible persons received fair punishment)[11], which have repeatedly been incorporated in the Public Defender's reports[12], do fit quite well into the context, which is not quite unexpected.

5 The Way to the East and the Disease of "Rating-o-mania"

Deregulation, such as reducing the number of licenses and permits, has limited the legal grounds for the government's interference with businesses. Cutting tax rates significantly eased the tax burden for businesses (Khaduri, 2006). Reducing the import-tax base for agricultural produce and construction materials as well as annulment of the import taxes for other goods have made Georgia much more competitive.

The new Labour Code is revolutionary. By limiting the rights of employees, it has substantially broadened those of employers. Although this may encourage businesses to develop, it also leaves employees unprotected.

These sorts of reforms resemble the Southeast Asian (Hong Kong, Singapore) and Anglo-Pacific (Australian, New Zealand, US and Canada) economic models. It should come as no surprise, therefore, if Georgian reforms track the way to the East and, in

[11] From the very outset of the post-revolutionary development in Georgia, the status of human rights have been associated with quite a lot of difficulties (e.g. Devdariani, 2004). See also: One Step Forward, Two Steps Back. Human Rights in Georgia After the "Rose Revolution," *Human Rights Information and Documentation Center*, 2004. http://66.116.100.86/humanrights.ge/eng/files/REPORT.pdf.

[12] See: "Public Defender Tells Saakashvili to Make Positive U-Turn," Civil Georgia, Tbilisi, January 14, 2008, online at http://www.civil.ge/eng/article.php?id=16894.

fact, we are moving towards the American-Canadian model along the path through Southeast Asia and Australia-New Zealand.

The Kazakh capital invasion of the Georgian economy fits this model only formally in that Kazakhstan is an "Eastern" country. Both Kaztransgaz and its only shareholder, Kazmunaigaz, however, are government-owned companies (Men'shikova, 2006). (Kazakhstan's government-owned holding company, Samruk, in turn, is the only shareholder of Kazmunai'gaz[13]). Despite this, the entry of the Kazakh government-owned companies into the Georgian market is, for some reason, labelled as "privatisation" (probably in order to create a reform image for the government),

Another circumstance that should be discussed separately is that all reforms described here are associated with a kind of disease, which conventionally could be labelled as "rating-o-mania."

After the publication of the World Bank's rating list, entitled "Doing Business", in 2006, according to which Georgia had made an impressive jump from the 112th to the 37th place amongst the world's nations, the Georgian Government announced that its next year's objective would be to push the country forward and to ensure that Georgia would be found amongst the 20 best nations of the world in the World Bank's next year's rating list and, in fact, it did assume the 18th place in 2007.[14] (Remarkably, the four best nations in the WB ranking include such non-European nations as Singapore, New Zealand, the US and Hong Kong). It is astonishing, that Georgia, the country where private ownership is not protected in any way, held the 18th place, which by itself deserves very little confidence.

The government should strive and provide for public wealth and prosperity and not for higher places in rating lists. As far as ratings are concerned, the government should pay more attention to the Economist Intelligence Unit Quality of Life Index[15] and the UNDP Human Development Index[16] by which, much to the people's disadvantage, Georgia ranks 87th and 96th respectively.

[13] See: "Ziat' Nazarbaeva vozglavil 'Kazmunai'gaz'" (Nazarbaev's Son-in-Law Takes Over "Kazmunai'gaz"), *Catalog SMI*, 27 June 2006, at http://www.catalogueofmass-media.ru/ index.php? dn =news&re=print&id=22922.

[14] See: *Doing Business. Economy Rankings*. Washington, DC: The World Bank Group, 2007, online at http://www.doingbusiness.org/EconomyRankings/.

[15] See: "The Economist Intelligence Unit's quality-of-life index," online at www.economist.com/ media/ pdf/QUALITY_OF_LIFE.pdf.

[16] See: "Statistics in the Human Development Report," online at hdr.undp.org/hdr2006/statistics

The "Doing Business" ranking is not of any particular value for Georgia as evidenced by Armenia's 39th place ranking. Furthermore, the 45th, 55th, 56th, and 74th places belong, respectively, to Hungary, Slovenia, the Czech Republic and Poland, all of which joined the EU and are developing rapidly. Similarly, neither Bulgaria's 46th place nor Romania's 48th closed their doors to the EU.

Ratings of this kind are useful for PR but are entirely worthless for serious economic analysis. Professional economists use statistics and not rating lists. As regards the former and as already noted above, Georgia has nothing of which to be proud.

6 In Lieu of a Conclusion: What Is the Value of the Way that does not lead to Europe?

The reader may well ask: Does anything that could be oriented to the West - to Europe - happen in the Georgian economy? The answer is: Yes, it does. The recent purchase of Bank Republic by the French Société Générale, for example, or the Bank of Georgia's offering at the London Stock Exchange show this to be true, but these examples are exceptions rather than the rule.

The way to the North only leads to Russia and it has nothing to do with the European choice because "Russia and the EU are already dancing together but not necessarily to the same tune."[17] Moreover, Georgia's joining the Liberal Empire exposes it to even greater danger than does staying in the CIS. Georgia may withdraw from the CIS whenever it wants. Withdrawal from Russia's Liberal Empire, if not impossible, would entail immense difficulties and costs.

Some economic measures taken by the government are senseless. The path to nowhere is the most dangerous.

The only more-or-less sound path is the Eastward one but what if, by taking this path, Georgia distances itself even more from the EU?

As the Earth is round, one may eventually arrive in Europe by travelling eastward from Georgia but we will approach Europe from a different side! It is worth noting that following the first five best nations in "Doing Business," which is the Georgian government's favourite rating list, the five next ones include Great Britain, Denmark, Australia, Norway and Ireland, which also supports the idea that we may reach the

17 See: Sapir (2003, p. 13).

European Union from the West and not from the East and which is a path that is obviously shorter for us...

Columbus also counted on the fact that the Earth was round. To reach India, he also took the opposite path and went to the West and not to the East. If he had not mistaken America for India, he would have continued travelling and, most probably, would have reached the Indian shore. I have an impression that, like Columbus, the Georgian government has not fully realised that it has started moving towards its goal in the opposite direction.

Georgia is not the first state to make its way to the EU by this strange route. The Baltic states, especially Lithuania and Estonia, reached their goals in the same way long before Georgia. Another indication that this path can be effective is the recently adopted European Neighbourhood Action Plan.

7 References

Albert, Michel, 1991. Capitalisme contre capitalisme. Paris: Editions du Seuil.

Anjaparidze, Zaal, 2006. "Georgian Government Questioned about Secret Funds," Eurasia Daily Monitor, The Jamestown Foundation, April 12, online at http://www.jamestown.org/edm/article.php?article_id=2370970.

Ascherson, Neal, 2004. "After the Revolution," London Review of Books, Vol. 26, No. 5, online at http://lrb.veriovps.co.uk/v26/n05/asch01_.html.

Åslund, Anders, 2005. "Putin's Decline and America's Response," Policy Brief, 41, August. Washington, DC: Carnegie Endowment for International Peace.

Balzer, Harley, 2005. "The Putin Thesis and Russian Energy Policy," Post-Soviet Affairs, Vol. 21, No. 3, pp. 210-225.

Bunce, Valerie, 1999. "The Political Economy of Postsocialism," Slavic Review, Vol. 58, No. 4, pp. 756-793.

Christiansen, Robert E., 2006. "Property Rights, Institutions and Growth: Will Georgia Make the Transition?," Presentation delivered at Cato Conference, International Monetary Fund, Resident Representative Office in Georgia, 27 October, online at http://www.imf.ge/view2.php?lang=1&view=239.

Chubais, Anatoliy, 2003. "Missia Rossii v XXI veke" (Russia's Mission in the 21st Century), Nezavisimaya gazeta, 1 October, online at http://www.ng.ru/printed/ideas/ 2003-10-01/1_mission.html.

Coppieters, Bruno, and Robert Legvold, eds., 2005. Statehood and Security: Georgia after the Rose Revolution. Cambridge: The MIT Press.

Crane, Keith, D. J. Peterson, and Olga Oliker, 2005. "Russian Investment in the Commonwealth of Independent States," Eurasian Geography and Economics, Vol. 46, No. 6, pp. 404-444.

Dugin, Alexander G., 2005. Eurasian Mission (Programme Materials of International Eurasian Movement). Moscow, Russia: ROF "Evrazia."

Fairbanks, Charles H. Jr., 2004. "Georgia's Rose Revolution," Journal of Democracy, Vol. 15, Vol. 2, pp. 110-124.

Farrell, Theo, 2005. "Strategic Culture and American Empire," The SAIS Review of International Affairs, Vol. XXV, No. 2, pp. 3-18.

Fioretos, Orfeo, 2003. "The Domestic Sources of Multilateral Preferences: Varieties of Capitalism in the European Community," in Peter A. Hall and David Soskice, eds., Varieties of Capitalism: The Institutional Foundations of Comparative Advantage. New York: Oxford University Press, pp. 213-244.

Gloveli, G., 2000. "Geopoliticheskaia ekonomia v Rossii" (Geopolitical Economy in Russia), Voprosy ekonomiki (Economic Affairs), No. 11, pp. 46-63.

Gularidze, Tea, 2003. "Russian Company Seals Controversial Takeover of Tbilisi Electricity Distribution," Civil Georgia, Tbilisi, August 2, online at http://207.218.249.154/eng/article.php?id=4694.

Hill, Fiona, 2004. Energy Empire: Oil, Gas and Russia's Revival. London, UK: The Foreign Policy Centre, online at http://www.brookings.edu/views/articles/Fhill/20040930.pdf.

Karumidze, Zurab, and James V. Wertsch, eds., 2005. "Enough." The Rose Revolution in the Republic of Georgia 2003. New York: Nova Science Publishers.

Khaduri, Nodar, 2006. "Georgian Economy," Central Eurazia 2005. Analytical Annual. Luleå, Sweden: CA&CC Press, pp. 147-152.

King, Charles, 2004. "A Rose Among Thorns," Foreign Affairs, Vol. 83, Iss. 2, pp. 13-18.

Krupnov, Yury, 2005. "Pochemu liberal'naia imperia v Rossii ne poluchitsia?" (Why the Liberal Empire in Russia Would Fail?), Vestnik analitiki (Analytical Herald), No. 2 (20), pp. 38-56.

Kupchan, Charles A., 2006. "Wilted Rose. Sukhumi Dispatch," The New Republic, February 6, online at http://www.boell.org/news/documents/Charles%20Kupchan %20Georgia% 20Report3.pdf.

Kurtz, Stanley, 2003. "Democratic Imperialism," Policy Review, Issue 118, April/May, pp. 3-20.

Matthew, H. C. G., 1973. The Liberal Imperialists. The Ideas and Politics of a Post-Gladstonian Élite. Oxford, UK: Oxford University Press.

Melikishvili, Alexander, 2006. "Response to Kupchan's Àrticle," Johnson's Russia List, February 10, online at http://www.cdi.org/russia/johnson/2006-40-19.cfm.

Men'shikova, Elena, 2006. "55 miliardov plius gazifikatsia vsey strany" ("55 Billion Plus Gasification of the Whole Nation"), Izvestia, Kazakhstan, online at http://www.izvestia.kz/news.php?date=15-12-04&number=9.

Miller, Eric, 2004. "Smelling the Roses: Eduard Shevardnadze's End and Georgia's Future," Problems of Post-Communism, Vol. 51, No. 2, pp. 12-21.

Nodia, Ghia 2005. The Dynamics and Sustainability of the Rose Revolution. In Democratisation in the European Neighbourhood, ed. Michael Emerson. Brussels: Centre for European Policy Studies, pp. 38-52, online at http://www.uquebec.ca/observgo/fichiers/57785_1267.pdf.

Papava, Vladimer, 2006a. "Comparative Advantages of the Central Caucasian Countries: Potential, Realised, and Missed Opportunities," The Caucasus & Globalisation, No. 1, pp. 87-96.

Papava, Vladimer, 2006b. "Corruption and Primary Accumulation of Capital in Transitional Economies," Ludwig von Mises Institute Working Papers, 10/17/2006, online at http://www.mises.org/journals/scholar/papava.pdf.

Papava, Vladimer, 2006c. "The Political Economy of Georgia's Rose Revolution," Orbis. A Journal of World Affairs, Vol. 50, No. 4, pp. 657-667.

Papava, Vladimer, and Frederick Starr, 2006. "Russia's Economic Imperialism," Project Syndicate, January, http://www.project-syndicate.org/commentary/papava1.

Papava, Vladimer, and Michael Tokmazishvili, 2006. "Becoming European. Georgia's Strategy for Joining the EU," Problems of Post-Communism, Vol. 53, No. 1, January/February, pp. 26-32.

Reiff, David, 1999. "A New Age of Liberal Imperialism?," World Policy Journal, Vol. XVI, No. 2, pp. 1-10.

Sapir, Jacques, 2003. "Russia's Economic Growth and European Integration," Post-Soviet Affairs, Vol. 19, No. 1, pp. 1-23.

Smith, Keith C., 2004. Russian Energy Politics in the Baltics, Poland and Ukraine. A New Stealth Imperialism? Washington, dc: The CSIS Press.

Socor, Vladimir, 2006. "Gazprom's 'Pure Commerce' in Georgia," Eurasia Daily Monitor, The Jamestown Foundation, 9 November online at http://www.jamestown.org/edm/article.php?volume_id=414&issue_id=3917&article_id=2371625.

Torbakov, Igor, 2003. "Russian Policymakers Air Notion of 'Liberal Empire' in Caucasus, Central Asia," Eurasia Insight. Eurasianet, 27 October, online at http://www.eurasianet.org/departments/insight/articles/eav102703.shtml.

Trenin, Dmitri, 2005. "Reading Russia Right," Policy Brief, 42, October. Washington, DC: Carnegie Endowment for International Peace.

Voropaeva, Liubov', 2005. "Gruzia daiot 'dobro'?," (Georgia to give 'OK'?), Gazeta SNG, 28 February, online at http://www.gazetasng.ru/article.php?id=1739.

Welt, Cory, 2004. "Georgia: Consolidating the Revolution," Russia and Eurasia Program. Center for Strategic and International Studies, April 6, online at http://www.csis.org/media/csis/pubs/ci.consolidatingrevolution.04.06.05.pdf.

THE NATIONAL ECONOMIC MODEL AND GLOBALIZATION (THE CASE OF GEORGIA)

TEIMURAZ BERIDZE

1 Introduction

The question formulated in the title of this article became a focus of social studies in the late 20th and early 21st centuries,[1] which was due, in the first place, to the disintegration of the bipolar world and the emergence of new sovereign states (CIS, Eastern Europe). Their natural response was a desire, on the one hand, to create new "independent" economic systems and, on the other, to enter the world economic sphere. Initially, this desire appeared to be contradictory, because integration into the world economy implied the need to "surrender" (in some measure) their economic sovereignty.[2] But it should be noted that economic sovereignty is always relative ("economic independence" in the strict sense of the term is utterly unrealistic), because otherwise it implies a closed (oriented toward maximum self-sufficiency) and, consequently, inefficient economic system. The degree of relativity of economic sovereignty is higher than that of political sovereignty, which follows from the very phenomenon of economic relations (satisfaction of material needs based on a division of labor and an exchange of its results), and also from the specific nature of human economic activity implying a removal of national state barriers and the creation of a nationwide and then a worldwide market. This suggests that unity (a contradictory unity) of the international and the national is something of a norm to which economic development "tends." At the same time, the absence or weakness of one of these factors is a symptom of an incomplete or uneven realization of the overall economic potential.

1 See, for example: J. Stiglitz, Globalization and its Discontents, W.W. Norton, New York, 2002.

2 The problem of economic sovereignty was examined in: T.A. Beridze, Economic grounds of Sovereignty: new aspects of economics and politics interaction, "Siakhle" Publishers, Tbilisi, 1995. A fruitful analysis of this problem is given in: H. Hinsley, Sovereignty, 2nd ed., Cambridge University Press, Cambridge, 1986; J.H. Franklin, "Sovereignty and the Mixed Constitution," in: The Cambridge History of Political Thought, 1450-1700, ed. by J.H. Burns, Cambridge University Press, Cambridge, 1991, pp. 309-328.

Assessing Georgia's current socio-economic developments is not an easy task: the aspects are varied (politics, economics, society, religion), while the period was far from easy (elections added political and economic tension). This is a strict confirmation that the economy and politics are intertwined and that their interaction plays an important role in public life.

2 Regarding the Definition of "National Economic Model"

The concept of "national" is used here as an attribute feature of a state entity, in contrast to "national" as an inherent attribute of a certain ethnos. In the latter case, one can only speak of national sovereignty where the actor is a nation (ethnos). A confusion of the concepts of "state" and "national" or their identification can have negative consequences. The trends observed in the territory of the former Soviet Union (Azerbaijan, Georgia, Moldova) are a good example of such confusion.

As regards the concept of "model," its most general definition is a simplified system used to simulate some aspects of the real economy.[3] More specifically, one can say that an economic model (paradigm) is a certain mode of organization of socioeconomic life, a type of relationship between economic agents in the production, distribution, exchange and consumption of goods and services.

A general requirement here is that an economic model of society should ensure sustainable economic growth and, as a consequence, reduce unemployment and inflation rates, as well as ensure appropriate distribution of income.

The concept of "national" used in reference to "economic model" implies, in the first place, effective use of absolute (to a lesser extent) and relative (to a greater extent) advantages in the course of economic activity.

These priorities are conditioned by a number of factors, such as the availability of natural and human resources, a convenient geo-economic position (Azerbaijan and Georgia are good examples), etc.

A "national economic model" is a reflection of the natural form of the state's economic existence, since it underlies "relative" economic independence (absolute economic

[3] See: J. Black, A Dictionary of Economics, Second Edition, Oxford University Press, Oxford, New York, 2002. 5 See: T.A. Beridze, op. cit., pp. 24-25.

independence does not exist), given that economic sovereignty is always relative and implies, first and foremost, independence in making and implementing decisions.

When speaking of a "national economic model," we do not mean a "model" reflecting the content, the very type of organization of the "producer-consumer" relationship (market or non-market), but the peculiarities (a number of distinctive features) of a given country's development. No "model" can be conceived or, most importantly, implemented in practice unless these peculiarities are taken into account.

What kind of economic development "model" can be chosen by the states of the Central Caucasus? Clearly, this question is not a simple one. The answer to it, in our opinion, could be as follows: first, organization of the economy according to market principles; second, reorientation of its structure in accordance with the existing realities (world market conditions, technical and technological ties); and third, due regard for traditions and other noneconomic circumstances.

Here we want to raise the question of “small countries”/ “small economies”. According to the political point of view there are no “small” and “big countries”; international law gives us a theoretical basis for such an assertion. But according to the economic point of view we can talk about “small economies”. What is to be understood as “small economy”: Territory? Population? GDP? GDP per capita? A positive balance of payments? Or something else? It is not an easy question, because it is composed of many indicators. The composite indicator construction was created and adopted by the UN: GDP per capita, longevity and education. Still, I think that this question is still open and thus deserves the continuation of the study. At any rate, this composite indicator must include economic (technical) and social aspects, and not include some demo-geographic aspects.

As regards the first factor, everything appears to be quite clear, just as the basic principles of organizing a market-based society, although the experience of the Central Caucasian countries gained in the course of market reforms shows that a number of mistakes have been committed in the process of this transition.[4]

The second factor - reorientation of the economic structure - is, on the one hand, a result of the operation of the first factor and, on the other, a part of current practical politics with all its pluses and minuses.

4 See, for example: V. Papava, International Monetary Fund in Georgia: achievements and errors," Imperial" Publishers, Tbilisi, 2001.

What is the structural composition of the Central Caucasus economies? Despite the region's small area, one cannot say that the structures of these economies are of the same type. In the case of Azerbaijan, it is an orientation toward energy resources and industry; in the case of Georgia, it is agriculture, small-scale industry, transport, communications and tourism; and in the case of Armenia, it is agriculture and small-scale industry. One can say that all of this is the structural effect of these countries, which should be both the cause and the consequence of the operation of their economies; it should underlie the construction and implementation of a "national economic model."

Strictly speaking, the "structural effect" concept needs to be further specified by economists. Its content could be spelled out as an optimal combination of economic sectors and subsectors ensuring maximum economic efficiency. For example, Georgia's GDP growth is mainly a result of the rapid growth of the financial sector and nor industry and this combination is responsible for the GDP's growth (around 12% in recent years). In our opinion this is not the optimal incentive for economic developments.

That is why the definition of "structural effect" is closely connected with the concept of a given state's comparative advantage.[5]

How can the "structural effect" be measured? There is no simple answer to this question. Evidently, it is a matter of definiteness, the share of key (core) sectors in the total effect, with possible changes in the content of the "structural effect" in view of progress. The general point is the optimal use of comparative advantages through global trade.

One-off injections into a country's economy (say, in the form of foreign investment) cannot reflect the "structural effect," although they can produce changes over the long term.

5 A sufficiently complete analysis of the comparative advantages of the countries of the Central Caucasus is given in the following articles: V. Papava, "Comparative Advantages of the Central Caucasian Countries: Potential, Realized, and Missed Opportunities," The Caucasus & Globalization, Vol. 1 (1), 2006, pp. 87-96; S. Zhukov, O. Reznikova, "Eco nomic Interaction in the Post-Soviet Space," The Caucasus & Globalization, Vol. 1 (1), 2006, pp. 96-109.

3 Globalization: A Challenge to National Economic Models

Globalization[6] in the broad sense of the word implies the interdependence of states and nations in the performance of the socioeconomic function. A distinction should be drawn between the political, economic, technological and cultural components of globalization. Although there are no clear dividing lines between them (for example, economic decisions taken at the supranational level often have a negative political effect and vice versa), let us take a brief look at their content.

The second problem arising in this context is how to measure globalization, especially its economic component. In our opinion, there should be different approaches to various aspects of globalization depending on the content of the phenomenon, i.e., qualitative definiteness is the key to quantitative definiteness.

- The political component of globalization implies the interdependence of states based on common political interests (for example, the creation of the European Union); establishment of regional associations as a means of resolving political problems, etc.
- The economic component is based on a community of economic interests (trade and provision of services) aimed at maximizing the economic effect.
- The technological component implies technical progress (primarily in transport and communications) and contributes to the interdependence of countries. Moreover, the very process of globalization initiates an active search in the field of information and communication technologies.
- The cultural component is based on the multilevel development of cultures, religions and traditions.
- The political component of globalization in the Central Caucasus varies from country to country. On the one hand, Azerbaijan and Georgia are oriented toward Western values, and on the other, Armenia takes a pro-Russian stand. Such a choice makes some concrete alterations to the political actions of the region's countries; it is also bound to have an effect on their economic decisions. Another

6 The globalization problem is investigated in many books and articles (see, for example: The Globalization of World Politics. An Introduction to International Relations, Second Edition, ed. by John Baylis and Steve Smith, Oxford University Press, Oxford, 2001; "Tired of Globalization," The Economist, November 2005; D. Held, "Democracy, the Nation-State and the Global System," Economy and Society, Vol. 20 (2), 1991, pp. 138—172; P. Krugman, "Growing World Trade: Causes and Consequences," Brookings Papers on Economic Activity, No. 1, 1995, pp. 327-377.

important aspect of this problem is the question of the territorial integrity of Azerbaijan and Georgia.[7]

A natural question arising in this context is whether globalization (its political aspect) helps to resolve the problems of territorial integrity. The answer lies in the effectiveness of the activity of international institutions and organizations, and also in the appropriateness of actions taken by the parties concerned.

The basis for the solution of this problem is the following: the concept of "national," as pointed out above, should be inherent to state entities, in contrast to "national" as an inherent attribute of a certain ethnos. In the latter case, one can only speak of national sovereignty where the actor is a nation (eth-nos). A confusion of the concepts of state and national sovereignties or their identification is unwarranted. This is evident, for example, from the fact that some national state and territorial entities proclaim their sovereignty, confusing national sovereignty and state sovereignty, which can and must be inherent only to the state. Another concept used in this context is "popular sovereignty" (sovereignty of the people), where the people are the key actor. The legal nature of the sovereign rights of the above-listed actors differs. In civil society, it is the population—as the citizens of a given state—that constitutes a nation. That is why the concept of "national statehood," in contrast to social, has a general civil meaning in international practice, i.e., in this context the concepts of "nation" and "state" appear as synonyms.

It should also be noted that in cases where the emergence of nations on the whole happened at the same time as did the formation of centralized states, nations naturally took a state form and developed into independent nation-states (England, France, Italy). In the east of Europe, on the other hand, the formation of centralized states, accelerated by the need for self-defense (against invasions by the Turks, Mongols, etc.), preceded the emergence of nations. That is why in such cases the latter did not and could not develop into nation states, but formed several multinational states (Austria, Hungary, Russia).

The economic component of globalization should be analyzed with due regard for global contradictions:

- Western integration contrasts with Eastern disintegration (to a certain extent). Suprastate (supranational) political integration in the West is accompanied by disintegration of states in the East;

7 According to experts, Georgia annually loses 20-30% of its GDP due to its territorial integrity problem (to some extent this also applies to Azerbaijan).

- Economics and politics are increasingly internationalized in the sense of a common understanding of problems and decision making. At the same time, the importance of regional associations tends to increase;
- European integration, nation states and regional associations—all these realities will continue to influence the political orientation of Europeans.

Above, we mentioned the indicators of globalization. On the economic side, one can speak about trade and its influence on the dynamics of the gross domestic product; about the dynamics of foreign investment (including direct investment); about the creation of multinational companies, etc.

If we look at the GDP, for example, we will find that in 2005, compared to the previous year, it increased by 26.4% in Azerbaijan, by 9.3% in Georgia, and by 14% in Armenia (in the past ten years, the overall GDP growth in the countries of the Central Caucasus averaged about 5%). It should also be mentioned that whereas, due to the 1998 global financial crisis, Georgia's GDP increased by 3.1% in 1998 and 2.9% in 1999, the respective figures for Azerbaijan were 10.0% and 7.4%, and for Armenia 7.3% and 3.3%.

As regards commerce, we are naturally dealing here, in the first place, with foreign trade (exports and imports). According to WTO data, a higher level of foreign trade operations means more effective economic development, faster GDP growth and lower transaction costs.

It is also difficult to overestimate the role of foreign investment, such as investment in the construction of the Baku-Tbilisi-Ceyhan (BTC) oil pipeline, which largely accounts for GDP growth in Georgia and Azerbaijan.

Plans of a global nature can only be implemented through the establishment of companies: multinational (MNCs) and transnational (BTC and TRACECA projects).

The difficulties in Georgian- Russian relationaships are mainly related with the "frozen" conflicts in breakaway regions of Abkhazia and South Ossetia. These tensions reached the culmination in the 5 days War in August. Georgia accuses Russia of imperialism, while Russia criticises Georgia for nationalism and pursuing an anti-Russian foreign policy. However the conflict has an economic dimension too. Many observers consider the economic policy of Russian Federation towords the former Soviet Republics including Georgia as an agressive, so called "liberal empire" tool to restore its influence. The active participation of the russian enterprises in georgian privatization and investment processes (energy sector), opening of their subsidiaries in industry and banking sectors (Vneshtorgbank) are seen as the economic measures interrelated with political interests.

Finally, we can say with certainty that globalization has an institutional component as well. It is the functioning of international economic organizations and institutions (like the World Bank, the International Monetary Fund or the European Bank for Reconstruction and Development). They have made and continue to make a tangible contribution to the development of the Central Caucasian countries, especially at the early stages of the transition period.

The technological component[8] of globalization is largely the result of scientific and technological progress, while technological determinism is the main driving force behind social development. New technologies are the key element in the globalization of culture, transport and communications; they reduce so-called transaction costs, thus promoting the growth of the economy and enhancing its efficiency.

The cultural component is quite important today, because it implies the need to take into account the multivariant and multivector nature of the world community (otherwise it would be featureless and uninteresting). There are many examples in today's world in which cultural circumstances and the factor of tradition have a significant positive effect on economic progress.[9]

States lacking self-sufficiency are more dependent on globalization and have more to gain from it.

Does globalization lead to general well-being or, on the contrary, intensify the struggle for survival? There appears to be no simple answer to this question, but it is a fact that the world has become "smaller" and that countries and organizations are more dependent on their "neighbors," competitors and partners than before.

4 Regional Economic Integration as a Way to Join the World Economy

It is no accident that the problems of the regional economy and regional economic integration have come on the agenda. This is dictated by the inner logic of economic research: the region is a major element in the self-organization of economic life. Soviet regional studies developed as a science dealing with the location of production and creation of territorial complexes. Later on, it began to study such areas of the regional

[8] See, for example: D. Archibugi, J. Michie, "The Globalization of Technology: A New Taxonomy," in: Technology, Globalization and Economic Performance, ed. by D. Archibugi and J. Michie, Cambridge University Press, Cambridge, 1997.

[9] See, for example: M. Porter, H. Takeuchi, M. Sakakibara, Can Japan Compete? Basic Books and Perseus Publishing, New York, 2000.

economy as housing construction, urban development, communal property, real estate and land management. In Western economic literature, the region (in the economic sense of the word) is seen as a territorial unit that "resolves" certain socioeconomic problems.

Our analysis is concerned with regional economic integration of independent states as subjects of political power. In this case, the territory of the Central Caucasus is regarded as a single region.[10] Within the framework of a general methodological analysis of this question, we leave aside the region's taxonomic rank, although it is obvious that each rank of this kind has its own specific features.

In our opinion, the effect of regional economic integration can be achieved in two ways: by integration within the region (Central Caucasus) and then (time of integration depends on two factors: – political – stability in the region, economic – speed of economic reforms) integration into the world community. This is the optimal alternative and effective, primarily from an economic standpoint.

As regards the first way, let us emphasize that, in principle, economic cooperation between the countries of the Central Caucasus is possible and necessary. In fact, this is how things stand in practice, although for well-known non-economic reasons the potential for such cooperation remains partly "untapped."

The second way is clear enough: the three republics should take part in global economic relations, but the question here is whether this is more effectively achieved through regional integration or through multi-speed and multi-vector steps. The answer here is obvious: the first option is more effective in view of economic factors.

Here we can refer to such concepts as "comparative advantage" and "structural effect." The "structural effect" is more peculiar for "small" and "transitional" countries, because of the non-stability of the economy and the short term change in demand. Due to these circumstances, the involvement of the Central Caucasian countries in the global economy through regional economic integration will be much more effective, because the realization of their economic potentials in this case will mean lower costs and, accordingly, will have a more significant effect.

[10] The basic methodological premises of this approach are given in E. Ismailov's article, "New Regionalism in the Caucasus: A Conceptual Approach" (see: E. Ismailov, op. cit.). Based on historical analysis, the author rightly believes that the historical process "shows the ... trends toward regional integration. In other words, the Caucasus is bound to form a single and integral entity of globalization." (ibid., p. 13).

5 Conclusion

The question formulated in this article: "The National Economic Model alternative to Globalization?" is by no means a rhetorical one. The answer to it, in our opinion, is as follows: this model will be an alternative if national economic interests and the comparative advantages of states in the interdependent world are not taken into account. At the same time, it will not be an alternative if the globalization process equally ensures economic growth in our interrelated, globalized world. This means, as we see it, that economic globalization will help each country to realize its comparative advantages, while regional economic integration—as a factor of gradual entry into the global economy—will come into full play.

Globalization should provide opportunities for using the economic potentials (in other words, the comparative advantages) of the Central Caucasian states. The existence of certain noneconomic problems (political problems in the first place) is further evidence of the truth about the interconnection between economics and politics.

IMPACT OF RUSSIAN SANCTIONS ON THE GEORGIAN ECONOMY

ERIC LIVNY, MACK OTT, KARINE TOROSYAN

1 Introduction

Russia, Georgia's largest neighbor and trading partner, has started imposing economic sanctions on Georgia beginning in December 2005, first, by restricting the exports of Georgian agricultural products to Russia, continuing with a ban on wine and mineral water imports in March and May of 2006, and severely limiting land and air transportation between the two countries. Amid concerns about possible military development and referenda in secession-prone provinces, as well as the harassment of Georgians residing in Russia, there has been surprisingly little discussion among Georgians about the realized and potential impacts of the sanctions on their economy.

The objective of this study was to analyze the economic impact of the restrictions Russia imposed on the movement of Georgian goods, production factors and people (hereinafter the "embargo regime"). To encompass the full gamut of potential impacts, the study comprised three main components:

- Overall analysis of economic growth, changes in the GDP structure, inflation, unemployment and capital.
- Analysis of developments in particular sectors of the Georgian economy: those affected directly by the Russian sanctions (e.g. agriculture, transportation), and those affected indirectly (e.g. trade and services, construction).
- We also analyzed data on migration and remittances which have direct implications for the Georgian labor market, household spending and investment patterns.

The data used in this report are from the official website of the Department of Statistics under the Ministry of Economic Development of Georgia (www.statistics.ge). We also used official National Bank of Georgia data (www.nbg.gov.ge) and household survey data reported by the CRRC (www.crrc.ge).

To measure the potential impact of Russian sanctions, for the whole economy and particular sectors, we begin our analysis by capturing the time trend present in the data for a range of years until December 2005 (the pre-embargo period). We try to account for seasonality, "bad year – good year" dynamics in agriculture, and quadratic trend, when relevant. We then extrapolate that trend to the post-embargo period and compare predicted with actual values of variables of interest. A major question for us has been to understand what part of the "gap" can be directly or indirectly attributed to the Russian embargo, to identify the causal links, and suggest possible policy measures to remedy its negative impact.

The major findings of the study are, perhaps, encouraging for Georgia. The country has clearly survived the Russian bear, as demonstrated by its growth performance in 2006 (between 8.5 and 9%) and 2007. As one would expect, those industries that depended upon the Russian markets – wine, bottled water and some agricultural products – have been harmed. However, this negative impact has been offset by expansion in other sectors. Expansion has been particularly strong in the services sector, e.g. financial intermediation, transportation, storage, etc. Thus, the major impact of the Russian embargo was on the speeding up of the pace of restructuring in the Georgian economy away from agriculture and established Soviet-era brands to other sectors of the economy. While the aggregate effect on GDP and other macro aggregates has been negligible, the social consequences of such a forced restructuring may be quite painful and long-lasting.

The structure of the paper is as follows. In Section A, we provide a literature review. The macroeconomic impact of sanctions is explored in Section B. We continue, in Section C, with the sectoral analysis. Finally, in Section D, we focus on issues concerned with Georgia's external economic relations: international trade, migration and remittances. Policy recommendations arising from the study are discussed in the concluding section.

2 Literature Review

The imposition of trade and transit barriers by Russia has been by far one of the largest negative shocks that the Georgian economy has recently experienced (along with energy price increase). Due to the recent nature of the phenomenon of interest, the body of literature trying to understand the impact of Russian sanctions on the Georgian economy is rather limited. This issue has been discussed in IMF staff papers (for 2006 and 2007), mission statements and other county documents. Another important source of information on the impact of sanctions on the Georgian economy is the reports issued by the National Bank of Georgia. Below, we review the mentioned sources and compare their findings on the impact of sanctions with our results.

According to the "Statement by IMF Staff Mission to Georgia" (IMF, 2006), economic growth in Georgia in 2006 has been hindered by the loss of Russian export markets, but remains robust. The mission estimates that Russia's economic embargo will increase the external current account deficit by about $250-$300 million in 2007. The expectation is that this will be financed by higher foreign exchange proceeds generated by the large inflow of foreign direct investment (FDI) projected for the years 2006/07, as well as by a very substantial increase in tourism revenues. Despite Georgia's strong underlying economic performance, the mission expects that the Russian embargo will lead to moderately slower growth of 6-7 percent in 2007. The recovery of the economy from this shock will rest largely on its ability to generate rapid export growth in the medium term.

A more recent staff review by IMF (IMF, 08/2007) finds that recent economic performance in Georgia has been stronger than expected: despite a sharp slowdown in export growth, declining output in the agricultural sector, other factors—including the significant private capital inflows—cushioned the impact of this slowdown, allowing the Georgian economy to continue to grow. Also, despite concerns that remittances would fall because of the tensions with Russia, remittances have in fact continued to increase, more than doubling in the first five months of 2007, compared to the same period in the previous year.

Our findings largely support IMF statements. We do observe a drop in export performance and increase in imports beyond what would be expected based on long-term trends. There is no evident drop in remittance inflows from Russia into Georgia, while there is a drop in outflows of money from Georgia into Russia. Overall, while certain sectors, and most noticeably agriculture, drop below the trend, the overall level of production in Georgia in 2006 does not seem to react very dramatically to the imposition of sanctions.

The assessment by the National Bank of Georgia is that the Russian sanctions tested the robustness of the changes that the Georgian economy has been going through during the last few years. The fact that Georgia withstood the embargo regime proves its economic independence from Russia. For the sake of comparison, they state that the 1998 Russian economic crisis represented another strong exogenous shock for the Georgian economy and that this shock had a tremendous negative impact on the economic growth rates and financial stability not only in 1998, but also in the following years. However, in 2006, the Russian "shock", despite its significance, only slightly influenced the economic growth and had virtually no impact on financial stability. NBG explains such a limited influence on the economic growth by two main reasons. First, the continuous reduction in Georgia's economic dependency on Russia is driven by the introduction of the visa regime, closure of borders, and other, similar events. As Russia's share in Georgian exports shrunk, the impact of the Russian sanctions turned out to be less negative than it would have been a few years earlier. The second reason is the improved business environment in Georgia. Ultimately, the structural adjustment has helped deal with the negative exogenous shock in a more effective manner.

The agricultural sector, which represents a large share of the GDP, experienced a rather dramatic contraction in 2006. NBG notes that the Russian trade embargo may have been an important factor, however, they argue that this contraction had more to do with unfavorable climatic conditions and sector specifics. Our findings do not support this point of view. While we agree that some part of the drop in agricultural production was due to sector-specific factors, the effect of Russian sanctions on the agricultural sector was very substantial.

3 Russian Embargo Macro Effects and Non-Effects on the Georgian Economy

The Russian embargo was imposed on selected products at the end of 2006/I, most notably on Georgian wine, bottled spring water and agricultural commodities—fruits, nuts and vegetables. Anticipating the negative impact of Russia's sanctions, by the end of 2006, the National Bank of Georgia (NBG) had reduced its forecasts for economic growth by 2-3%-11.5% to 8.5%-9%. Clearly, there are specific and visible industries that have suffered, notably wine, but our analysis shows that the aggregate effects are not large.

3.1 GDP Trend Analysis

In this section we look at aggregate quarterly GDP data for Georgia for the period of I. 1996-IV.2005 (total of 40 quarters). Our goal is to capture the time trend present in the data up through December 2005 (pre-embargo period), then to extrapolate that trend to the period of I.2006-II.2007 (post-embargo time periods) and compare our results with actual GDP in each of these quarters. For our analysis we continue to use quarterly GDP for Georgia in constant (1996) prices, million GEL, published by the Department of Statistics under the Ministry of Economic Development of Georgia.

To capture the general time trend in GDP series as well as seasonal fluctuations we have included a quadratic time trend and a series of seasonal dummy variables. All variables are statistically significant and jointly explain 96% of variation in GDP series. We check for serial correlation by using the Breusch-Godfrey test for AR(1). There is no evidence of serial correlation in the regression and we use results from the ordinary least squares estimation of the model.

Figure 1: Predicted and Actual Quarterly Values of GDP (I.1996-II.2007), Million Constant GEL

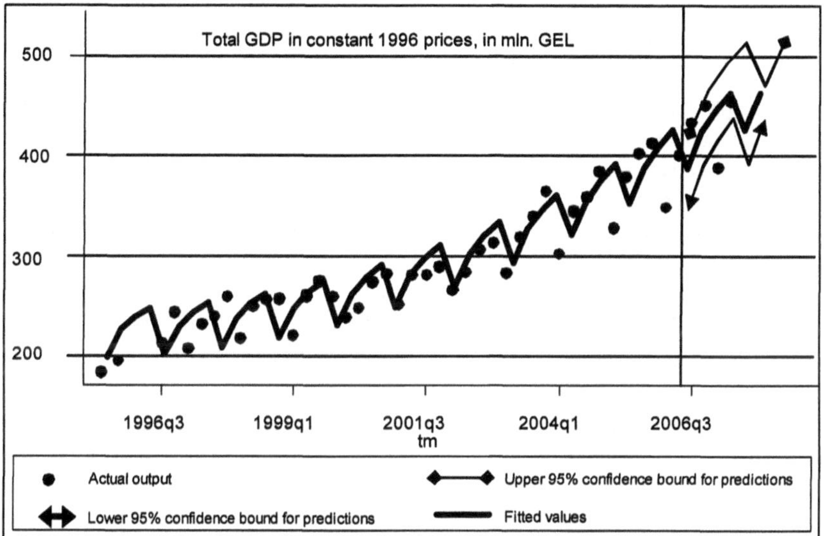

Note: vertical line corresponds to the last quarter of 2005

To add some confidence to our estimates we have calculated a 95% confidence interval around our predicted quarterly GDP values (see Figure 1). It turns out that in all the quarters except for the first quarter of 2006, actual GDP was within the confidence interval. The level of actual GDP in the first quarter of 2006 is well below its predicted level, reflecting the immediate impact of the sanctions on the Georgian economy. The same pattern is observed in the first quarter of 2007. GDP data in the other quarters do not show any obvious impact from the embargo. Interestingly, our results do not show a reduction in the growth rate of real GDP following the imposition of sanctions.

3.2 Inflation And Unemployment In Georgia

In the first quarter of 2006, inflation decreased from 5 % to 4 %, but from April onwards inflation started to increase and reached its peak of 14.5 % in August. From August till January inflation dropped to 8.8%. The increase in inflation was driven by the increased price of natural gas, bird flue, production costs increase and general expectations of inflation by the public because of the Russian embargo. The decrease in inflationary pressures in the second half of 2006 can be explained by the decline in world oil prices, appreciation of GEL with respect to USD and the tightening of its monetary policy by the NBG. The NBG increased reserve requirements: the minimal obligatory reserves requirement in national currency increased to 2% as of August 20, and to 4 % as of December 10 (Inflation Report of the NBG, 3rd Quarter, 2006, p.13).

Accordingly, there is no evidence that the NBG attempted to use its monetary policy to soften the effect of the embargo on the economy, despite the temptation to do so. During this period, many producers, e.g. in the agricultural sector, were confronted with the need to carry larger inventories, sell at lower prices, and perhaps absorb higher transportation costs to compensate for the Russian blockade. At the same time, the money supply did grow in 2006 due to a significant credit expansion to the private sector by commercial banks from 99 million GEL in December 2005 to 151 million in December 2006. Credit expansion peaked in May 2006, reaching 231 million GEL

Credit expansion to the agricultural sector is worth special consideration. The volume of credits to this sector has a strong seasonality. As a rule, the volume of credit goes up in fall and spring, which is not surprising. This seasonal pattern was largely maintained during 2006. However, there was an unusual spike in credit in December 2006 and January 2007, which does not fit the general trend (the fall spike in agricultural loans is typically followed by a lull in the following winter months).

The increased level of lending has affected the dynamics of monetary expansion: reserve money expanded sharply in excess of trend in November 2006, with most of the increment—not all—remaining in the system. However, it is important to note that the increase in reserve money can also be attributed to the activities of the Bank of Georgia on the London Exchange in November 2006.

Unemployment edged up slightly in 2006/III, but remained at or below the rate for the full year 2005; thus unemployment figures do not by any means reflect a downturn in the economy.

4 Sectoral Impact of Sanctions

4.1 Agriculture, Forestry, Fishing

The relative share of this sector in GDP has been historically high; with a decreasing trend over time: in 1996 this sector accounted for 34% of the overall GDP, while in 2006 it was 19%. Even with its decreasing share, agriculture is still the largest sector of production in Georgia, and as such plays a very important role for the whole economy of Georgia.

Trade in agricultural products with Russia was rather high before the imposition of sanctions. Exports of Georgian agricultural products to Russia were restricted on December 19, 2005.

In order to estimate a potential level of agricultural output in 2006, we fitted a quadratic time trend with seasonal adjustment through the quarterly time series on agricultural output for 1996-2005 (for the total of 40 quarters). We also added a dummy variable for odd years to capture the "bad year – good year" pattern observed in agriculture. Test for AR(1) shows no evidence of serial correlation in the model. In all post-sanctions quarters except for III-IV.2006 agricultural production is below the lower bound for the predicted output of the confidence intervals.

Our estimate of underproduction in agriculture is almost 12% of the predicted output in this sector and is a significant part of GDP in I.2006-II.2007. It is important to note that this underproduction is not due to a bad year in agriculture, since we have controlled for that factor. Hence, we conclude that the estimated drop in agricultural production is due to Russian sanctions. Another important observation is that the level of production has not increased in 2007 – in the first two quarters of 2007 we see a very significant underproduction. The situation has not improved despite steps aimed at buttressing the agricultural sector. For example, after the Russian embargo on Georgian wine, the Georgian government induced local businesses to support the wine sector; the policy was not official, though. This fact has been mentioned many times in interviews, including by the President of Georgia (see, for example, "News Georgia" official website[1]).

To summarize, despite efforts to support the agricultural sector, including credit expansion by commercial banks, as of II.2007 the sector has not recovered from the shocks of 2006 and needs further assistance. Thus, we find it important to introduce various farmer support programs. One such program is the agricultural development project[2], presented on 1st of August 2007 by the Ministry of Agriculture, the Ministry of Finance, the World Bank representative office in Georgia and other donor agencies. The project aims at providing long-term loans for farmers, refining businesses and agribusinesses. 10.4 mln USD have been distributed among 5 commercial banks, which have taken responsibility to service the loan-providing process. The duration of

1 http://www.newsgeorgia.ru/geo1/20061227/41861203.html
http://www.newsgeorgia.ru/geo1/20060908/41804562.html,
http://www.newsgeorgia.ru/geo1/20071015/42071871.html

2 Ministry of Agriculture of Georgia, official web-page: http://www.maf.ge/?class=94

the credit line to commercial banks is 10 years with a 4-year preferential period and an interest rate determined by LIBOR 6-month deposit rate plus 2%.

4.2 Manufacturing

Another very important sector in the Georgian economy is manufacturing. The relative share of this sector in overall GDP was 11% in 1996 and continued to be the same in 2006.

Overall, we estimate "overproduction" in the manufacturing sector at 64 million GEL for 6 post-embargo quarters. Thus, prima facia, the Russian sanctions did not have a visible negative impact on the Georgian manufacturing sector. Moreover, the increased level of output in manufacturing helped balance the negative effect of sanctions on other sectors and the overall GDP.

The fact that the manufacturing sector in Georgia has not been hit by the Russian sanctions is both encouraging and surprising. It is encouraging since manufacturing is one of the most important and quickly developing sectors in the economy and the fact that it did not contract due to Russian sanctions is a sign of its maturity and economic strength. However, it is surprising given that parts of the sector were directly hit by the sanctions regime. These include mineral water, wine, some heavy equipment (locomotives), where we expected to see a large decrease in output.

To better understand the situation, we conducted a series of interviews with several large producers of manufactured products that had strong ties to Russia. There are a few factors that could plausibly explain the lack of negative impact on the manufacturing sector. These facts include:

- Even though the sales of some products to Russia were drastically reduced, the production level was kept high, with firms increasing inventories. Such a temporary increase in inventories could be justified given the prospect of finding alternative markets and/or restoration of full economic ties with Russia;
- some output was redirected to other markets (Ukraine, Armenia, etc.);
- companies started development and production of new products.

4.3 Construction

This is a quickly developing sector in the Georgian economy: the relative share of construction in total production rose from 2.5% in 1996 to 9% in 2006.

Using the results of our time trend analysis, we estimate the expected level of construction for each quarter of I.2006 - II.2007. During all six quarters except for III.2006 there was underproduction in this sector compared to the predicted levels. This deviation is particularly strong in the last quarter of 2006 and the second quarter of 2007, when the actual level of production is below the lower bound of the confidence interval. Overall loss in production in this sector is estimated to be 159 million constant 1996 GEL. However, such a large drop in construction activity in 2006-07 is due to the fact that a large-scale construction project in the energy sector came to an end. Hence, there is an obvious direct link between the Russian sanctions and the drop of output in construction.

4.4 Trade and Repair Services

The relative share of trade services and repair services varied from 12% to 15% during 1996-2006. Estimation results for this sector are rather interesting. As it turns out, the level of production in this sector exhibited exceeded the predicted level throughout the post-embargo period. The overall "surplus" is estimated at 130 million constant 1996 GEL, also helping to balance the overall drop in the GDP level caused by the Russian sanctions.

4.5 Transport and Storage

The relative share of transport and storage varied from 4.5% to 9% during 1996-2006. This sector shows a sizeable positive deviation from the predicted trend in all six post-embargo quarters. Aggregate excess in production in this sector is 175 million constant 1996 GEL. This significant economic growth in the transportation sector in 2006-07 has been achieved as a result of better exploiting Georgia's transit possibilities. It is worth noting that the sector could have developed at a faster rate had the sanctions regime not been in place. In particular, motor and air transport lost potential revenues from cargo and passenger transportation to and through Russia.

4.6 Financial Intermediation

The relative share of this sector has rapidly grown from less than 1% of GDP in 1996 to over 8% in 2006. This rapid expansion is expected to continue in the future, further increasing the share of financial intermediation in overall GDP.

The estimation technique for this sector is slightly different from the previous cases, since we found evidence for AR(1) processes in the model. To correct for serial correlation, we performed feasible generalized least squares estimations using the Cohrane-Orcutt technique. During the quarters of interest we have found significant overproduction (production above the historical trend) in this sector. The overall level of overproduction is estimated to be 144 million GEL (constant, 1996). Financial intermediation is yet another sector that has been experiencing rapid growth, helping to fill the gap in production caused by the Russian sanctions.

4.7 Conclusions

As we have seen above, aggregate production in all six post-embargo quarters is very close to the predicted trend, contrary to what may have been expected. Based on our estimates, total GDP losses for 2006 and the first two quarters of 2007 are about 36 million 1996 GEL, which is not statistically different from the trend.

Looking at the disaggregated data, however, we find that during this period Georgia went through a major structural change whereby contraction in one sector – agriculture – was cancelled out by above-trend performance in other major sectors, in particular services and manufacturing.

Whether this trend will be sustained into the future will depend on many factors, including Georgia's political stability. It seems likely that as Georgia further integrates into the global economy, positive developments in the services sector will continue, further reducing the relative weight of agricultural output in GDP.

5 Impact on External Economic Relations

The Russian embargo has had a significant impact on Georgia's in- and outflows of goods, capital and labor. In this section we analyze the time trends in Georgian exports and imports, as well as remittances, with a special focus on money transfers to/from Russia.

In addition to being Georgia's largest trade partner, Russia was a very important country of destination for Georgian migrants. Not surprisingly, Russia has also been the largest source of remittance inflows into the Georgian economy.

5.1 Analysis of Georgian Exports

The most direct effect of the Russian embargo was on Georgia's exports. The impact is not as strong as one would have expected, however. In this section we look at the aggregate monthly export data for Georgia for the period of 01.2000-03.2007 (total of 87 months). Overall, Georgia's exports continued to expand in 2006. However, the Russian embargo has slowed down the rate of export growth.

To understand the extent of this negative impact, we apply the same statistical procedure to extrapolate the time trend present in data until December 2005 for the post-embargo period (after January 2006) and compare our results with the actual volume of exports in each month. We start by fitting a quadratic time trend and drop the linear or quadratic term depending on their significance level. A dummy variable for each month is included if it is individually significant and has a t-statistic above 1. We use the Breusch-Godfrey test for AR(1) and when we find evidence of serial correlation we perform the Feasible Generalized Least Square (FGLS) estimation. The data used in this section was obtained from the National Bank of Georgia. Trade values are reported in thousands of current USD.

Using data on aggregate exports from Georgia for the time period at hand we can easily infer the pattern of trade and its changes over time. We check for serial correlation in the model and find no statistical evidence of such. Since data exhibits slight heterogeneity, we are using the heteroskedasticity-robust estimation technique that adjusts standard error estimation. The estimation procedure is based on 72 observations of the aggregate exports volume covering January 2000 to December 2005; data for 2006 and the first 3 months of 2007 is not included in the regression. Our goal is to study the pre-embargo pattern of exports; predict, based on our findings, the pattern for 2006-2007; and compare it with the actual level of exports for the post-embargo period. If there are differences in predicted and actual trade levels, they could be attributed to the Russian sanctions. The benefit of our approach is that it allows us to calculate the total value of export losses compared with the potential level. Our results show that export volumes are significantly below its predicted level for almost all months after December 2005 (below the lower confidence bound).

5.2 Georgian Imports

We will now look at imports to Georgia from its main import partners. Imports to Georgia continued to grow at a high rate, very slightly outperforming the time trend. We estimate a relatively modest increase in the volume of imports, which could be partially attributed to higher world prices of oil and oil products, natural gas and other commodities such as wheat and sugar. However, during the same period, Georgia has also dramatically increased its imports of FDI and construction-related goods such as steel rods, trucks, computer and telecommunication equipment. Overall, imports may have been boosted by stronger aggregate demand and real appreciation of the Georgian Lari.

Import dynamics from Russia show a larger variation in the post-embargo months, but there is no clear deviation from the long-run trend. The bulk of Russian imports consists of two commodities: natural gas and wheat. According to figures released by the NBG, in the first 9 months of 1996, Russia provided Georgia with the lion's share of natural gas and wheat consumption (96% and 80% of total imports to Georgia, respectively).

5.3 LABOR MIGRATION AND REMITTANCES

Our primary focus in this section is on the impact of Russia's embargo and political standoff on the flow of migration and remittances between Russia and Georgia. We extend our analysis to Armenia and Azerbaijan as useful "benchmarks". Our analysis is based on data from the Caucasus Research Resource Center's (CRRC) Annual Household Surveys, which collect detailed data on household demographic, economic and other characteristics from regionally representative sample of households in three republics of the South Caucasus[3].

Only about 2% of the entire pool of individuals recorded by the survey is reported to be in migration. This share does not change significantly from year to year, and might reflect rather well the true share of people that were indeed abroad for each given year.

Most people emigrating from Armenia are headed to Russia (80%), mostly for work purposes. For Azerbaijan, Russia is an even a more important migration destination in relative terms (about 90% of all migrants from Azerbaijan go to Russia). However, it is worth noting that Azerbaijan has the lowest numbers of migrants for all years and all destination countries: about 1/2 of the total number of migrants from Georgia and only

[3] Data that are currently available cover 2004, 2005, and 2006 and provide information for almost 16,000 families from urban as well as rural areas of the region. Tables A1 and A2 in the Appendix provide more information on regional data coverage for all years in which the survey was implemented.

1/3 of that from Armenia. Interestingly, the list of other migration destinations for these two countries is rather short.

The picture is quite different for Georgia. First, migration destinations for Georgians are much more diverse than for Armenians and Azerbaijanis. Second, being the most popular destination country for Georgian migrants, Russia hosts less than 40% of all migrants from Georgia as compared to 80% and 90% of migrants from Armenia and Azerbaijan, respectively. Further, of those Georgians who are in Russia less than 50% are there for work purposes, the rest state other reasons, primarily family reunification and education.

This sharp difference in the migration pattern for Georgia could be a result of the visa regime put forward by Russia since 2000[4]. According to these data, Georgia has been diversifying away from Russia and towards new migration destinations. Apparently, migration to Russia has become more costly and, possibly, less beneficial, given the negative attitude towards Georgians. Hence, the net benefit from migrating to Russia has become lower compared to other countries, which caused many people to switch to alternative destinations. Among these, Ukraine is a new and increasingly attractive country for Georgian emigrants alongside such European countries as Greece and Germany.

The embargo regime will further reduce the net benefit from migration to Russia (due to further increases in the cost of migrating to this country, and a reduction in benefits because of the higher level of intolerance towards Georgians in Russia). Hence, we expect a further diversification away from Russia and towards other countries. With time, as richer migrant networks emerge in the "new" destination countries, they will become even more attractive for Georgians. The work motive is much weaker for migrants from Georgia. This makes Georgia relatively immune to Russian sanctions in terms of loss of income from remittances.

5.4 Remittances

In this section we will focus on the trend in remittances to and from Georgia, with a special emphasis on remittances to and from Russia. Figure 2 shows the amount of remittance inflows as whole and separate inflows from the 3 largest source countries –

[4] Unlike Georgia, Armenia and Azerbaijan have visa-free access to Russia, which explains the major role Russia plays as a migration destination for these two countries.

Russia, USA, and Greece. Starting in 2003, the flow of remittances from Russia started to grow at a very high rate, while remittances from the other countries shown in the figure have shown only a modest increase (compared to Russia).

Figure 2: Total Monthly Inflow of Remittances Into Georgia, Billions of Current USD

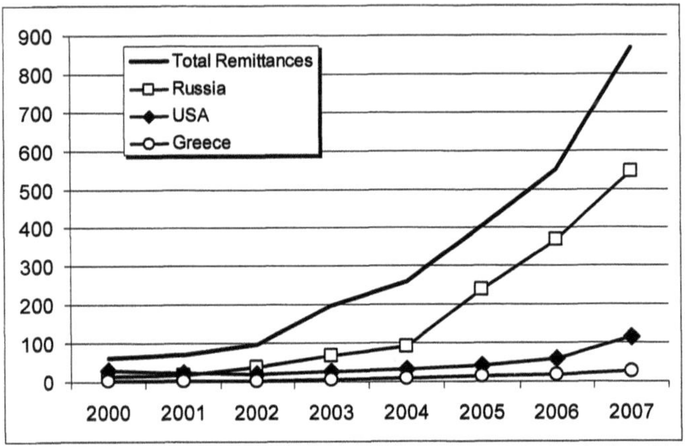

Source: National Bank of Georgia at http://www.nbg.gov.ge/.

The large volume of remittances from Russia can be explained by the fact that Russia is still the major migration destination for Georgia, and in CRRC data the share of Georgians living in Russia was 40% - 4 times larger than in the second most popular destination – Greece. Even though the list of other countries that host Georgians is rather diverse, according to CRRC data their individual shares in overall migration flows are still small compared to that of Russia.

To follow possible changes in monthly remittance inflows into Georgia (and particularly from Russia), we perform the same type of analyses of these time series as was done above for exports and GDP flows.

Remittances flows into Georgia exhibit an interesting pattern: there has been what seems like a significant change in the pattern starting in 2004 (which coincides with the Rose Revolution in Georgia). There has been a sharp increase in the growth rate of remittance inflows, as well as a sharp increase in the volatility of these flows compared to earlier years (2000-2003). This pattern is present for aggregate data on remittance flows into Georgia, as well as for these inflows from individual source countries. This phenomenon may be attributed to some reforms in the banking sector. In any case, it needs further investigation to understand the nature of this sharp increase in monetary transfers and factors behind the higher volatility of these transfers over time.

Time trend regression results for money transfers from Russia to Georgia are arrived at, using 72 pre-embargo observations. As before, we use a heteroskedasticity-robust estimation of standard error. We allow for an intercept and slope change for data after January 2004. All variables in the model are highly significant and explain 91% of total variation in remittance inflows from Russia.

Based on our estimation results we predict remittance inflows from Russia and compare these predictions with actual inflows to detect any statistical difference. The drop in overall remittances into Georgia from Russia is relatively mild (only about 5% of the potential level), and it is well within the limits of the confidence level for our predicted volume of remittances.

Now, we will turn to the analysis of remittances flows out of Georgia and into Russia. The technique that we apply here is identical to the one used for remittance inflows from Russia, but the outcome is somewhat different. There has been a sharp drop in remittances outflows going to Russia. This pattern starts right after the arrest of Russian military officers on September 27, 2006 and a series of sanctions that followed this incident, including the suspension of issuing Russian visas to Georgian citizens.

Data on remittance outflows suggest that after October 2006 money transfers from Georgia to Russia dropped on average by 7.3 million USD per month. The drop in remittance outflows to Russia is partially due to the departure of around 2000 families of Russian military in the last quarter of 2006. This pattern in cash outflows means that some part of the money that was otherwise going to Russia stays in Georgia and, as one could speculate, is used in local economic transactions.

6 A Summary of Findings and Policy Recommendations

The most controversial and surprising finding from our study is that the Russian embargo has had no impact on the dynamics of the main macroeconomic aggregates: GDP, inflation, and employment. This finding is surprising in as much as Russia has been a key trade partner for Georgia. It provided Georgia with the bulk of essential inputs e.g. wheat and natural gas; it was also a major supplier of components for the manufacturing sector. As a market for Georgian exports, Russia was an important consumer of agricultural products (fruits, vegetables, nuts and walnuts), wine, spirits and bottled water as well as ferroalloys. Finally, Russia has been the largest destination for Georgian migrants and the largest source of remittances into Georgia.

At a more disaggregated level, the findings are perhaps not less perplexing. While, as expected, the sanctions delivered a major blow to Georgian agriculture (we estimate losses, compared to the potential level, of 240 mln. constant 1996 GEL), a few other sectors of the economy appear to be outperforming the long term trend. This is the case with manufacturing (despite the loss of export markets for wine and bottled water) and the services sectors, with financial intermediation leading the way. The losses in agriculture are easily attributable to the Russian embargo. Could it be the case that the Russian embargo – ironic as it is – facilitated growth and development in these winning sectors?

We believe that this may indeed have been the case. The imposition of the Russian embargo provided Georgia with excellent public relations in the rest of the world, leading to a surge in interest, tourism, and foreign direct investment. Politically, Georgia became strongly associated with the US and the West, becoming a target for technical assistance programs and projects. Technical assistance efforts by Western nations have been instrumental in propelling Georgia to top rankings in the widely cited Doing Business report; Georgia's credit rating was adjusted to B+/B in 2006 (Standard & Poor's, December 2006).

An interesting finding of our study is the limited visible impact of the Russian embargo and transport blockage on the labor migration from Georgia to Russia. Apparently, the visa regime that had been put in place for Georgian citizens before the recent escalation in sanctions had weakened the pull factor of labor migration to Russia. Therefore, the subsequent blockade did less damage than it would have under normal circumstances.

Another paradoxical but positive finding of our study concerns the flow of remittances from and to Russia. We do not observe any sizeable decrease in the flow of remittances from Russia. Prima facie, this suggests that economic activity by Georgian migrants in Russia has not gone down, at least not in the short run. It would be important, however, to analyze the structure of remittances from Russia (size and number of transfers) in order to better understand the dynamics that were set in motion by the Russian blockade. There are at least two complementary explanations that would be consistent with a downscaling of Georgian labor migration to Russia.

First, while the number of Georgian labor migrants in Russia goes down, more Georgians are considering return migration due to harassment by the Russian authorities, and are therefore remitting larger amounts of money in order to start a new business or purchase property in their home country. Second, as travel by physical persons became more costly in 2006, migrants may be forced to increasingly remit money through the banking system, bringing up the officially recorded level of remittances.

Very importantly, and consistent with the above logic, we do see a sharp drop is the outflow of remittances to Russia. Monthly bank transfers to Russia began to decline in October 2006, simultaneously with the imposition of further restrictions on Georgians travelling to Russia.

7 Policy Recommendations

During the last decade, starting in 1996 according to our data, the Georgian economy has been shifting away from agriculture towards manufacturing and the services sectors. While agricultural output continued to grow in absolute value, its relative share in GDP has declined steadily. It appears that the Russian embargo and the subsequent "westernization" of the Georgian economy have sped up this process of structural change. The losing sectors were those depending on the Russian market, including such signature brands as Borjomi and semi-sweet red wines. While these structural changes were to some extent inevitable and perhaps desirable, their pace was accelerated by the Russian embargo and would take a special effort to accommodate.

In particular, Georgia would have to devote attention to restructuring the agricultural sector and facilitating the process of social and economic adjustment in the rural regions. The social consequences of a forced restructuring may be quite painful and long-lasting because of a lack of alternative employment options, with the important exception of tourism. Redirection of agricultural products to other markets will require major marketing efforts, including investment in international branding and quality assurance. To facilitate this process, the government – together with relevant business leaders and associations – may need to take steps to induce competition and quality control in the food processing industry which suffered the most from the Russian embargo. Many Georgian products can be re-oriented to highly competitive EU and North American markets should such measures be taken.

To help the small agricultural producers, the government may want to promote cooperation among small producers in the procurement of inputs, credit, storage, transportation and marketing. Such cooperation through regional and/or national associations is a key feature of agriculture in e.g. Israel and Alsace (Northern France). Though thus far cooperation has not been very successful in Georgia's agriculture, it may be tried again under the new, post-embargo circumstances.

The Russian embargo has created a strong impetus for Georgia to diversify its trade and industrial production away from the Russian market. We consider this to be a positive development which will already pay off in the medium term. That said, to fully integrate itself in the global economy, George would need to make a massive investment in the road and railway infrastructure to reduce the costs of transportation to

Armenia, Azerbaijan and Turkey. Georgia must also reduce other barriers to trade through e.g. simplification of customs procedures (still quite cumbersome according to the IFC Business Environment report, 2006).

The slowdown in migration and concomitant capital flows to Russia left Georgia with a larger and potentially better educated and industrious workforce, more money for local investment, as well as increased domestic demand for locally produced goods and services. While restrictions on outmigration may have strained the labor market, the growing Georgian economy was able to absorb most of the surplus labor, as is evident from the stability of unemployment figures. The government, however, has to deal with local pockets of unemployment in regions that suffered directly from the Russian embargo. The longer-term impact of restrictions on outmigration and remittances from Russia is less certain. It will to a large degree depend on the continued performance of the Georgian economy.

To summarize, the Russian embargo did more good than harm to the Georgian economy. It was introduced at a time when Georgia had the economic slack to cushion the negative impact of the embargo; it created strong incentives for the Georgian private sector to modernize and diversify; it catalyzed an increase in economic aid and technical assistance from the international community, triggering dramatic improvements in the business climate as reflected in analytical reports (e.g. IFC, WB) and international credit ratings.

We must conclude with a word of caution, however. Many of the positive shifts that occurred in the Georgian economy are a function of the excellent international public relations Georgia enjoyed in the wake of the Rose Revolution and the scuffle with Russia. The rapid economic expansion was assisted by the genuine efforts to help on the part of the West; at least to some extent, however, it was a function of speculation by international and local investors. Credit expansion during the last year or two – including consumer credit and (subsidized) credits to agricultural producers, may not be sustained if the economy is hit by a recession. The political uncertainty in the run-up to the presidential elections in early January 2008 and its aftermath will provide an excellent test of the strength of the Georgian economy and its ability to survive the loss of Russia.

8 References

Caucasus Research Resource Centers, http://www.crrccenters.org/

Georgia: Fifth Review Under the Three-Year Arrangement Under the Poverty Reduction and Growth Facility and Request for Waiver of Performance Criteria - Staff Report; Press Release on the Executive Board Discussion; and Statement by the Executive Director for Georgia", IMF Series: Country Report No. 07/107, March 09, 2007

Georgia: Fourth Review Under the Three-Year Arrangement Under the Poverty Reduction and Growth Facility and Request for Waiver of Performance Criteria - Staff Report; Press Release on the Executive Board Discussion; and Statement by the Executive Director for Georgia", IMF Series: Country Report No. 06/395, November 09, 2006

Georgia: Sixth Review Under the Three-Year Arrangement Under the Poverty Reduction and Growth Facility and Request for Waiver of Performance Criteria - Staff Report; Press Release on the Executive Board Discussion; and Statement by the Executive Director for Georgia", IMF Series: Country Report No. 07/299, August 31, 2007

National Bank of Georgia, http://www.nbg.gov.ge

Online Magazine- Civil Georgia, "Estimating Impact of Potential Gas price Hike", Giorgi Sepashvili, 11.03.2006;

Statement by IMF Staff Mission to Georgia", IMF Press Release No. 06/276, December 15, 2006

Statistical Department of Georgia, http://statistics.ge/

TAX REFORM IN GEORGIA - WHICH ALTERNATIVES SHOULD BE PURSUED?

ELGUJA KHOKRISHVILI, DAVID NARMANIA

1 Introduction

After 90 years of soviet hegemony large economic, political and social changes were made in former Soviet Union countries to support the transformation process. A key role in the reform agenda is given to public finances, especially tax reform. During the last decade several researches and recommendations have been made in scientific literature regarding a consumption-based tax reform. The advantages of a consumption based tax system are based on the elimination of ineffective elements existing in the traditional tax system of Western countries and on the reduction of incorrect allocation effects of resources provoked by it[1]. Income taxes distort the spend/save decision, leading taxpayers to prefer current over future consumption. Consumption taxes are thought not to have this effect. This suggests that a consumption based tax system is closer to the policy makers' notion of a fair and efficient system of capital income taxation and, hence, probably a better candidate for a tax reform in transformation countries than competing reform strategies. Stable economic growth is the essential basis for a country's successful transformation to a market economy. After the adaptation of the first tax code in Georgia in 1998, several changes were made to a new tax code which became effective on January 1, 2005. According to the varieties of taxes and their administration, Georgia more or less adopted the tax systems existing in Western countries. It becomes obvious that the chance of developing a rational, sustainable and adjusted tax system for transition countries was missed. A good portion of the tax reform process in Georgia has been carried out without an explicit evaluation of how well the different proposals would perform against standard objectives including revenue performance, economic neutrality, tax burden distribution, simplicity and administrative feasibility. Short-changing the preparation stage inevitably led to a continuous ad hoc patching of the system, creating confusion among tax

[1] For a good first reading on a consumption-based tax reform see Rose (1991), for proposals on such a reform in the UK see Kay and King (1986) and Institute for Fiscal Studies (1978), in the U.S. see Hall and Rabushka (1995) or Bradford (1986).

administrators and taxpayers alike and creating uncertainty for domestic and national investors.

In this article, we will describe how the tax system codification has been developed and which results have been reached by these reforms up to now. In order to answer the question, what a rational and effective tax system should be like, it is necessary that the offered alternatives implement the functions awarded to taxes in a market economy.

In the first section, an appraisal of taxing principles will be given. The early and present situation of a tax system is shortly discussed in the second and third sections. The analysis of the tax reform will be based on these basic criteria, according to which some taxes are especially deemed to be more expedient in a democratic and legal state. The following corresponds with the concept of an adequate tax system for a market economy: correspondence with the market economy's conditions, free of inflation effects, easily to administer and fair taxation. Especially for Georgia, a tax system is desirable, which will work successfully by taking into consideration the small resources of taxpayers and the administration[2].

2 Guiding Principles for Taxation in Transition Countries

In order to evaluate an existing tax system and to develop alternatives for reform, criteria for a "good" tax system are needed. These criteria can be inferred from the basic goals of fiscal policy and the associated functions of taxation, as well as knowledge of the tax incidence (tax burden). The demand for a criteria-oriented tax reform is of particular importance for the former socialist countries of Eastern Europe, which usually have no experience with taxes as the main source of financing the government budget in the new framework of democracy and a social market economy.

Tax is an obligatory, unconstrained money contribution to the state budget, which is paid by the taxpayer, in the form of an inevitable, non-equivalent and gratuitous payment. Tax revenues are used to finance the budget (fiscal function). Herewith, taxes have regulative function, because they decrease the purchasing power of the private sector and transform it into the state's purchasing power. The regulative

[2] See Rose, M. (1999): Einführung marktorientierter Einkommensteuersysteme in Osteuropäischen Reformstaaten, in: Smekal, C., Sendlhofer, R., Winner, H. (Hrsg.), Einkommen versus Konsum. Ansatzpunkte zur Steuerreformdiskussion, Heidelberg, pg. 167.

function is also shown in cases in which taxes are used to decrease the chance of or stop a market failure. The social-political function of taxes is the distribution of primary revenues from the market process. Finally, taxes have an informational function, because the tax is considered as a "price" paid by each individual for state services[3]. Various political programs in democratic states, with their appropriate "prices", are expressed as tax burden and they compete with each other to obtain the electors' majority. From this, taxes should execute informational function in democratic states[4]. A "good" tax system should ensure the execution of the above mentioned general functions of taxes[5], but the execution of these purposes is related to undesirable lateral effects, of which one of the most important is excess burden. Over the last decade, scientific literature has paid more attention to understanding undesirable lateral effects provoked by taxation. For this reason, the rational tax system to be established in Georgia must not only achieve the goal of establishing steady sources of revenue for the state, but must also satisfy the criteria of efficiency fairness, harmonisation and simplicity[6].

Efficiency: The efficiency criterion pertains to the extent to which the tax system collects the necessary revenue without otherwise affecting economic behavior. The concept of 'neutrality' is used to describe taxes which do not result in taxpayers altering their economic behavior. Neutrality is a reality with respect to individuals' most basic decision: once you have earned income, you have to decide whether to spend it or to save it for future consumption. An income tax that does not influence the savings decision (and this means: the growth rate of an economy) is called *intertemporally* neutral. When the individuals save they have to choose between different investment alternatives. *Investment or intersectorally* neutral income taxation will not influence the investor's decision where to invest. There will be no tax discrimination between industry sectors or types of assets. Investment in a corporate business is regarded as one

3 See Stiglitz, J. (1989), Finanzwissenschaft, München. pg.417.

4 There is a great discrepancy between the information function and the political reality given here, because in most cases in real politics the government tries to declare incorrect "prices" of a political program to the electors or hides them. See Stiglitz/Schonfelder (1989).

5 Criteria are based on general functions of taxes. These criteria belong to: Fiscal productivity, administrative effectiveness, tax burden effectiveness, etc. We try not to discuss separate criteria in this article. For detailed information see: Musgrave, R. A./Musgrave, P. B./Kullmer, L. (1993)

6 See Rose, M. (1999): Recommendation on Taxing Income for Countries in Transition to Market Economies, In: Rose, M. (Hrsg.), Tax Reform for Countries in Transition to Market Economies, Stuttgart, pp. 27ff.

alternative among others, like bank deposits, self-employment or some small business[7]. *Intersectoral* neutrality also includes neutrality with respect to reinvestment decisions: there must be no tax discrimination between profit paid out as dividends and profits reinvested in the same taxable entity. The last income taxation should be neutral with respect to financing decisions. This means, that the decision between debt and equity should not be influenced by taxation. A neutral income tax system is also prerequisite for not influencing economic behaviour through taxation[8].

Fairness: From the Georgian perspective, direct taxes must satisfy this criterion in that they do not impose upon certain types of income a higher or lower burden than is imposed on others. This requires that the profits of enterprises of different legal forms and capital income should not bear a higher tax burden than income from employment or self-employment[9].

Harmonization: The existing tax system should be constituted with a feasible basic structure for the realization of a harmonized system for the taxation of personal income and of company profits. It does not make sense to have, as an objective, a system that is harmonized with respect to the tax base, if the taxation of the incomes of natural persons is out of step with the taxation of the profits of corporate bodies. Such harmonization, however, is a "sine qua non" because at the end of the day a company's profits are its owner's income[10].

Simplicity relates to how well taxpayers, policy makers and administrators understand the system and can comply with it. Compliance costs on taxpayers should be minimized and taxpayers should be able to readily understand the tax consequences of their actions[11].

7 See Kruschwitz, L./Husmann, S./Schneider, D. (2002) „Investitionsneutrale Steuersysteme vor dem Hintergrund der Kontroverse um Einkommen oder Konsum als geeignete Steuer-bemessungsgrundlage". In: Diskussionsbeiträge des Fachbereichs Wirtschaftswissenschaft der Freien Universität Berlin. Nr.2002/16, pp. 2ff.

8 See Wagner, F.W./Wiessel, H. (1995): Entscheidungsneutralität der Besteuerung als Leitlinie einer Reform der Einkommensteuer, Wirtschaftwissenschaftliches Studium Heft 2, pp. 65-70.

9 See Petersen, H.-G. (2004): Redistribution and the Efficiency/Justice Trade-off. Finanzwissenschaftliche Diskussionsbeiträge, Nr. 42, Potsdam 2004, pp. 1-20.

10 See Petersen, H. - G. (2003): Zu einer Fundamentalreform der deutschen Einkommensteuer: Das Einfachsteuermodell des Heidelberger Steuerkreises. Finanzwissenschaftliche Diskussionsbeiträge, Nr. 34, Potsdam 2003, pp. 5ff.

11 See Rose, M. (1999): Recommendation on Taxing Income for Countries in Transition to Market Economies. pp. 29-30.

The goal underlying the criterion of neutrality in decision-making is the formation of efficiently functioning markets. In the early stages of development it is of central importance for the functioning of a free market economy that the tax system be relatively neutral in order to, as far as possible, avoid the undesired effects of distortionary taxes. Roughly speaking, this means that ideally the tax system should not prompt anyone to act differently from how he would have under no-tax conditions, where he only obeys market signals. The theory of optimal taxation tells us that complete neutrality cannot be achieved because first-best taxation via lump-sum taxes is not possible. Instead, the relevant question is what kind of tax system is closest to second-best. In other words, what are the implications of the optimal tax theory for the tax policy which should be oriented to an optimal tax system?

3 History of Development

In 1991, after the declaration of independence, the Georgian tax system was based on taxes which existed during the times of the Soviet Union. In the following years, VAT and excise tax were established. Taxes of the period mentioned above were regulated by separate laws[12]. Taxes were established by separate regulations created by the Cabinet of Ministers, where administrative and organizational competences were determined in a very general matter. There was no united and regulated tax legislation, and tax culture was low.

In 1995 independent Georgia adopted the first constitution and a law of the National Bank, and in 1997 the first tax code was implemented. The adoption of the constitution evoked great changes in the country's economic and public life. Due to the disorganization caused by the planned economy and the transition into a Western market economy, structural changes of state revenues and the private sector were inevitable. The approval of a new tax code in September 1997 served to rationalize the tax system, bringing all tax legislation together in a single legislative document made up of 29 separate legal acts. The code came into force progressively and in January 1998 it was adopted. The code also broadened the tax base by eliminating some exemptions, clearly defined taxpayers' rights and obligations, and reduced various excessive tax penalties introduced during an earlier period of high inflation. The main taxes and rate structure of the Code are summarized in table 1.

[12] See Narmania, D./Khokrishvili, E. "Georgian Tax System (Part. 1), magazines "New Economist", Advertising redaction, Tbilisi 2006, pp. 10.

Table 1: Main Taxes and Taxes Rates of the 1997 Georgian Tax Code

Single rate of VAT at 20 percent with a low registration threshold 24,000 GEL. Exports zero rated.
Single rate of corporate tax at 20 percent with allowable deductions broadly in line with international practice
A personal income tax with bands at 12, 15, 17 and 20 percent, with very modest progressivity as most incomes were taxed at the top marginal rate.
Social tax on payroll of 28 percent (tax code did not govern payroll taxes payable to the medical insurance or employment founds).
Withholding tax of 10 percent on interest and dividends.
Taxes on property and land, vehicle ownership and inheritance tax.
A limited number of ad valorem excises (alcoholic beverages, tobacco products, gasoline and a few minor excises).
A wide range of natural resource taxes.
Seven local taxes including a turnover tax of 1-2 percent depending on region.

Source: Tax Code of Georgia, 1997

As a rule, the successful execution of a tax reform is determined by the effective usage of tax legislation in practice. If we take a look at the tax practice of recent years, we can see that we had a quite different picture in practice. Especially the practice of taxation of enterprises was selected with Georgian taxation peculiarities: On one the hand, tax bodies confiscated material assets and products in order to cover the tax debts of entrepreneurs. Products received from enterprises with debts were handed in by the tax inspectors at a higher price (in most cases due to an arrangement arrived at by way of corruption) than the price which existed on the market. In this way, tax debts were deducted based on illegally high prices. On the other hand, tax payments in Georgia were often 'negotiated' between tax payers and tax inspectors. In this context, tax inspectors have resorted to the practice of requesting advanced payments against future tax liabilities, in order to meet their monthly and quarterly revenue targets. The stock of advance payments as of the end of 2001 amounted to GEL102 million (or 1.6 percent of GDP). While the bulk of these advance payments pertains to legitimate advances (e.g., profit taxes and outstanding VAT claims), it also includes such negotiated advanced payments. Taxpayers go along with such a scheme in exchange for lower tax assessments and the ability to draw on these funds at will. On their part, tax officials are able to prevent sanctions which may include their dismissal, and to supplement their bribes. These practices not only tend to undermine revenue collection but also make revenue flows unpredictable even within a month's horizon[13].

Figure 1: Tax Revenues to GDP 2000-2006

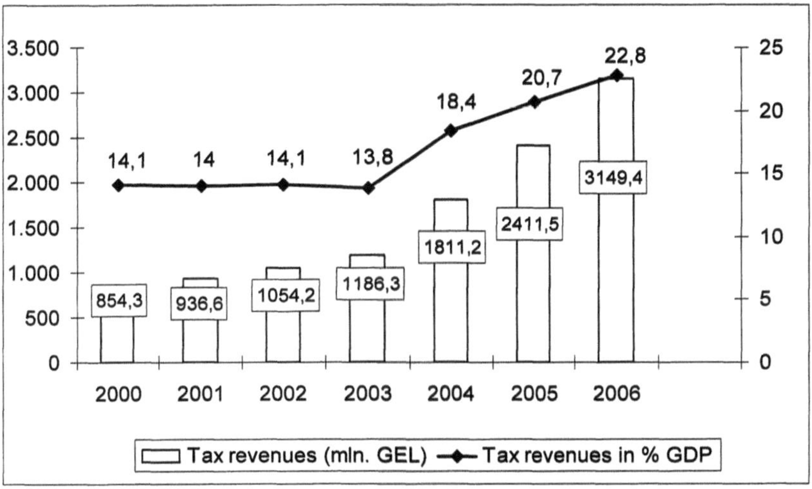

Source: National Bank of Georgia, Bulletin of monetary and banking statistics, N 12, 2002

Over time, the proliferation of innumerable amendments, mostly providing exemptions and tax reductions on VAT and excises, has seriously undermined the consistency of the code. The history of tax policy changes in Georgia since the adoption of the tax code demonstrates an unhealthy focus on short-term measures and a complete disregard for the integrity of the system. New taxation schemes, mostly initiated by parliament and driven by influential interest groups[14], are too often introduced at any stage of the budget cycle without proper assessment of their fiscal impact or indication

[13] See Georgia: Recent Economic Developments and Selected Issues, IMF Staff Country Report Nr. 211, Washington DC. 2001, pp. 12-20.

[14] The IMF has also documented how state capture has operated in the area of taxation, leading to exemptions, special treatments, and a general perception that the law is manipulated (with the cooperation of parliament) and not implemented in a fair manner. (See IMF- Georgia Recent Economic Development and Selected Issues, November 2001).

of their duration[15]. For example, between July 1997 and April 2001, 113 amendments to the code were passed, of which 58 percent related to VAT and excises. The development of tax revenues in 2000-2006 is presented in Figure 1, and the shares of tax revenues in GDP in 1997-2001 are given in Figure 2, both of which make clear that among the GUS countries, Georgia's revenue to GDP ratio is very small.

Figure 2: Tax Revenues to GDP 1997-2001

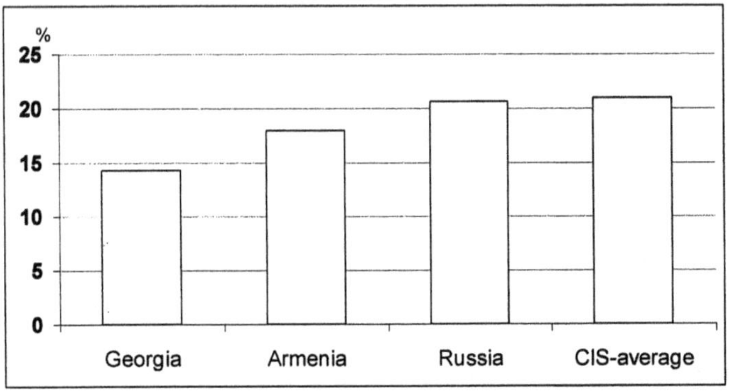

Source: IMF (2001): Armenia, Georgia, Kyrgyz Republic, Moldova, and Tajikistan: External Debt and Fiscal Sustainability, February 6, 2001; MoF, Republic of Georgia.

Low level of tax revenues on its side impeded the realization of state incomes. The lack of state resources hampered the formation of effective tax administration and miserable supply of public goods decreased the willingness of citizens to pay taxes. Or to put it in Holmes' (1996) words: "The resource-strapped state cannot provide public goods and is therefore less and less able to extract resources through either voluntary compliance or threats of coercion"[16]. The understanding of such effects will help us understand why

[15] Several studies have document the problems of state capture and corruption in Georgia. Some studies have ranked the country amongst the worst in the Former Soviet Union (FSU). See Hellman, J., G. Jones and D. Kaufmann, 2000, 'Seize the State, Seize the Day: State Capture, Corruption, and Influence in Transition," World Bank Policy Research Paper 2444 (http//www.world bank.org/ wbi/goveernance/).

[16] See Holmes, Stephen, (1996): Cultural Legacies or State Collapse? Probing the Post communist Dilemma. In: Michael Mandelbaum (Ed.): Post communism: Four Perspectives. NY: Council on Foreign Relations, pp. 22-76.

most transformation countries had such a low level of tax revenues in the first years and why they are so difficult to improve[17]. Special treatments lead to distortions, abuses, increased compliance and administrative costs, as well as taxpayer inequities and resentment. In addition to complicating tax compliance and administration, such a short term approach has made the tax policy framework unstable both for the tax administration and for tax payers. Under these circumstances it is not surprising to find that the number of registered tax payers in Georgia was low by international standards. Georgia's tax system had, for the most part, not provided the desired level of stability in tax institutions. Continuous changes in the tax structure had contributed to the increase of administrative and compliance costs, facilitated tax evasion and discouraged economic activity.

4 The Current Tax System

A reformed tax code was implemented on January 1, 2005. The number of taxes was reduced from 21 taxes to only 7 taxes and the rates of some taxes and taxation objects, also of distributing mechanisms and numbers of amounts between the various levels of budgets were reformed. General state and local taxes are applied in Georgia. General state taxes are common to the whole territory of Georgia and the identical rates determined by the tax code apply to every territorial unit of Georgia[18].

The social and income tax rate were also reduced significantly (see table 2). Still, it should be mentioned that according to the new tax code the property tax includes the property tax of old enterprises and physical entities, land tax and the tax regarding owners of auto transport means[19].

[17] See Daianu, D./Vranceanu, R. (2000): Public Finance and Low Equilibrium in Transition Economics: The Role of Institutions. William Davidson Institute, Working Paper No. 319; Johnson, S./Kaufman, D. (1998): Corruption, Public Finances and the Unofficial Economy. World Bank, Policy Research Working Paper No. 2169, Washington, D.C.

[18] See Georgian Law – "Georgian Tax Code", Article 6-10 dated on December 22, 2004

[19] See Narmania, D., "Tax Policy", lecture course, Georgian Institute of Public Affairs (GIPA), 2007, page 4.

Table 2: Former and current tax rates in Georgia

State taxes	Tax rate	Tax base
Income tax* (Art. 165-168)	12%	The taxable income is determined as the difference between gross income per calendar year and the deductions defined by the Tax Code
Corporate tax** interest, dividends (Art. 169-172)	20% 10%	Profits are defined as gross income minus all authorized deductions specified in the Tax Code. on interest and on dividends
VAT (Art. 220-224)	18%	Supply/import of goods and services, customs value of all taxable goods imported into the territory of Georgia. Single threshold of 100,000 GEL. Export of goods is zero-rated
Social tax*** (Art. 266-270)	20%	Levied on wages and salaries paid by employers.
Excise (Art. 253-265)	Various specific rates	Taxable transaction, for example alcoholic beverages, tobacco products, gasoline, et cetera.
Local taxes	**Tax rate**	**Tax base**
Property tax Enterprises	max. 1%	The balance sheet residual value of assets is calculated according to the average value of assets at the beginning and at the end of the calendar year.
Physical persons (Art. 271-276)	0.8 – 0.2 %	of the inventory value of the building or structure .
Gambling tax**** (Art. 277-280)	Various specific rates	Income earned by sale of lottery tickets, gambling tables and machines used for gambling purposes, etc.

Notes:* From January 2008 onwards the income tax rate amounted to 25%; ** From January 2008 onwards the tax rate was reduced to 15%;
*** From January 1, 2005 it was 20%; but as of January 1, 2008 it was cancelled;
**** From January 1, 2007 it became the excise.

5 The Personal Income Tax

5.1 DESCRIPTION

Income tax payers are resident and non-resident natural persons. Residents pay taxes on their worldwide income, non-residents on their income from domestic sources. The definition of income includes monetary income and any kind of compensation from employment and non-employment sources. Income tax is paid by physical entities on a joint revenue basis, which is determined within the calendar year. The tax rate is 25%. Dividends and other capital revenues are taxed with a 10% rate at the tax source and are not entered into the total revenue. The income tax system in Georgia and in many Western countries is supposed to follow the principle of comprehensive income taxation (known as the Schanz/Haig/Simons system)[20]. Comprehensive income is defined as the net growth in the taxpayer's personal wealth during one taxable year. Technically, under the Schanz/Haig/Simons system, each individual should be taxed annually on the sum of all their annual income from labour and capital regardless of source[21]. In the Georgian Tax Code the determination of joint revenue is determined in accordance with the synthesis principle (enumeration of revenue varieties, Article 174-176) and with the net principle of revenue varieties[22] (all expenses which are related to this revenue will be deducted according to the net principle, i.e. deductibility of expenses incurred in the generation of revenues, Article 177). Revenue received from employment is taxed at the place of issue. Individual entrepreneurs (self-employed activity) are obliged to pay the appropriate amount quarterly. Physical persons, whose annual revenue does not excel 35,000 GEL, are not obliged to turn in the annual tax declaration. Distinctive for Income tax in Georgia is that deduction of personal expenses (expenses spent on education, health insurance contributions) and tax-free

[20] The traditional concept of income taxation has its origin in the famous article "Der Einkommensbegriff und die Einkommensteuergesetze" by Georg von Schanz. See von Schanz, G. (1896). In the American tax literature, the birth of the traditional concept of income taxation is linked with the names of Henry C. Simons und Robert Murray Haig. see Simons, H. C. (1938) and Haig, R. M. (1921[1959]). see also Musgrave, R. A. (1990).

[21] Revenues from employed activity, profit from self-employment activity (selling of agricultural products, rent received for property), rates, dividends, capital revenues, granting, inheritance. See: Holmes, K (2001): The Concept of Income. A Multi-Disciplinary Analysis, Amsterdam. p. 60.

[22] All expenses which are related to this revenue will be deducted according to the net principle, i.e. deductibility of expenses incurred in the generation of revenues. Article 177, Georgian Tax Code (2005)

amount are not allowed. Thereby, the tax burden of physical entities is higher in comparison with the entrepreneur's. This does not conform to the horizontal fairness criterion. This requirement follows from the subject net principle[23] that living wages will not be taxed (the deduction of a tax-free allowance to guarantee subsistence consumption and the deduction of further personal cost, e.g. health insurance contributions). Every kind of minimal taxation which does not pay heed to untaxed living wages is in conflict with this criterion. Besides, expanses for education are an investment into the human capital, which will ensure or increase revenues; this is why they are unequally treated in comparison with equity and financial investment. Such unequal taxation of various forms of investments does not conform to the horizontal fairness criterion.

Self-produced agricultural produce produced in a private enterprise before its industrial processing is free from income tax, if the total annual revenue does not exceed 100,000 GEL. The tax code also frees individual entrepreneurs of several groups from taxes, if they produce without employed labour in the area which is defined by law. Besides, the following are free of income tax: alimonies, gifts and inheritances, grants, state stipends, state benefits and state pensions (see article 168). An income of up to 3,000 GEL per year earned by the following physical persons is not subject to taxation: handicapped persons, single mothers, persons who adopted a child, etc. (see article 168).

5.2 Assessment of the Status Quo

In the traditional income tax system[24], the fairness principle is oriented towards the annual income burden and assumes that equal taxation means equal yearly burdens. It is well known, however, that the traditional system of income and profit taxation imposes a heavy tax burden on enterprises. In the traditional income tax system, the fairness principle is oriented towards the annual fairness concept and assumes that

[23] According to subjective net principle taxpayers' personal expenses are allowed to be deducted from joint revenues.

[24] The traditional concept of income taxation has its origin in the famous article "Der Einkommensbegriff und die Einkommensteuergesetze" by Georg von Schanz. Cf. von Schanz, G. (1896). In the American tax literature the birth of the traditional concept of income taxation is linked with the names of Henry C. Simons und Robert Murray Haig. Cf. Simons, H. C. (1938) and Haig, R. M. (1921[1959]). Cf. also Musgrave, R. A. (1990).

equal taxation means equal annual tax burden. It is well known, however, that the traditional system of income and profit taxation imposes a heavy tax burden on taxpayers[25]. This system depends exclusively on annual incomes. The previous history of the backgrounds of capital formation does not play any role. Therefore, capital formation is usually made from taxed income. In the following periods this capital itself forms a new tax base and the interest payments (or profits, dividends, rent, etc.) on that capital are taxed again. Capital itself and capital income is consequently several fold burdened. During inflation the double taxation effect on the taxpayer is higher. Figure 3 demonstrates this so-called avalanche effect of capital income taxation in Georgia with a simple example[26]: Let us assume an income tax rate of 25 percent – which will be adopted from next year onwards. An employee or an entrepreneur saves 10,000€ and invests that amount in his company or on the capital market at an interest (net) rate of 5% for 35 years. The assumed inflation is 10 percent. Without any taxation his interest earnings would grow to 55,160.20€ (see Figure 3) and would be available for his consumption in his old age.

In the case of the Georgian income tax, savings are accumulated from the taxed income, so that at the assumed wage tax rate of 25 percent only 7,500€ can be invested for that 35-year period[27]. Because of the tax-reduced investment amount, the interest payment for the first years is no longer 500€, but only 375€. Despite the original 25 percent burden, the gross interest payment of 375€ is taxed again at 10%, as has been mentioned above. Consequently, the savings only grow by 337.50€ and during inflation by 105€. Over the following 34 years, income tax has to be paid on the annual interest income. The effective tax burden, including the original paid amount, is 33 percent (with inflation 79 percent) after the first year. The disposable amount for old-age consumption is reduced to 35,505.11€ (with inflation 12,200.84€). Compared to the

[25] This theorem of tax incidence is well-known in the literature of taxing income from capital. Unfortunately the theorem is often described as "double taxation". The correct description, however, should be "double burdening" because the pre–burdened return on savings is taxed only once. See Petersen, H.-G. (2003): Social Justice and Ability to pay in a Dynamic Perspective. In Finanzwissenschaftliche Diskussionsbeiträge, Nr. 35, Potsdam 2003.

[26] The calculation is based on Petersen's (2003) model and inflation is added in this model, see Petersen, H-G., Rose, M. (2003): Zu einer Fundamentalreform der deutschen Einkommensteuer: Das Einfachsteuermodell des Heidelberger Steuerkreises. Finanzwissenschaftliche Diskussionsbeiträge, Nr. 34, Potsdam 2003.

[27] At the time of 10 % inflation and 5% real interest rate according to Fisher's formula i = p + r (1+p), where i=Nominal Interest Rate; r = Real Interest Rate; p = Inflation, the nominal interest rate will be 15.5 %. Present capital income taxation with 15.5% nominal interest rate brings us to [10 %(1-t)-p] / (1+p) = 1.4% real after-tax market rate.

55,160.20€ in the situation without any income tax, the effective lifetime tax burden on the interest income is 38.9 percent and under inflation it is 89.1% (see chart 4) – much more than the annual 25 percent flat rate. Figure 3 provides a striking example of how the multiple taxation, which arises from the Georgian taxation of interest and enterprise profits under inflation, causes the effective tax rate to skyrocket in the course of a lifetime[28]. During inflation, the *intersectoral* and *intertemporal* distortions become more important. As all cash flows including the market rate of interest contain a compensation for inflation, taxing them nominally can result in very low or even negative real rates of return.

Figure 3: Capitalincome (€) in the Georgian Income Tax System, Tax Burden of Interest

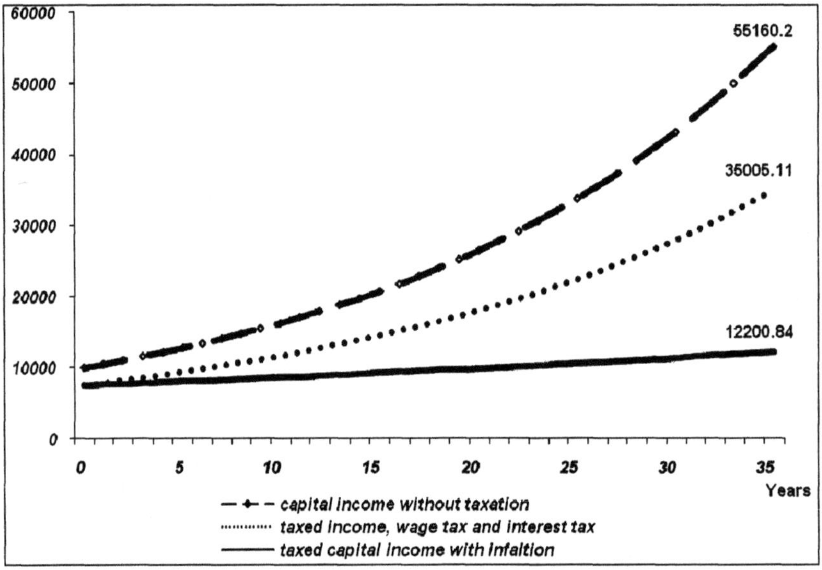

Source: Petersen (2003); own calculations

[28] On the concept of lifetime income, see Petersen, H-G., Rose, M. (2003).

Figure 4: Lifetime Burden on Interest Income of the Georgian Income Tax

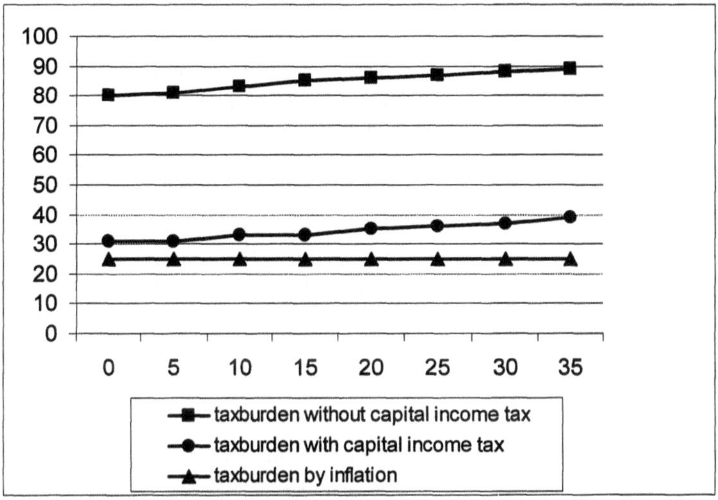

Source: Petersen (2003); own calculations

Economic development in Georgia will depend essentially upon whether there is a significant increase in real *capital formation* as a result of investment in the enterprise sector. In view of the current low level of citizens' incomes, there are only very limited possibilities of raising the necessary savings capital by foregoing consumption. In this case it is more important than in many developed economies to ensure that savers and investors are not at a tax disadvantage against the immediate consumer. The traditional concept of income taxation, however, a concept that is characterised by its comprehensive discrimination of savers and investors, can therefore not be considered as an option for the future income tax law in Georgia[29].

[29] Rose, M. (1999): Einführung marktorientierter Einkommensteuersysteme in Osteuropäischen Reformstaaten, in: Smekal, C., Sendlhofer, R., Winner, H. (Hrsg), Einkommen versus Konsum. Ansatzpunkte zur Steuerreformdiskussion.

6 The Corporate Income Tax

6.1 Description

Profit tax payers are Georgian and foreign enterprises, the latter with respect to their gross income from Georgian sources[30]. Georgian enterprises' profit is taxed at a rate of 15%[31/32]. Exceptions are profits, percent and dividend received from oil and gas operations, which are taxed at a 10% rate (i.e. profit received due to execution agreements determined by the Georgian law "about oil and gas", if these agreements were made till January 1, 1998). Basis is the profit, defined as the difference between gross income and the deductions stipulated by the tax code. Business income below the profit tax is calculated using an accrual method similar to the financial accounting profit. Current tax payments are made three times a year, with an annual adjustment by April 1 of the following year. Varieties of deductions[33] are determined by the tax code at the time the profit tax is calculated. Their general content is the following:

- outlays connected with the receipt of income;
- interest costs shall be deducted in the amount not exceeding 24% of the annual rate for the corresponding period of time and corresponding ratio;
- depreciation charges (annually, in varying degrees from 5 to 20%) and repair of fixed assets;
- losses can be carried forward for a period of up to five years.

Some kind of revenues, like the sale of objects for religions purposes, budget-funded and charity organisations except for profits from economic activities, self produced agricultural produce produced in a private enterprise before its industrial processing, are free from profit tax.

[30] *Georgian enterprises* are enterprises whose place of activity is Georgia, or whose place of management is in Georgia. A *foreign enterprise* is an enterprise whose place of activity is not in Georgia, or which is not managed from Georgia.

[31] From January 1, 1998 till January 01, 2008 the rate of profit tax amounted to 20%.

[32] According to the tax code, sole proprietorships are burdened by the personal income tax, and partnerships, limited liability companies and corporations are burdened by the corporate profit tax.

[33] t is firmly determined that profit and income taxes will not be deducted, also sanctions paid in the budget, expenses spent on entertainment and these expenses which are not related to economic activity, also expenses related to receiving revenues, which is free from the payment of profit and income tax. See Kruschwitz, L./Husmann, S./Schneider, D. (2002), a. a. O. ,pp. 4-6.

6.2 Assessment of the Status Quo

Under real life conditions it is practically impossible to design a comprehensive income tax that is neutral with respect to investment decisions. This is as true for the Georgian as for the German profit tax. The reason is that capital is valued at its historic value whereas economic income calculation is based on the present value of future cash flows[34]. This means, that a traditional profit tax will always influence capital allocation to industry sectors. In a comprehensive income tax – and Georgia should not be an exception here – interest expenses from a business loan are deductible and the creditor's interest income is taxed as regular income. This means that the traditional income tax system has a tendency to privilege debt over equity[35].

The Georgian profit taxation mainly follows the guidelines of the traditional system and therefore an obstacle to the efficient performance of business enterprises. A typical solution, which can be found in many tax codes, is a more or less selective adjustment of assets or equity values.

7 A Proposal for Neutral Income and Business Taxation in Georgia

Given the distortionary effects on economic decisions of the Georgian income tax system the question must be raised what alternative is at hand. Tax theory has developed a system that is much less of a hindrance to the economic performance of enterprises. A specific feature of this system is a new concept of fairness which is based on the notion of equal tax burden for equal lifetime incomes. Therefore, it takes into consideration the fact that, by virtue of the former taxation of the income used for its formation, actual capital income already bears a tax burden without ever having been taxed. For a country – as Georgia – trying to gain from tax competition there are options for it to be able to offer better tax environments for investments than those in other countries which are bound to the traditional concept of taxing personal income

34 The ideal traditional system of income taxation serves as a reference system for a traditional income tax, as, in accordance with the taxation of net accrual to wealth, it takes into consideration the true economic depreciation. In practice only a rough estimate of the true economic depreciation is possible. Tax depreciation frequently precedes true economic depreciation. See Kaiser, M. (1991), a. a. O., pg. 18.

35 Rose, M. (1999): Steuerfreiheit investierte Gewinne? in: Wirtschaftsdienst, 79. Jg. (1999), Heft 1, pp. 35-38.

and business profit[36]. The *Heidelberger Steuerkreis*[37] has developed an "easy tax proposal" (The easy tax) that, on one hand, integrates income and corporate taxes into one law and, on the other hand, secures equal treatment of wages and capital income as far as possible. The conflict between efficiency and justice is reduced to an absolute minimum[38]. The legislation is based on mostly transparent laws that the average taxpayer can easily understand and, therefore, comply to. Both tax declaration and collection are regulated in the simplest way. The Easy Tax has two specific forms of tax collection: personal income tax and profit tax. Taxable income comprises three sources: wages, self-employment income, and retirement income. Expenses for vocational education are subtracted. The citizen's existence minimum of consumption is exempt from taxation through a corresponding tax allowance over the entire lifespan (see table 3).

The profits of small corporations, which are corporations with a small number of shareholders, are taxed as income from self-employment. Profits of S corporations, considered pass-through companies[39] in the Easy Tax draft law, are considered distributed to shareholders and are taxed as personal income. Pass-through companies are companies of any legal form where the enterprise is managed in large measure by its owners (company members). Here one can also refer to companies that are managed personally. The integrated profit tax also follows a simple calculation scheme that is shown in table 4. Naturally, the profit taxation of companies cannot be as simply designed as the one of an average employee, but there are fundamental possibilities of simplification that pay special attention to the liquidity needs of small and

36 See Petersen, H-G., Rose, M. (2003): Zu einer Fundamentalreform der deutschen Einkommensteuer: Das Einfachsteuermodell des Heidelberger Steuerkreises. Finanzwissenschaftliche Diskussionsbeiträge, Nr. 34, Potsdam 2003, pp. 3-12.

37 The members of the "Heidelberger Steuerkreis" are Joachim Lang (Köln), Hans-Georg Petersen (Potsdam and DIW), Manfred Rose (Heidelberg), Bernd Rafellhüschen (Freiburg and Bergen). The permanently updated draft law and additional information can be found at www.einfachsteuer.de;

38 Since 1994 the Republic of Croatia had an income tax system which was neutral with respect to investment and financing decisions as well as inter-temporally neutral. After the general elections in 2001 major changes have been made to the tax code, which have destroyed most of the consistent principles of this system. It was easy to handle, generally accepted among taxpayers and, most important, it generated considerable tax revenues. For details see Schmid, Wissel and Stöckler (1996) and Keen and King (2001).

39 The concept of a pass-through company is based on the model of the US S-corporation, in which the profits of a company, managed by a juristic person are taxed at the shareholder level.

medium enterprises in transformation countries. Corporate profits are calculated with emphasis on liquidity according to the modified cash accounting, in order to assure that enterprises do not get financial difficulties because of tax payments and that fiscal administration cannot drive companies into bankruptcy.

Table 3: The Base of the Personal Income Tax

Tax Base of the Personal Income Tax
Income from employment + Income from self-employment + Income from old age security - Expenses for professional training - Loss carryover
Market-derived income - Personal allowances (for all household members, payments into the public or private health insurances)
= Income to be taxed

Source: Petersen, H.-G./Rose, M. (2003).

For the equal treatment of wages and capital income over the course of a lifetime, the above-mentioned avalanche effect must be avoided[40]. Two different methods could be applied. Their effects on capital income taxation are equal, but they would influence the periodic distribution of tax revenues. In the interest adjustment method, a standard market interest rate is subtracted from all capital income. For all sources of capital income (interest, profits, rents, and so forth) a basic rate of return – for instance, the interest rate for a two year government bond – remains tax free as remuneration for the

[40] See Petersen, H-G., Rose, M. (2003), pp. 3-12.

abnegation of consumption[41]. Consequently, only capital incomes above this basic rate of return (also called the protective interest rate) are taxed, whereas a steady tax base on capital income remains. The protective interest rate avoids the avalanche effect and in the lifetime perspective, the equal treatment of wages and capital income is assured[42]. The interest adjusted profit taxation also meets all the other neutrality requirements for businesses[43]. Of particular importance is the fact that the interest adjustment neutralizes the type of chosen depreciation procedure, since depreciations determine the size of the equality capital and, therefore, the deductible interest adjustment. For transformation countries it is important that the taxation of virtual profits, threatening enterprises especially in times of high inflation, no longer exists.

Table 4: The Tax on Enterprise Profit

The calculation of the Profit of the enterprise
Profit from current market activities of the businesses belonging to the enterprise, calculated according to the principle of cash accounting + Profits from the sale, withdrawal or liquidation of the businesses of the enterprise, as well as from the reorganisation or a change in the type of tax liability of the business person + Allocated profits of related enterprises = Profit of enterprise - Loss carryover and protective interest thereon
= Profit to be taxed (tax base)

41 In Belgium, enterprise profits will be taxed on an interest-adjusted tax base from 2007 onwards.

42 See Petersen, H-G., Rose, M. (2003), pp. 4-6.

43 For details see Rose, M./Zöller,D.(2008): Die Heidelberger Einfachsteuer: Eine optimale Besteuerung von Einkommen in der Marktwirtschaft. In: Sustainable Fiscal and Social Policy in Georgia, Working Papers of the German-Georgian Working Group for Fiscal and Social Policy. pp. 224-272.

8 Conclusion

The prime objective for any country attempting to choose an optimal tax system is to ensure that the tax law adopted secures the highest level of economic welfare possible for its citizens. The level of economic welfare of Georgia is fundamentally determined by the economic performance of its enterprises in whatever organisational form (sole-proprietorship including all kinds of self-employment, partnerships and corporations). It should and will be of the greatest interest to Georgia that the economic performance of enterprises develops as freely as possible and that the tax system does not constitute a hindrance thereto, i.e. it should as far as possible guarantee decision neutrality[44]. The authors have analyzed the main elements of the Georgian income and profit tax with respect to their effect on economic decisions. Like other traditional income tax systems in OECD countries it heavily affects economic decisions. Some major distortionary effects of the Georgian income tax system can be easily replaced through minor reforms, which would turn it into a consumption-based income tax system. Unfortunately, the public discussion of the Georgian tax system has a rather different focus. As economists, we must keep in mind that taxpayers ask for low taxes (for themselves) but they do not care about neutral taxes. A neutral tax system is a typical public good. Apart from some strange economists, nobody will miss it. But society pays a price for a non-neutral tax system in form of less growth and a distorted industry structure.

[44] See Rose, M. (1997), Eine konsumorientierte Neuordnung des Steuersystems für mehr Entschei-dungsneutralität, Fairness und fuTransparenz, in. Buhler, W. (Hrsg.), Steuervereinfachung, Festschrift für Dietrich Meiding zum 65. Geburtstag, Heidelberg, pp. 233-252.

9 References

Bauer, Thomas: Die Grundzüge des Sowjetischen Steuersystems; in: Intertax 5/1978, pp.182-194.

Belsky, V.: Das Steuersystem der Sowjetunion, Veröffentlichungen des Instituts für Finanzwissenschaft und Steuerrecht Nr.81, Wien 1973. Berlin-Heidelberg, 29-42.

Boadway, R./Bruce, N. (1984): A general proposition on the design of a neutral business tax. Journal of Public Economics 24, pp. 231-239.

Bowles, Roger A.: Tax policy, tax evasion and corruption in economies in transition, in: Feige, E.L. / Ott, K. (Hg.): Underground economies in transition, Aldershot 1999, pp.67-86.

Cangiano, M.: „Accountability and Transparency in the Public Sector: The New Zealand Experience", IMF Working Paper, November 1996.

Gabitzashvili, B: Messung der Schattenwirtschaft in Georgien. In: Sammelband wissenschaftlicher Arbeiten des wissenschafts- und Forschungsinstitut für Statistik. 2001, Tbilissi.

Hedtkamp, Günter: Das Steuersystem im Dienste der Sowjetischen Staats- und Wirtschaftsordnung, in: Finanzarchiv, N.F. 2/1960, S. 181-200. Heidelberg, pp. 167-195.

Hellman, J., G. Jones and D. Kaufmann, 2001, 'Seize the State, Seize the Day: State Capture, Corruption, and Influence in Transition," World Bank Policy Research Paper 2444, (http-//www.worldbank.org/wbi/goveernance/).

Hellman, Joel, Mark Schankerman (2000): Intervention, Corruption and Capture: The Nexus Between Enterprises and the State. Economics of Transitions. World Politics 50 (2) 203-234.

Hermann Clement, Michael Knogler (1998), Wachstum in schwierigem Umfeld – Wirtschaftslage und Reformprozesse in Ostmittel- und Südosteuropa sowie den GUS Ländern, Veröffentlichungen des Osteuropa – Institut München Nr. 242.

Holmes, K. (2001): The Concept of Income. A Multi-Disciplinary Analysis, Amsterdam.

Homburg, S. (1997): Allgemeine Steuerlehre, Vahlen, München.

IMF: Staff Country Report Nr. 211, Washington DC. 2001, pp. 12-20.

IMF: Armenia, Georgia, Kyrgyz Republic, Moldova, and Tajikistan: External Debt and Fiscal Sustainability, February 6, 2001.

IMF: IMF Approves US$141 Million PRGF Loan for Georgia, Press Release No. ¼ January 12, 2001.

IMF: Georgia: Recent Economic Developments and selected Issues, IMF Staff Country Report; November 2001.

IMF: Georgia: Selected Issues and statistical Appendix, IMF Country Report Nr. 03/347, Washington DC. 2003.

IMF: Outcomes of the Georgia Tax Administration Modernization Project, 17/08/2001; published on the internet under: http://imf0018. imf.org/eca.nsf/049 80307d447

Keen, M. and J. King (2001): The Croatian Profit Tax: An ACE in Practice, IMF staff paper.

KNS-Steuerreformgruppe (1991): Empfehlungen zur Reform des ungarischen Steuersystems, Heidelberg. pg. 14.

Kruschwitz, L./Husmann, S./Schneider, D. (2002) „Investitionsneutrale Steuersysteme vor dem Hintergrund der Kontroverse um Einkommen oder Konsum als geeignete Steuerbemessungsgrundlage". In: Diskussionsbeiträge des Fachbereichs Wirtschaftswissenschaft der freien Universität Berlin.

Lang, Joachim (1993): Entwurf eines Steuergesetzbuches. Schriftenreihe des Bundesministeriums für Finanzen, Heft 49. Bonn, sowie Lang, J.: The Concept of Tax Code; in: Rose Manfred (Hrsg.): Tax Reform for Countries Transition to market economies, Stuttgart 1999, pp. 185-195.

McLure Jr.,Charles E.:The New Tax Code of Kazakhstan, in : Asian Development Bank (Hg): Fiscal Transition in Kazakhstan, Singapur 1999, pp. 83-135.

McLure, C.E.Jr. (1988): The 1986 ACT: Tax reform's finest or death throes of the income tax? in: National Tax Journal, vol. 41 (1988),no. 3, S. 303-315.

McLure, C.E.Jr. (1991): A Consumption-based direct tax for countries in transition from socialism, The World Bank: Policy research and external affairs working papers, no.751, Washington, D. C. 1991.

Ministry of Finance (1998), Public Finance in Georgia, No. 61: Modified Acts on Taxes: Act on Company Tax, Tbilissi.

Ministry of Finance (2005): Nr.2002/16, pp.1-4.

Musgrave, R. A. - Musgrave, P.B. - Kullmer, L. (1994): Die öffentliche Finanzen. Mohr-UTB

Musgrave, R. A. (1990), "On Choosing the 'Correct' Tax Base – A Historical Perspective", in: M. Rose (ed.), Heidelberg Congress on Taxing Consumption, Berlin-Heidelberg, 29-42.

Newbery, D.; Stern, N. (1987): The Theory of Taxation for Developing Countries. Taxation and Development, Chapter 7 by Newbery, D., Oxford University Press. pg. 117.

Petersen, H.-G. (1993): Finanzwissenschaft I., pp.285ff. 3. Aufl., Stuttgart, Berlin, Köln.

Petersen, H.-G. (1997): Das Neuseelandexperiment: ist das die zukünftige Entwicklung des deutschen Sozialstaates? In: Echhard Knappe, Albrecht Winkler (Hrsg.): Sozialstaat im Umbruch. Herausforderungen an die deutsche Sozialpolitik. Frankfurt/New York: Campus 1997. pp. 119.149.

Petersen, H.-G. (2003): Werte, Prinzipien und Gerechtigkeit: Zu einem dynamischen Verständnis von Leistungsfähigkeit. In: M. Ahlheim, H.-D. Wenzel und W. Wiegard (Hrsg.), Steuerpolitik – Von der Theorie zur Praxis. Festschrift für Manfred Rose, Springer, Heidelberg et al., pp. 59 –100.

Petersen, H.-G. (2004): Redistribution and the Efficiency/Justice Trade-off. Finanzwissenschaftliche Diskussionsbeiträge, Nr. 42, Potsdam 2004, pp. 1-20.

Petersen, H.-G. (2006): Konsumorientierte Besteuerung als Ansatz effizienter Besteuerung. In: Steuer und Wirtschaft, Nr. 3/2006, pp. 266 – 277.

Petersen, H-G., Rose, M. (2003): Zu einer Fundamentalreform der deutschen Einkommensteuer: Das Einfachsteuermodell des Heidelberger Steuerkreises. Finanzwissenschaftliche Diskussionsbeiträge, Nr. 34, Potsdam 2003.

Rose, M. (1992): Reform der Besteuerung des Sparens und der Kapitaleinkommen: zur Neuregelung der Zinsbesteuerung aus der Sicht einer Konsum- und damit Marktorientierten Neuordnung des Gesamtsystems. Heidelberg: verl. Recht u. Wirtschaft Betriebs – Berater: Beilage: 1992/5.

Rose, M. (1997): Eine Konsumorientierte Neuordnung des Steuersystems für mehr Entscheidungsneutralität, Fairness und Transparenz, in. Buhler, W. (Hrsg.), Steuervereinfachung, Festschrift für Dietrich Meiding zum 65. Geburtstag, Heidelberg, pp. 233-252.

Rose, M. (1999): Steuerfreiheit investierte Gewinne? in: Wirtschaftsdienst, 79. Jg. (1999), Heft 1, pp. 35-38.

Rose, M.(1999): Einführung marktorientierter Einkommensteuersysteme in Osteuropäischen Reformstaaten, in: Smekal, C., Sendlhofer, R., Winner, H. (Hrsg), Einkommen versus Konsum. Ansatzpunkte zur Steuerreformdiskussion.

Rose, M. (2002): Ein einfaches Steuersystem für Deutschland, in: Wirtschaftsdienst, Nr.8, 1994, pp.423ff, und zum Konzept der konsumbasierten Besteuerung: Rose, M.: Reform der Einkommensbesteuerung in Deutschland, in: Schriften des Betriebs-Beraters, Bd.122.

Rose, M./Zöller, D.(2008): Die Heidelberger Einfachsteuer: Eine optimale Besteuerung von Einkommen in der Marktwirtschaft. In: Sustainable Fiscal and Social Policy in Georgia, Working Papers of the German-Georgian Working Group for Fiscal and Social Policy. pp. 224-272.

Rosen, H. S. (1992): Public Finance, 3rd ed., Homewood, IL

Sandmo, A. (1976): Optimal Taxation - An Introduction to the Literature, Journal of Public Economics 6, pp. 37-54.

Schwinger, R. (1992): Einkommens- und konsumorientierte Steuersysteme, Heidelberg.

Shah, A.; Whalley, J. (1995): The Redistributive Impact of Taxation in Developing Countries. In: Khalizadeh-Shirazi, J.; Shah, A. (Hrsg.): Tax Policy in Developing Countries, paper presented at the World Bank Conference on Tax Policy in Developing Countries, Washington D.C.: 166−187.

Simons, H. C. (1938), Personal Income Taxation: The Defi nition of Income as a Problem of Fiscal Policy, Chicago.

Stiglitz, J. (1989): Finanzwissenschaft, München.

Stiglitz, J. E. /Schönfelder, B. (1989): Finanzwissenschaft. R. Oldenbourg Verlag, München-Wien.

von Schanz, G. (1896), "Der Einkommensbegriff und die Einkommsteuergesetze", Finanzarchiv 13, 1-87.

Wagner, E. (1989): Besteuerung und Kapitalbildung als internationales Optimierungsproblem, in: Hax, H./Kern, W. (Hrsg.): Zeitaspekte in betriebswirtschaftliche Theorie und Praxis, Stuttgart 1989, pp. 279-295.

Wagner, F. (1999): Warum eine Einkommensteuer konsumorientiert sein muss, in: Smekal, C. u. a. (Hrsg.): Einkommen versus Konsum, Heidelberg, pp. 15-35.

Wagner, F.W./Wenger, E. (1996), Theoretische Konzeption und legislative Transformation eines marktwirtschaftlichen Steuersystems in der Republik Kroatien, in: Sadowski, D./Czap,H./Wächter,H. (Hrsg.) Regulierung und Unternehmenspolitik, Wiesbaden, pp. 399-415.

Wagner, F.W./Wiessel, H. (1995): Entscheidungsneutralität der Besteuerung als Leitlinie einer Reform der Einkommensteuer, Wirtschaftwissenschaftliches Studium Heft 2, pp. 65-70.

Wiswesser, R. (1996): Einkommens- und Gewinnbesteuerung bei Inflation, Frankfurt a.M., pg.165.

Zodrow, G. R. (1997): Reflections on the consumption tax option, in Head, J. G./Krever, R. (ed.):Taxation towards 2000, Melbourne 1997, pp. 45-79.

AGRICULTURE IN GEORGIA - COMMERCIAL SECTOR OR SOCIAL SAFETY NET?[1]

KERSTIN KÖTSCHAU, EKA SEPASHVILI, NINO NARIMANIDZE

1 Introduction

After the breakdown of the Soviet Union many CIS countries started to implement reforms to transform the national economy toward a market economy. The reform process in the agricultural sector was initiated by land reform and farm restructuring. Formerly state-owned land was distributed to the population. In doing so, the former Soviet countries had two goals in mind. First, they wanted to boost the efficiency of agricultural production (Adams, 1995, p. 7) through the creation of a market environment (Daviddi, 1994, p. 4). Second, they aimed to avoid starvation and to fight poverty during transition (Deininger, Feder et al., 2003, p. 139; Adams, 1995, p. 7).

The agricultural transition in Georgia was made up of two stages of land reforms. While the first stage aimed to avoid poverty, the second stage was more focussed on establishing a commercial agricultural sector.

The aim of this paper is to analyse the success of the Georgian land reform by means of the two goals of Georgian land reforms. This will be done by a descriptive analysis using secondary data as well as data from household surveys conducted in the fall of 2007 in Shida Kartli.

The paper starts with an overview of the Georgian land reform and its achievements on the macro-level in the second chapter. It will continue with an analysis of the mentioned household data with regard to land endowment, household income and poverty. The paper will end with a conclusion and outlook for prospective political measures.

1 This paper has been developed during a research project funded by Volkswagen Foundation.

2 Rural Transformation in Georgia

2.1 LAND REFORM

In Soviet times, the agricultural sector was characterized by strong dualistic structures. Large agricultural enterprises, so-called kolkhoses and sowchoses, cultivated most of the land. Besides these enterprises every rural household kept some land for its own purposes, so-called household plots; in Georgia, around 0.25 ha per family (Thiesenhusen, 2002, p.1; Ebanoidze, 2002). All land was owned by the state and was given to the households only for their own cultivation. At the end of the Soviet Period 87% of all agricultural land belonged to kolchoses and sowchoses, while auxiliary family plots added up to 5.5% (Didebulidze, 1997; Thiesenhusen, 2002). In the early 1990s Georgia had 1,300 collective and state agricultural enterprises which cultivated around 3 million hectares of agricultural land. Nearly 1 million ha were arable land or land for perennials, the remaining 2 million ha were mostly pastures and meadows (Stanfield, 2002, p.2 Kneib, 2002). Averagely, 255 workers were employed on collective farms (UNECE, 2001). Additionally, about 200,000 ha were cultivated by households (auxiliary plots) (Stanfield, 2002, p.2).

Land reform in Georgia was conducted in two steps. The first step was the distribution of land at the beginning of the 1990s, the second step started in 2005 with privatising the remaining state-owned land to the population.

2.1.1 First Stage of Land Reform

The first stage of land reform started one year after the declaration of independence in 1991. The reform began in times of civil war and political unrest (Lerman, 1996, p.1;, p.1f; Didebulidze, 1997, p.40). During this time the Georgian government noticed that it would not be able to provide governmental social security programs. So, the aim of the land reform was to protect the population from growing poverty in the country (Thiesenhusen, 2002, p.1; UN- Economic and Social Council, 2001, p.4; Ebanoidze, 2002).

The first stage of land reform began with the implementation of the Government Resolution 48 in January 1992, the so-called "land privatisation decree". According to this resolution a private reserve of 850,000 ha was established, which amounts to less than 30% of the total agricultural area (Ebanoidze, 2002, p. 135). Following Resolution 48 and its amendments (Arsenashvili, 2000, p. 1; Ebanoidze, 2001, p. 1) every farmer received up to 1.25 ha per household in Low Lands and 5 ha in Highlands. Rural dwellers who had not been engaged in agriculture received up to 0.75 ha, while urban dwellers received up to 0.25 ha of agricultural land (Thiesenhusen, 2002, p.2, Bullard, 2000, p.9, UN- Economic and Social Council, 2001).[2]

All land was given to the citizens free of charge (APLR, 2003, p. 11; Koechlin, 1999, p. 13). The distribution of agricultural land was managed by a land reform committee elected by each village (Ebanoidze, 2003, p. 126), however, land allocation to the citizens was mainly arranged by those who managed the former collective farms. (Bullard, 2000, p. 9). Thus the privatization was very homogeneous within a Sakrebulo (village) (Koechlin, 1999, p. 8), but differed between Sakrebulos in terms of the size of the distributed parcels as well as in terms of the completeness of the process of land allocation (Shuker, 2000, p. 13).

Following the principles of equity each family was meant to receive the same quality of land. Former large land parcels of the same soil quality were split into several plots. Thus, each household received up to four or five land parcels, usually in different locations (UN- Economic and Social Council, 2001, p. 4).

The privatisation process at the beginning of the 1990s started spontaneously and lacked necessary legal guarantees. In order to rectify this situation, local reform committees were authorized to issue a receive-delivery act, which was the main document for granting ownership of agricultural land until 1996 (Decree # 503 as of June 28, 1993) (Ebanoidze, 2002).

Until 1996 nearly 4 million land parcels with an amount of 930,000 ha were distributed to 1.4 million persons. Most of them did not obtain the receive-delivery act due to an unwillingness or inability to pay for it. Initial registration hampered and hindered the land market's development (Ebanoidze, 2002). As the registration of land transfers has only been regulated since the "Law on Land (Immovable Property) Registration" in 1996, transfers of ownership rights were not accurately registered before 1996. The existing system was not able to meet the requirements of recording subsequent transactions or new laws adopted from 1996-1999. In order to speed up the initial

[2] The size of the parcels received by the population depended on several aspects including the former employment status, land endowment of the villages (sacrebulos) (Thiesenhusen, 2002, p.2) as well as military aspects (distance to country borders, etc.) (Shuker, 2000, p. 24, APLR, 2003, p. 12) or when agricultural land was very fertile (in this case the local authorities often decided to distribute less than the aspired amount of land) (APLR, 2003, p. 12). In the conflict area Abkhazia only 8.5% of all agricultural land was privatized in April 2001, in South Ossetia a mere 1.6% (Kneib, 2002, p. 15).

registration of ownership rights and to issue registration certificates to the beneficiaries, a number of international organisations supported the country in creating land cadastre and registration systems (Ebanoidze, 2002) such as KfW Entwicklungsbank and USAID.

Although called "Land Privatisation Decree", all distributed land remained in state ownership and was given to individuals only in inheritable lifetime use. Actual transfer of land ownership from the state to the individual households only became possible after the adoption of the "Law on Private Ownership of Agricultural Land" in March 1996, more than four years after the distribution of the land had begun (Lerman, 1996, p. 1; Thiesenhusen, 2002, p.1). According to this law all resolutions granting agricultural land ownership rights to the population of Georgia were considered legitimate (Ebanoidze, 2003, p. 127).

Instead of further privatizing agricultural land the government released "The Law on the Lease of Agricultural Land" in 1996 to give the population the chance to lease the remaining state-owned land. The land rent was the same amount as the land tax (Shuker, 2000). It was intended to create a market-oriented agricultural sector controlled by larger leaseholders (Terra Institute, 2005, p. 4). The lease price was intended to be at least equal to the land tax rate and could be higher (Tsomaia, Ebanoidze et al., 2003, p. 186).

However, the lease of land was associated with corruption. Most of the fertile land parcels were allocated to influential state officials who had no interest in farming and sub-leasing the land (Ebanoidze, 2002, p. 138). Not infrequently land was leased for speculative purposes.

In total, around 25% of Georgian agricultural land or 54% of arable land were distributed between 1992 and 1998 when the first stage of land reform was completed (Thiesenhusen, 2002, p.2). Nearly four million land parcels with a total amount of around 760,000 ha of agricultural land had been distributed to more than one million households (Ebanoidze, 2003, p. 127; APLR, 2004, p. 4; UN- Economic and Social Council, 2001, p. 5).

2.1.2 Second Stage of Land Reform

The second stage of land reform came into force with the "Law on Privatization of State-Owned Agricultural Land" in July 2005. The law and its amendments aimed to develop the commercial agricultural sector and to boost land market development by privatising additional agricultural land. The remaining state-owned land was intended to be privatized by selling it. Aim of this stage of land reform was the development of the Georgian agricultural sector and of the land market in general (APLR, 2006, p.3).

State-owned agricultural land parcels which had not been leased out should be privatized through special auctions. Leased state-owned agricultural land parcels shall be privatized through direct sale. Agricultural land parcels that failed to be sold through special auctions and the parcels that were formerly leased shall be privatized through open auctions. Agricultural state-owned land parcels that are not leased or/and formerly leased agricultural land parcels are subject to privatization through direct sale under competition. It must be mentioned that the rules and conditions for direct sale under competition are provided by the Law of Georgia on Privatization and Alienation of State and Local Self-government Properties. As per July 31, 2006, 1,944 leased land parcels with the total area of 16,991 ha have been privatized. (APLR, 2006, p.5)

According to figures published in the official Agricultural Yearbook 2007, 58% of the cultivated agricultural land is in private ownership, while 40% is rented from the state. Renting land from other private persons plays only a minor role with a mere 1% of the total operated area (Georgian Department for Statistics, 2008).

2.2 Change of Agrarian Structure

The result of the land privatisation process was a huge change in the agrarian structure. As large enterprises were almost entirely deconstructed, the agricultural production is now greatly based on very small households, which either produce for some commercial purpose or mainly for subsistence. The agricultural land is highly fragmented; some refer to the land as "over-fragmented" (Lerman, Csaki et al., 2004a). According to the Agricultural Census conducted in 2004, the average total farm size accounts for 1.2 ha. 40% of all farms are less than 0.5 ha in size, 70% less than 1 ha (Georgian Department for Statistics, 2004).

The share of private farms and households in total animal production increased from 85% in 1995 to almost 100% today. And while 13% of the total sown area was cultivated by private farms in 1990, its share increased to 49% in 1995 and further to 98% in 2005 (Georgian Department for Statistics, 2006a). Almost the entire agricultural production in Georgia and the supply of domestic products to the Georgian population depends on very small household production.

The dismantling of large agricultural enterprises as well as the emergence of small family farms has lead to a significant change in land use. The agricultural area sown to annual crops dropped from 700,000 ha in 1990 to nearly 300,000 ha in 2007[3]. This drop includes a significant drop of area sown to fodder crops from 329,000 ha in 1990 to 10,000 ha in 2007 (Georgian Department for Statistics, 2008), which was in accordance with a substantial decrease of animal production[4].

Although plant production could recover somewhat since the middle of the 1990s (Lerman, 2005), unfavourable weather conditions in 2006 (Georgian Department for Statistics, 2007) set back the level of agricultural production, resulting in a sown area of a mere 297,000 ha. Correspondingly, agricultural production decreased significantly until 2007. Particularly intensive crops like wheat or barley lost importance, while maize production even increased somewhat. Georgia, which was the main exporter of vegetables in Soviet Times, had to face a sharp reduction of the vegetable production of 57% to 190,300 tons from 1990 to 2007 (Georgian Department for Statistics, 2008; Georgian Department for Statistics, 2007). Grape production decreased by 67% to 227,300 tons in 2007. Grapevine has been privatized to the population at the beginning of the 1990s. As the beneficiaries often could not afford the expenses to maintain the grapevines (due to low prices or missing output markets; vine factories often stopped producing in the 1990s) or the household needed the area for production of staple food for subsistence, many vines were destroyed and replaced by arable land. Thus, not only agricultural production decreased but values built over years were destroyed, resulting in a lack of production capacities.

When the governments started to privatize agricultural land one has expected the land market to develop quickly and thus the agricultural sector to become more efficient by transferring land to the most productive users. However, in Georgia land transfers only became possible in the second half of the 1990s after legitimating private land ownership in 1996 (Lerman, 1996, p. 1).

3 Since the bad harvest in 2005 agricultural production could not recover and resulted in an even lower sown area than prior to the transition process from 534,000 ha in 2004 to the mentioned 297,000 ha in 2007 (Georgian Department for Statistics, 2008).

4 Cattle inventories in large agricultural enterprises have almost entirely disappeared - from 1.3 million head in 1990. However, the number of cattle on family farms has increased by 200,000 heads to 1 million but they are mostly held in an extensive manner. Pig, sheep and goat as well as poultry production have significantly decreased (Georgian Department for Statistics, 2008).

Table 1: Agricultural Land Transactions in Georgia (1998 – 2005)

	1998-1999	2000	2001	2002	2003	2004	2005
Buying and Selling	1,485	1,143	3,729	6,471	5,158	10,429	14,268
Mortgage	15	106	567	1,324	2,216	3,323	3,327
Inheritance	1,414	2,521	1,882	2,614	4,099	9,793	7,835
Gift	63	107	175	285	555	1,358	1,549

Source: APLR, 2006

Close to 2 million agricultural land parcels were registered until the end of 2003. In the same year 5,158 land sales had been registered (see table 1), which amounts to less than 0.3% of all registered agricultural land parcels (Terra Institute, 2005, p. 46). Thus, the land market had not been very active until then. Although the land market development further improved up to 14,268 land sales in 2005, there is still a need to extend land market activities.

2.3 Agricultural Production

The agricultural sector has traditionally played an important role for the whole Georgian economy. In 1990 nearly 30% of the GDP was delivered by the agricultural sector (see table 2). At the same time, one quarter of the employees were engaged in agriculture. Although the agricultural production significantly decreased during the first years of transition, the importance of the agricultural sector for the whole economy even increased. While the industrial and other sectors began to recover since the middle of the 1990s and economic growth started, the agricultural sector could not keep up with this economic development. The share of agriculture in GDP dropped from 42% in 1995 to 9% in 2007. One may argue that the decreasing agricultural share in GDP indicates economic development. However, the low agricultural production in Georgia shows that the decreasing share of agricultural production in GDP does not result from more efficient agricultural but from the low level and decreasing agricultural production. The value added in the agricultural sector is the lowest in Georgian economy and could not improve since 2003 (UNDP, 2008, p. 36).

Despite the decreasing share of agriculture in GDP the share of this sector in total employment more than doubled since 1990. Today, more than 50% of the employees are engaged in the agricultural sector (see table 2). This may have statistical reasons as according to the Law on Employment every household which owns at least one ha of agricultural land is counted as self-employed in the agricultural sector, even if the household is more strongly engaged in non-agricultural work (UNDP, 2002, p. 15). Nevertheless, these figures indicate the low labour productivity in the agricultural sector and underline the still existing "buffer role" of the agricultural sector.

Table 2: Agricultural sector- Share on GDP and employment (1990 – 2007)

	1990	1995	2000	2003	2005	2006	2007
GAO, in current prices, m GEL	-	-	1,633	2,262	2,330	2,131	2,310
Share of GDP (%)	29.7	41.7	20.2	19.3	14.8	11.2	9.4
Number employed, in 1,000 head	695	530	957	996	948	966	910
Share of total employment (%)	25.1	30.6	52	54.9	54.3	55.3	53.4

Source: Own table, based on Georgian Department for Statistics, 2008 and Georgian Department for Statistics, 2006a)

2.4 Rural Poverty

Poverty is a phenomenon which had been officially neglected during Soviet times. After the breakdown of the Soviet system and the political and economic reconstruction, poverty[5] significantly increased in all CIS countries, including Georgia. Due to weak databases at the beginning of the 1990s poverty figures are only provided from the middle of the 1990s onwards. And even from that time poverty data should be treated with caution due to inadequate poverty measurements (IMF, 2006, p. 5) as well as changing methods, a fact which does not allow a time series from 1995 until today (Georgian Department of Statistics, 2001). However, general observations can be conducted.

In most CIS countries rural poverty sharply increased from the beginning of the transition process. In most CIS countries the rural population was more strongly

[5] Poverty is a multi-faceted and multi-dimensional phenomenon with a huge variety of definitions and conceptions. A commonly agreed definition of poverty does not exist. This paper is not able to fulfil the requirements of adequate poverty description and analysis. To receive more information about poverty concepts and definitions please refer to Hemmer and Wilhelm, 2000 as well as Sautter and Serries, 1993.

affected by poverty than the population living in urban areas (World Bank, 2000). The situation in Georgia is different.

As mentioned above, poverty figures are only available from the middle of the 1990s. A time series is available from 1998 until 2005[6]. Figure 1 clearly shows that in the mid-1990s the rural population was less hidden by poverty than the one in urban areas or small towns.

Figure 1: Poverty in Georgia (1998 – 2006)

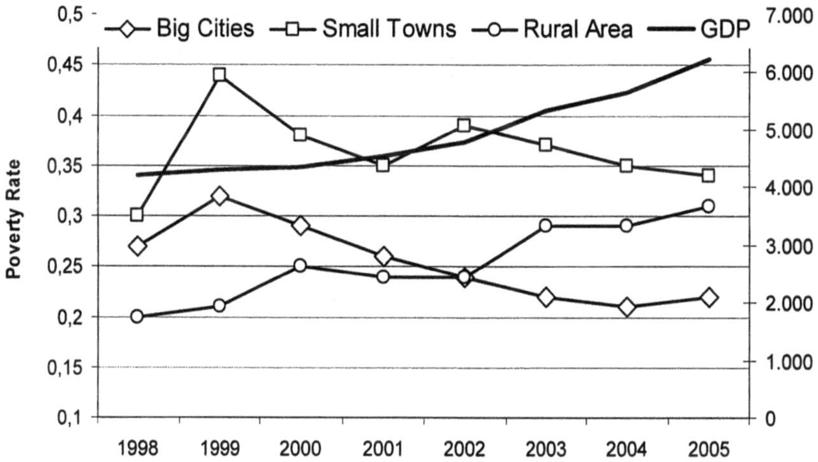

Source: Georgian Georgian Department of Statistics, 2001; Georgian Department for Statistics, 2006

As intended by the land reform of 1992, the agricultural sector seemed to be the social safety net for the rural population. From 1997 to 2004 the share of non-cash income in the total per capita income measured by the average monthly per capita income decreased from 45% to 23% (Georgian State Department of Statistics, 2006). Although the income from selling agricultural products somewhat increased during the same period, it seems to have been inefficient to protect the rural population from starvation.

6 Old official poverty figures are available until 2005. The government has implemented a new methodology to measure poverty, starting from 2005. Thus, there is no consistent time series available from the mid-90s until today (Georgian Department of Statistics, 2001).

From 2002 on, the poverty rate in rural areas exceeded that in urban areas, which decreased since 1999.

The urban population's income has been strongly connected to the economic growth. When Georgia started to recover after the Russian financial crisis of 1998, the GDP increased year by year. Corresponding to the economic growth the urban poverty rate has decreased. The agricultural sector, however, seems to be decoupled from this development. The economic growth does not have positive impacts on the income situation in rural areas. Unfavourable weather conditions like the unusual flood in 2005 (UNDP, 2008, p. 35) can have a negative influence on agricultural production from which the agricultural sector only slowly recovers. An example is the use of fertilizer or manure after one year of low harvest. Thus, after the very low harvest in 2006 10% less ha were cultivated using 25% less fertilizer and 15% less manure (Georgian Department for Statistics, 2008).

Based on a macro-level analysis it seems that the intended goal of the first stage land reform succeeded. In the middle of the 1990s poverty in rural areas was below that in urban areas. As agricultural production was the main income source in rural areas and still is, cultivating land seemed to keep households out of poverty. However, the level of agricultural production has significantly decreased and could not recover until now. The second wave of land reforms, which started in 2005, has not shown any positive results until now.

3 Rural households - A Micro-analysis

3.1 Survey Design

Within a research project by the Center of international Development and Environmental Research (ZEU) of Giessen University, financed by the Volkswagen Foundation, a household survey was conducted in Georgia in the fall of 2007. The survey took place in four villages in the project region Shida Kartli. The villages were selected based on criteria of agricultural production and distance to larger towns (access to labour markets, input and output markets, credit markets). In each village 50 households were selected for interviews by random sample. The questionnaire included questions regarding household structures, land endowment and agricultural production, income sources and access to input and output markets.

One of the villages (G-I) has very urban structures with up to 3,870 inhabitants and access to non-agricultural labour in the industrial sector. G-III is located close to G-I and the population has, thus, access to non-agricultural labour in G-I. G-II and G-IV are typical villages in rural areas with long distances to regional towns and, thus, to access to labour, input, output, and credit markets.

Table 3: Village Characteristics

	G-I	G-II	G-III	G-IV
Inhabitants (in number)	3,870	1,300	3,185	1,470
State-owned agricultural land (in ha)	1.7 (leased out)	150 (leased out)	10.51 (leased out)	112 (leased out)
Privately owned agricultural land (in ha)	150	485	818	680

Source: Household Survey 2007

3.2 Land Endowment

Between 0.2 and 1.5 ha of land, averagely, are owned by the surveyed households (see table 4). The large differences occur because of the different amounts of agricultural land belonging to the villages. The minimum size of land parcels does not fall below 0.2 ha, the maximum size amounts to up to 0.8 ha.

Table 4: Average Land Endowment

	Total	G-I	G-II	G-III	G-IV
Total land owned (in ha)	1.2	0.2	1.5	0.9	1.3
Min size (in ha)	0.2	0.1	0.2	0.2	0.2
Max size (in ha)	0.5	0.1	0.5	0.6	0.8

Source: Household Survey 2007

This land is the result of the first stage of the land reform and some land transfers until today. The second stage land reform did not succeed in the project villages. Only 2 households have indicated to have bought land since 2005. One of them stated that it had leased the land before it had bought it.

Secure land ownership rights are a precondition for land market development and access to credits. Although Georgia distributed agricultural land in 1992 and privatized the land in 1996, a mere 33% of all households indicated that they have officially registered the land, up to 71% thereof in urban areas, 6-12% thereof in remote areas. Main reasons for not registering the land have been financial reasons as well as lack of information. Only one household stated that the distance to the registration office was the main reason. Considering the results of similar household surveys in Moldova in 2008 where land was distributed to the population in the late 1990s but with simultaneous registration, nearly all of the households indicated to hold ownership certificates for the land. This leads to the conclusion that the fact that land distribution and land registration do not go hand in hand in Georgia results in a relatively low registration rate. Land owners are required to devote greater effort and money (at least for transportation and opportunity costs during the registration procedure) to register the land. Additionally, the Georgian government should have given better information about land registration to the population, namely about the procedure as well as the motivation for and benefits of registered land.

As already shown in Chapter 2, the land market was developed very slowly and only in recent years. Less than 10% of the households indicated to have been transferred (buying, selling, leasing) agricultural land. The main reasons for not buying or leasing land were financial reasons with 65% or 58%, respectively, as well as no need for more land with 21% or 16%, respectively (see table 5).

Table 5: Reasons for not Leasing or Selling Land

	Georgia	G-I	G-II	G-III	G-IV
Not Leasing Land					
N	153.0	30.0	38.0	36.0	49.0
Lack of money	57.5	36.7	52.6	58.3	73.5
No need for more land	16.3	16.7	18.4	25.0	8.2
No land available	13.1	23.3	18.4	-	12.2
Not Buying Land					
N	153.0	30.0	45.0	34.0	48.0
Lack of money	65.0	53.3	60	58.8	81.3
No need for more land	20.4	13.3	31.1	35.3	4.2
No land available	4.5	-	4.4	5.9	6.3
Bad quality of the land	5.7	30.0	-	-	-

Source: Household Surveys 2007

However, in some villages no land was available for selling or leasing, especially in G-I, but also in the rural village G-II. This may be result from a lack of agricultural land (very probably in G-I), but also from missing land market structures and institutions.

Many households are not able to cultivate all of their land. Thus, 13% (G-I) – 54% (G-III) of the surveyed households stated to have at least let some agricultural land lie fallow. The main reasons were lack of inputs (41%), bad quality of the land (15%), and no irrigation possibilities (15%).

More than half of the population indicated that they do not intend to extend their cultivated area. From those who are interested in extending the cultivated area only one quarter indicated that the government (local or national authorities) was their most preferred trading partner. Most of the respondents declared to trade best with neighbours, friends or family members.

3.3 Household Income Structure

It is not surprising that agricultural production plays a major role in rural households' income. When analysing household income in Georgia one should consider that the Georgian government has tried to boost subsistence by the first stage land reform. Thus, it is not sufficient to refer only to monetary income but also to non-monetary income (subsistence).

Averagely 44% of the income (monetary and non-monetary) in all Georgian project villages comes from agricultural production, either from commercial plant or animal production, wage employment in agriculture or agricultural production for own consumption (see table 6). In those villages which are located in more remote areas the share of agricultural in total income even increases up to 62%.

Averagely 33% and in single villages up to averagely 48% of the income is non-monetary income[7]. Thus, subsistence plays a major role in most of the surveyed households. On average, the share of subsistence in all villages is more than double that of monetary income derived from agricultural production.

[7] Non-monetary income (subsistence) was calculated by multiplying the self-consumed products with the local market price of these. The amount of self-consumed products as well as local market prices has been surveyed in the questionnaire.

Table 6: Income Structure (in % of total monetary and non-monetary income)

	Georgia	G-I	G-II	G-III	G-IV
N	201/205	51-52	50	50	50
subsistence	32.7	22.5	47.5	22.9	38.2
self-employment in plant production	8.2	1.4	12.6	1.7	17.5
self-employment in animal production	1.1	0.0	2.0	1.2	1.2
self-employment in plant and animal production	9.3	1.4	14.7	2.9	18.7
agricultural wage employment	2.3	0.8	0.0	5.5	2.9
Agriculture, monetary and non-monetary income	44.3	24.8	62.2	31.2	59.8
non-agricultural self-employment	13.5	2.8	18.0	25.3	8.3
non-agricultural wage employment	22.3	47.6	7.3	21.0	12.5
Non-agricultural income, total	35.8	50.4	25.3	46.3	20.8
public transfers	15.6	18.5	10.5	16.7	16.7
private transfers from abroad	2.5	3.2	2.0	3.7	1.2
private transfers from Georgia	0.1	0.0	0.0	0.4	0.0
Private transfers, total	18.3	21.7	12.5	20.9	17.9

Source: Household survey 2007

The question is now why the households themselves consume most of their own produce instead of selling it. Undervalued output prices are the main impediment of selling products. More than 60% of those who indicated to sell at least some of the production stated this reason. Around 20% do not have any problems with selling products. And another 20% declared problems with transport, dilate payments, difficulties of finding a buyer or other problems to be the main barrier of selling products.

Income from wage employment in agriculture hardly plays a role in the income of the surveyed households. This is due to the fact that the large agricultural enterprises stopped working at the beginning of the 1990s and commercial agricultural structures could not be established again in the project villages until now. Due to the fact that farmers produce mainly for own consumption, hardly any non-household members are employed on these farms.

The share of non-agricultural income in total income is especially high in those villages located close to industrial enterprises. In G-I, for example, a juice, as well as a sugar factory have been established in recent years. Many of the households in G-I and G-III, which is located close to G-I, include at least one member who is employed in one of these enterprises. In the villages in remote areas non-agricultural production is also important, as an average of 20-25% of the households receive an income from this sector. Among others, non-agricultural employees are teachers or work in administration. Also, non-agricultural self-employment is an income source that is not to be neglected. Last but not least, transfer payments should be mentioned at this point. Considering that more than 50% of the households are headed by a person who is not employed but either a pensioner, veteran or disabled (18%), a housewife (10%) or unemployed (27%), public as well as private transfers are highly important for the households' income. But although they are important, they are rarely sufficient.

3.4 Poverty

Poverty is calculated according to the concept of the World Bank. The poverty line used for this analysis is 2.15 $ PPP (Purchasing Power Parity) per day. This line was developed for transition countries, as the 1 $/day line developed for developing countries would not suffice for clothing and heating costs, which are higher in transition than in typical developing countries (Alam, Murthi et al., 2005). The analysis is based on the disposable income minus the expenditures for production and taxes and is calculated per household equivalent[8].

When calculating poverty while regarding only monetary income, 43% of all surveyed households are considered poor (see table 7). There are major differences between the villages ranging from 25% in G-III to 67% in G-II.

Table 7: Poverty Rate (Head Count Index)

	Georgia	G-I	G-II	G-III	G-IV
N	195.0	48.0	49.0	48.0	50.0
2.15 $ PPP Mon	42.6	37.5	67.3	25.0	40.0
2.15 $ PPP MonNoMon	20.5	25.0	36.7	10.4	10.0
Difference	**22.1**	**12.5**	**30.6**	**14.6**	**30.0**

Source: Household Survey 2007

[8] The household equivalent has been calculated according to the new approach of OECD (OECD, 2008). Every household member has been converted into household equivalents according to this standard and all member equivalents within one household have been summed up. The household income has then been divided by the household equivalent.

The head count index measures the number of poor in relation to the whole population:

$$H = \frac{q}{n}$$

with H = head count index, q= number of poor and n= total number of population. (Hemmer and Wilhelm, 2000). As soon as non-monetary income comes into consideration the poverty rate decreases in all villages. The difference ranges from 13% in G-I, where agricultural production is not that important, to 31% in G-II, where agricultural production and especially subsistence is highly common. In other words: subsistence in the project villages saves 13-31% of the households from starvation. Of course this conclusion is somewhat poor as, if the institutional framework were sufficient and markets were undistorted, households could sell their products, receiving better income and thus achieving a higher living standard. However, under certain conditions subsistence is highly important for many households in order to live outside of poverty.

Considering monetary income, most vulnerable are those households whose household head is either unemployed (60% of the unemployed are poor), a pensioner, veteran or disabled (54%) or a housewife (80%). These groups are also the ones that mostly benefit from subsistence. Thus, the poverty rate for unemployed decreases to 30% when non-monetary income is included into the calculation; that of pensioners, veterans or the disabled is at only 24% and that of housewives at 40%. In national statistics these groups may be considered as individual farmers, as according to the Law on Employment every household owning at least 1 ha of agricultural land is counted as an individual farmer (UN- Economic and Social Council, 2001, p.15). Thus, although being engaged in agricultural production and may be defined as farmers by the official statistical department, many households do not have the strong connection to the soil as in evidence in western European countries. They feel more like an unemployed or a housewife than like a farmer.

4 Conclusion and Outlook

After the breakdown of the Soviet Union most CIS countries started to transform their nation's economy towards a market economy. The transformation process in the agricultural sector was induced by land reform and farm restructuring. Land reform in Georgia has been implemented in two stages. The first stage from 1992-1998 included the distribution of agricultural land to the population and the transfer into private ownership. Unlike most other CIS countries, Georgia used this measure mainly to protect the population from starvation and poverty. As no social safety nets existed, the population was to be enabled to take care of itself by cultivating its own land. The second stage of land reform started in 2005 and aimed to establish an efficient and commercial agricultural sector. Land was not, like before, given into private ownership for free but has been sold to the population via auctions or direct sales.

Today, the Georgian agricultural sector seems to be decoupled from overall economic growth. Agricultural production dramatically decreased during the first decade of transition and could not recover until now. Missing markets and access to credits, highly fragmented land structures as well as lack of management knowledge of the newly established farmers led to an agricultural sector in a desolate situation. While the idea of subsistence as a means of fighting poverty might have been successful during the middle of the 1990s when rural poverty was below that in urban areas, today the rural area is the one most ridden by poverty.

Still, subsistence is highly important for large parts of the population. In the villages surveyed in a research project financed by the Volkswagen Foundation, being engaged in agricultural production for one's own consumption keeps up to 31% of the households out of poverty. Mainly those without regular work, such as unemployed or housewives, profit from subsistence. However, many households cultivate only a part of their agricultural land. Other parcels lay fallow. The low agricultural productivity as well as the land market are not yet well developed and potential buyers of agricultural land are often missed. In order to protect the rural population from poverty, the establishment of an efficient agricultural sector is essential. The Georgian government should focus on strengthening institutional frameworks and developing a rural infrastructure. Access to markets (input as well as output markets) and to credits is essential for a commercial sector. Off-farm employment opportunities should be broadened so that people are not engaged in the agricultural sector because it is the only possibility for them to survive. As in other countries, such as Ukraine or Russia, labour outflow will be important to increase labour productivity.

5 References

Adams, M. (1995). Land Reform: New Seeds on Old Ground? Natural Resource Perspectives Paper. London, Overseas Develpment Institut (ODI).

Alam, A., M. Murthi, et al. (2005). Growth, Poverty, and Inequality- Eastern Europe and the Former Soviet Union. Washington, D.C., World Bank.

APLR (2006). Georgian Land Market Development Activity- Report for Quarter 3. Tbilisi.

APLR (2003). Georgia Country Paper/ Land Policy Action Plan. South Caucasus Regional Land Policy Conference. Tbilisi.

APLR (2004). Analysis of Real Estate Market Development in Georgia. Tbilisi, Association for the Protection of Land Owners Rights (APLR): 33.

Arsenashvili, D. (2000). Land Privatization in Georgia. A Land Privatization Index, Minsk, Belarus, The National Land Agency of the State Committee for Land Resources, Geodesy and Cartography of the Republic of Belarus and the Land Tenure Center, University of Wisconsin-Madison.

Bullard, R. K. (2000). EC Food Security Programme- Second Land Reform and Privatization Programme. Brussels, Réseau Europèen de Sécuritè Alimentaire: 46.

Daviddi, R. (1994). Privatization in the Transition to a Market Economy. Sam Domenico di Fiesole, EUI: 39.

Deininger, K., G. Feder, et al. (2003). Land Policy to Facilitate Growth and Poverty eduction. Land Reform, Land Settlement, and Cooperatives- Special Edition: 5-18.

Didebulidze, A. (1997). Problems and Prospects of Agriculture and Rural Development in Georgia. Tbilisi, UNDP- Georgia.

Ebanoidze, J. (2001). Agrarian Reform in Georgia. Access to Land, Bonn.

Ebanoidze, J. (2002). Integrating land issues into the broader development agenda- Country Case Study for the Regional Workshop on Land Issues in Central and Eastern Europe and the CIS. Final Report Georgia Land Market Development Project. USAID: 129-175.

Ebanoidze, J. (2003). "Current land policy issues in Georgia." FAO Land reform- Land Settlement and Cooperatives(3): 125-140.

Georgian Department for Statistics (2004). Agricultural Census 2004. Tbilisi.

Georgian Department for Statistics (2006). Food Security Report 23. Tbilisi.

Georgian Department for Statistics (2006a). Agriculture 2005, Georgian Department for Statistics. 2006.

Georgian Department for Statistics (2007). Food Security Situation Bulletin No. 27. Tbilisi: 39.

Georgian Department for Statistics (2008). Agricultural Yearbook 2007. Tbilisi.

Georgian Department for Statistics (2001). Food Security Situation Bulletin No. 3. Tbilisi, Ministry of Economic Development of Georgia.

Hemmer, H.-R. and R. Wilhelm (2000). Fighting Poverty in Developing Countries. Frankfurt, Peter Lang.

IMF (2006). Georgia: Poverty Reduction Strategy Paper Progress Report. Washington, International Monetary Found.

Kneib, W. (2002). Georgia: Study on the Formulation of Guidelines to Support Structural Change in Georgian Agriculture as a Basis to Extend the Land Consolidation Component of the Project by a Soil Atlas. Frankfurt am Main, KfW: 82.

Koechlin, J. M., Jegat (1999). Survey on Land Privatisation Process in Georgia, European Commission Food Secururity Programme: 56.

Lerman, Z. (1996). Land Reform and Private Farms in Georgia. Washington, DC, The World Bank: 27.

Lerman, Z. (2005). The Impact of Land Reform on Rural Household Incomes in Transcaucasia and Central Asia. Rehovot, The Center for Agricultural Economic Research: 19.

Lerman, Z., C. Csaki, et al. (2004a). "Evolving farm structures and land use patterns in former socialist countries." Quarterly Journal of International Agriculture 43(4): 309-335.

OECD (2008). What are equivalent scales? 2008. Internetsource: http://www.oecd.org/LongAbstract/0,3425,en_2649_33933_35411112_1_1_1_1,00.html (Date of access : November 1, 2008).

Sautter, H. and C. Serries (1993). Inhalt und Methodik von Armutsanalysen. Bonn, BMZ.

Shuker, I. (2000). Georgia: An Update of Agricultural Developments. ECSSD Environmentally and Socially Sustainable Development Working Paper: 34.

Stanfield, J. D. (2002). Options for State Owned Agricultural Land in Georgia. Mt. Horeb, Terra Institute: 11.

Terra Institute (2005). Final report: Georgia Land Markets Development Report, USAID Georgia Land Market Development Activity. Mt. Horeb, Terra Institute: 374.

Thiesenhusen, W. C. (2002). Discussion of the Draft Law prepared for the Second Phase of Land Reform and Privatization in Georgia (Prepared by the Association for the Protection of Landowners' Rights (APLR)). Mt. Horeb, Terra Institute: 24.

Tsomaia, E., J. Ebanoidze, et al. (2003): The Other Agricultural Land Reform in Georgia: State Leasing of Land to Private Farmers. Prepared under Cooperative Agreement 114-A-00-01-00111-00 between Terra Institute and the Association for the Protection of Landowners' Rights (APLR), and USAID/Caucasus. Online: http://www.terrainstitute.org/georgiareport.html (Date of Access: Nov. 20, 2008),

UN- Economic and Social Council (2001). Land Administration Review Georgia.

UNDP (2002). Human Development Report- Georgia 2001-2002. Tbilisi, UNDP: 105.

UNDP (2008). Human Development Report 2008- The Reforms and Beyond. Tbilisi, UNDP.

World Bank (2000). Making Transition Work for Everyone- Poverty and Unequality in Europe and Central Asia. Washington, D.C., World Bank.

III. ENVIRONMENT AND LAND USE CHANGE

AGRICULTURE AND LAND USE CHANGE IN GEORGIA

ALEXANDER DIDEBULIDZE, TENGIZ URUSHADZE

1 Introduction

In the 21st century, Georgia, as any other country, should define the priorities and essential vector of its agricultural sector, including the food industry. Namely, it is undoubted that this development will be caused by the general condition of the economy, the policy of governmental structures and the geopolitical situation.

After the restoration of national independence in 1991, Georgia experienced one of the deepest recessions among the countries of the former Soviet Union. A subsequent civil war and internal conflicts, extreme inflation and problems due to the socio-economic adaptation as a consequence of the disintegration of Soviet production and distribution systems forced Georgia's rural population to introduce subsistence agricultural production systems. Agriculture underwent a deep crisis, the total cost of the agricultural production was reduced, and many food enterprises have stopped operating. The main reasons for this were uncompetitive raw materials, lack of financial and energy resources, uncertainty about demand and an increased competition from foreign products. As a result, Georgia faced the problem of famine from 1993-1998. Only with international assistance it became possible to provide the poorest portions of the population with a minimum of food products (table 1).

The political independence of Georgia must be strengthened by economic growth and its rate; production resources and the character of their use are essential in this respect. With the deficiency account of fuel and energy resources, land resources are one of the basic material riches of the country, and despite rather low indicators (0.12 hectares of the cultivated land per capita), they are distinguished by a high variety of soil types (some of them were first identified in Georgia) and high natural fertility.

Table 1: Food Aid to Georgia (Based on actually received deliveries, 1000 Tons)

	1993	1994	1995	1996	1997	1998	1999	2000
Programme Aid*	345.2	715.3	536.8	203.1	118.0	78.7	67.8	35.0
Project Aid**	2.9	37.2	50.8	47.5	31.8	31.1	19.2	26.9
Total Aid	348.1	752.5	587.6	250.6	149.8	109.8	87.0	61.9
	2201	**2002**	**2003**	**2004**	**2005**	**2006**	**2007**	
Programme Aid*	4.9	35.0	0	50.0	0	24.5	0	
Project Aid**	72.9	20.7	17.5	19.9	14.0	9.8	6.8	
Total Aid	77.8	55.7	17.5	69.9	14.0	34.3	6.8	

* Food for budgetary support and concessional sales; ** Food distribution to beneficiaries.

Source: UN (2007)

Georgia has a rich tradition in agriculture. Agricultural lands occupy roughly 43% of the country's total land area; approximately the same area is covered by woods. Vertical zones with almost all types of soil and climate characterize the country. Georgia is characterized by the main types of soil degradation (loss of organic matters and biological activity, physical degradation, soil erosion, water logging, salinization and alkalization, acidification, loss of chemical fertility, disappearance of soil, etc). Among main types of soil degradation the most significant is soil erosion caused by intensive degradation and desertification of land [Urushadze, T. / Urushadze, A., 2003]. Besides, soils are strongly contaminated with various toxic substances and radionuclides, and are eroded. Agricultural land in West Georgia is basically subject to water erosion and in East Georgia to wind erosion that is caused by uncontrollable wood cuttings due to the unstable power supply of the mountain regions (in West Georgia 33.8% of the agricultural land is eroded: 16.3% of it weakly, 12.3% moderately, 5% strongly; in East Georgia 29.4% of the agricultural land is eroded: 16.5% of it weakly, 10.5% moderately, 2.4% strongly); the whole country requires continuing care to prevent soil erosion.

Georgia's soil cover is extremely variable and includes almost all major soil types of the world. In the former Soviet Union, Georgia was the only republic with widespread subtropical soil landscapes, suitable for subtropical crops, accounting for the heightened interest of Soviet soil scientists in its soils. Practically all leading Russian and Georgian soil scientists contributed to their studies, beginning with the Report of the Dokuchaev Trans Caucasian Statistical Committee in 1899. The law of vertical soil zonality was based on these studies in the Caucasus Mountains. Subsequently the red, yellow and other subtropical soils were studied by the Russian scientists B.B. Polynow, S.A. Zakharov, I.P. Gerasimov, V.A. Kovda, S.V. Zonn and the Georgian scientists D. Gedevanishvili, M. Sabashvili, M. Daraselia and T. Urushadze. S.A. Zakharov first described what became known as the calcareous cinnamonic soils of the drier subtropics; T. Urushadze also first described new soil types – yellow-brown forest and black-brown forest. According to the 1:500 000 soil map the brown forest soils (with the illuvial B_t horizon) are the predominant soil type.

Unfortunately, the level of soil science in Georgia has recently decreased. It has been explained that many soil scientists do not know about new approaches in world soil science. In this respect it was very important to realize the German-Georgian project "Cadastre and Land Register", co-financed by KfW. For some years, German specialists, with participation of Georgian colleagues, created the "Atlas of Georgian Soils". This document was prepared according to world standards. They were translated into Georgian and published as "World Reference Base for Soil Resources" and "Textbook of the Field Investigation of Soils". Detailed soil investigations were carried out. All of these activities were important steps in the field of the popularization of soil science and now soil science occupies a place of honor among environment sciences. The example of close cooperation between German and Georgian scientists in soil science was the joint project which examined the spatial distribution, the amounts and the mobility of heavy metals in the Mashavera valley. The project was realized under the leadership of Prof. Peter Felix-Henningsen and its results are of ecological, economic and sociological importance [Felix-Henningsen et al., 2007].

2 OBJECTIVES

The role of agriculture in the development process of Georgia continues to be the topic of a debate in the development policy of the country. Over the last years, international and bilateral organizations have substantially reduced their food aid to Georgia (table 1), and this is one of the indicators of stabilization of the national economy, but a considerable share of the population is living in deteriorating conditions. The unbalanced diet of the poor poses a threat to the overall health of the population and creates the risk of human resource deterioration. More then 200 thousand internally displaced persons have limited access to employment, live in inadequate conditions and receive inefficient social assistance [UNCT, 2004]. From this point of view, agricultural production is important for the food security of Georgia, because it is a source of income for the majority of the rural poor.

Since the Rose Revolution in November 2003, the new policy conducted by the government was characterized by a strong reform process towards the Western standards of a market economy. Georgia quickly integrating into the global economy with the clear strategic orientation towards joining NATO (Georgia has signed the NATO Individual Partnership Action Plan, NATO IPAP in October 2004) and towards the European Union in the framework of the European Neighborhood Policy Action Plan (ENP AP, signed in October 2006). At the same time, unresolved territorial conflicts, involving Abkhazia and former South Ossetia, remain a source of tension with the Russian Federation, the former main economic partner, accelerating the economic separation from this country. The Georgian economy's performance has improved in the last 4 years:

- real gross domestic product (GDP) growth has increased from an average of 7.4% in 2000–2003 to 9.3% in 2004–2007;
- the current account deficit, although widening, has been largely financed by the surge in foreign direct investment inflows accompanied by rising foreign exchange reserves that crossed the 1.7 billion USD threshold in 2007;
- external debt sustainability has strengthened, with improvement in key debt indicators [Ministry of Economic Development of Georgia, 2008].

However, this robust growth has not yet translated into a net job creation, nor is there a clear downward trend in poverty incidence. The unemployment rate remains high and more than one third of the population lives below the poverty line, with the rural poverty incidence estimated at 41.7% and widening rural-urban gaps [ADB, 2007]. In Georgia there is a large potential for increasing agricultural productivity based on land consolidation and functioning land markets, good prospects for raising exports of agricultural produce provided farmers are better connected to markets, but the transition to commercial agriculture is slow. Small and fragmented land plots have limited many farmers to subsistence agriculture, irrigation and drainage infrastructure assets have deteriorated, and most crops have low yields. Under these conditions, factors favouring the development of agriculture, food and the processing industry of Georgia are:

- the most needed legislation for a market-oriented agro-food sector is in place;
- development and strengthening of private ownership and market relations in rural areas;
- high level of privatization;
- membership in WTO;
- good reputation of Georgian products;
- closeness to large markets;
- increase of organic production;

- Georgia's location as a transport hub;
- development of tourism;
- well-educated population, extensive network of agricultural researchers and institutions.

On December 7, 2007, the Parliament of Georgia amended the Law "On the Structure and Rule of Activities of the Executive Authorities of Georgia". According to the amendments, the National Agency of Public Registry (NAPR), instead of the Ministry of Agriculture of Georgia, shall be entitled to change the category (arable land, pasture, meadow) of agricultural land parcels. As a result, the process of agricultural land category change shall be simplified and the same body shall ensure the change of both agricultural land status and category. According to Georgian legislation, in particular to the "Law on Compensation of the Substitute Land Development Costs and Inflicted Damage in Allocating Agricultural Land for Non-Agricultural Purposes", NAPR is authorized to change the status of agricultural land into non-agricultural land (except for agricultural lands within the recreational territories, as well as within administrative borders of Tbilisi and Batumi, where the Ministry of Environment and Natural Resources is responsible for the change of land status). Amendments shall fill the gaps in the existing legislation, after the State Department of Land Management, together with the Ministry of Agriculture, has been authorized to change the category of agricultural land. Originally, there were a number of bureaucratic barriers impeding the process and usually it took rather a long time to change the land category. After the adoption of the mentioned legislative change, it will become possible to change the agricultural land category within a short period of time and under simplified rules. As of today NAPR has information on agricultural land parcels existing in Georgia and is able to direct the process in an efficient way [Gvaramia, 2008].

Climatic conditions are favourable for the production of a variety of agricultural products in Georgia. However, at the same time, Georgia is a highly disaster-prone country, and it has been frequently affected by serious natural disasters such as earthquakes, droughts, floods, landslides and avalanches; institutional disaster management capacities are very limited in terms of disaster prevention, mitigation, preparedness and response. In order to get reasonable yields, much of Eastern Georgia requires irrigation and much of Western Georgia requires drainage.

Under the former soviet system, Georgian agricultural production was concentrated on high value agricultural export products such as tea, citrus, wine, fresh and processed fruits and vegetable, essential oils and spices; one of the factors for the concentration of high value crops in Georgia was their high labour requirement. Only small amounts of cereal crops, meat, milk and sugar were produced in the former Georgian SSR. Most products were imported from other Soviet Countries where they could be produced more cheaply. Today the irrigation systems, perennial crops, and livestock farms are seriously damaged, activities connected with soil protection, veterinary services, selection and breeding are abandoned as well and the positions on the domestic markets and traditional export markets are lost. Since 1990 the production of the main agricultural products, as well as the livestock population has declined (table 2 and 3). As a result Georgia is far from self-sufficient in the production of food and still depends on import. Besides, the lion's share of imported food is of poor quality or contains pollutants harmful to and dangerous for health. The majority of the population uses poor food. Georgia relies on imports of grain, flour, dairy, and meat products to satisfy the domestic food demand. The situation on the consumer market is chaotic. Some businessmen take advantage of that and provide the market with food of extremely poor quality. The sharp decrease of the living standard, poverty that sparks off comparatively high prices of food and the unending process of growing poor even more complicate the situation. Also, the per capita consumption of major products has fallen. The government approach towards economic and social reforms implies alleviating current poverty by means of intense and sustainable economic growth. The government's readiness to target the national development policy, which is aimed at progressive integration into European structures and standards, is widely demonstrated.

There are about 0.16 million ha of natural meadows and 1.8 million ha pastures in Georgia. Most pastures areas (1.3 million ha) are for summer grazing, and are mostly characterized by salty soils with extremely low fertility. The total productivity of the summer pastures amounts to about 800,000 tons, while the figure for winter pastures is only 200,000 tons. Historically, this disproportion was partially equalized by the use of winter pastures outside Georgia (Dagestan, Azerbaijan), but currently it is impossible to use these pastures (war in Chechnya, transportation problems, closed border to Russia) and the former transhumant system collapsed [Didebulidze and Plachter, 2002] . Before its independence Georgia imported 400 thousand tons of feed grain, 75 thousand tons of soybean and other protein meals, 8 thousand tons of fish meal, 8 thousand tons of yeast and 7 thousand tons of vitamin and mineral premixes from other Soviet republics; today, this import has stopped.

Table 2: Instability of Agricultural Production (1000 Tons)

Production	1990	1996	1997	1998	1999	2000	2001	2002	2003	2004	2005	2006	2007*
Cereals and beans	666.3	652.2	901.9	597.8	780.5	420.5	713.6	672.2	754.1	679.3	702.9	326.8	413.1
O.w. Wheat	257.7	114.0	311.0	145.0	226.1	89.4	306.5	199.7	225.4	185.8	190.1	69.7	74.0
Barley	117.8	28.1	45.4	15.7	50.8	30.2	98.9	57.5	48.3	61.3	65.4	30.6	38.8
Oats	11.7	3.4	8.0	2.3	3.3	2.0	9.5	4.2	6.2	4.8	2.8	1.3	1.4
Maize	270.2	490.9	546.3	420.2	490.5	295.9	288.6	400.1	461.9	410.6	421.3	217.4	288.1
Beans	7.3	21.7	15	9.2	9.3	2.6	10.0	10.7	12.2	16.6	23.0	7.8	10.8
Sunflower seeds	7.7	4.0	31.4	22.8	40.5	2.6	41.8	21.2	25.5	22.3	22.3	12.3	16.1
Potatoes	293.8	285.6	353.0	349.8	443.3	302.0	422.2	415.3	425.2	419.5	432.2	168.7	226.6
Vegetables	443.2	423.9	513.9	380.0	417.0	354.2	396.0	405.6	430.1	400.5	436.7	179.7	181.9
Melons	38.8	31.5	27.0	32.2	108.2	80.0	83.9	125.1	125.0	109.5	119.6	37.8	73.7
Fruit and berries	591.2	361.0	299.4	279.0	296.0	250.0	200.0	172.0	260.0	160.2	264.9	153.3	229.0
Grape	691.0	311.3	309.1	238.5	220.0	210.0	150.0	90.0	200.0	180.0	250.3	162.5	230.0
Citruses	283.1	86.8	57.1	85.1	56.0	40.0	60.0	33.1	59.2	38.2	122.4	52.2	98.1
Tea leaf	501.7	34.0	33.2	47.2	60.0	24.0	23.0	24.0	25.5	20.0	22.8	6.6	7.5
Meat	170.3	117.8	120.7	104.1	100.5	107.9	102.4	106.9	108.9	109.2	109.5	100.3	86.9
Milk	659.4	530.3	600.2	634.7	660.3	618.9	710.0	742.1	765.1	780.4	787.7	619.4	639.6
Eggs X 1000	769.2	350.2	370.4	380.4	390.1	361.4	395.4	408.8	458.1	496.6	504.6	359.6	383.2
Wool	6.2	3.0	1.7	1.7	1.7	1.9	1.9	2.0	2.0	2.2	2.2	2.1	2.2
Honey	1.6	0.8	0.9	1.2	1.5	1.4	1.7	1.8	2.0	2.3	2.3	2.2	2.3

*Preliminary data;

Source: Department of Statistics (2007)

3 ANALYSIS

Since 1990 the production of the main agricultural products, as well as the livestock population has declined (table 2 and 3). Recalculating the population's cash income per capita (in 1990 – USD 900, and in 2006 – USD 680), the decrease results from the fact (table 2) that the production of annual crops in Georgia is extremely unstable. The basic cultures, for example cereals and beans, potatoes and vegetables, are considerably higher in odd years than in even years, and every year with a rich yield is followed by a year of less productivity. This testifies to the dependence of the average yields on climatic conditions. A sustainable growth in agriculture will be possible if the yields could be increased in all years. However, this will be difficult as Georgia suffers from unfavourable climate change.

Grains account for more than half the area sown, followed by vegetables. The area under annual crops treated by pesticides in 2006 amounted to no more than 52.3% of total land. Family holdings account for 94% of the production of grains, 100% of vegetables, and 100% of fruits, but a lower share of potatoes (78%, 2006). Of forestland, only a small area comes under "forest production", with the rest of the area classified as conservation and recreation forests.

Table 3: Livestock Numbers (as of end of year, 1000 heads)

	1985	1990	1995	2000	2001	2002	2003	2004	2005	2006	2007
Cattle	1646	1298	974	1177	1180	1216	1243	1251	1260	1164	1129
Of which: Cows	649	552	531	646	678	705	728	736	740	681	571
Horses	25	20	24	35	39	43	43	44	43	N/A	N/A
Pigs	1173	880	353	443	445	446	474	484	455	343	140
Sheep and goats	1980	1618	725	628	659	700	722	805	815	789	780
Poultry	24296	21760	13847	7826	8496	8905	9201	9836	7482	5332	6337
Bees (Families)	127	96	55	98	105	118	133	152	150	146	N/A

Source: Department of Statistics (2007)

Although the forest area has not decreased, illegal logging has contributed to the degradation of the composition and quality of forests. The production of livestock-products, as well as most livestock population (table 3) has declined since 1985-1990; grain-based livestock, also poultry and pig production has shrunk the most. The livestock sector should continue to be a supplier, mainly for domestic markets. The egg production increased in 2007, but the overall situation in animal husbandry was unfavourable due to continued outbreaks of bird flu, and a massive death of pigs took place; over 68 thousand pigs, usually allowed to roam freely, which contributes to the spread of the disease, died or were culled during 2007 as a result of African swine fever in Georgia. An institutional structure was established and a plan of activities adopted to combat this disease. The ongoing climate warming will necessarily cause an increase of the frequent natural disasters in Georgia.

Over the last foru years, the business environment in Georgia has undergone a significant change. Regulation of business entry has been streamlined, with:

- the number of business activities subject to licenses and permits reduced by 84%;
- licensing procedures simplified;
- the "one-stop shop" approach to business regulation adopted.

Furthermore, entry barriers in the form of minimum capital requirements have been lowered. The reform of business exit, however, is less advanced. As regards the regulation of ongoing business, operational licensing, permits, inspections and other compliance systems, including taxes and customs have become much more business-friendly. Contract enforcement has been strengthened by reforming court procedures for commercial disputes. The new Labour Code simplifies restrictions on the duration of contracts and overtime, as well as regulations regarding severance pay and labour dispute resolution.

The relevant key priorities for the development of the agricultural economy for 2008-12, as outlined in the new cabinet's program "Georgia without Poverty" [Parliament of Georgia, 2008], are:

- The average economic growth rate will be 9 percent annually over the next five years;
- income tax will decrease from the current 25 percent to 15 percent;
- the Fund for Stable Development, comprised of 25 percent of budget surplus and 25 percent of income received from privatization, will also be established. GEL 3 billion is expected to be accumulated in the fund which is designed to deal with unexpected economic challenges;
- Georgia's share of exports will amount to 40 percent of GDP, and the share of agricultural products will increase to 25 percent of overall exports;
- guarantees of private property will be secured through the constitution and better judicial practice;
- Georgia's sovereign credit rating will increase by at least four points over the next five years;
- poverty reduction will be a key focus, with the number of social program beneficiaries being halved over the next five years as a result of poverty reduction;

- more than 200 enterprises processing agricultural products will be created, 10 000 people will be employed and about 100 000 people will be involved in the process;
- an increase of the effectiveness of the agricultural sector and its commercialization will be reached; during the next 5 years, the export of agricultural products will equal 25% of the total exports, and the territory supplied with irrigation systems will be enlarged 2.5 times (from 130 000 hectares to 300 000 hectares).

The government's priorities with regard to agriculture are to develop an agricultural infrastructure, to reform the management of irrigation and drainage systems, develop viticulture, achieve convergence with sanitary and phyto-sanitary (SPS) standards and promote consumer protection through workable systems of food and product safety, providing high-quality goods and services to the national market. Expected results in the medium term include an enabling environment for agribusinesses and enhanced competitiveness of Georgian agricultural products on international markets, but the state budget expenditure for agriculture and forestry is low (The financing provided by the state budget for the Ministry of Agriculture amounted to 78.1 Mio GEL or 0.41% of GDP in 2008 [Law "On the State Budget of Georgia of 2008" of the Government of Georgia, 2007]). As a result, activities that are known to have high returns on investment remain underfunded: e.g., agricultural research and extension. Public expenditures that help strengthen relations between farmers and markets also need to be stepped up.

Georgia's foreign trade turnover was worth over USD 6.5 billion in 2007, a 40 percent increase compared to the previous year. Exports reached USD 1.2 billion, up 33 percent, whilst imports rose by 42 percent to a total of USD 5.2 billion. The share of trade with the Commonwealth of Independent States is 36 percent, down from 39 percent. The EU member states account for 28 percent. Trade with Turkey, Georgia's main trading partner, exceeded USD 900 million last year with the comparative figure of USD 646 million in 2006. With Ukraine soaking up most of the Georgian wine and spirits traditionally destined for Russia, it is now Georgia's second largest trading partner, replacing Russia with USD 669 million. Russia comes third with USD 632 million, which is 0.4 percent less than in 2006.

Agricultural products still form a significant part of Georgian exports (35.1% in 2005, only 26.7% in 2006). However, Georgia has a negative trade balance of agricultural products (table 4) and the country of agricultural net-export, in which the export of agricultural and food products exceeded the import 1.7 times, changed into the country of agricultural net-import; the largest share in agricultural imports belonged to cigarettes, wheat and meal, sugar and meat. For instance, provided that international cereal prices have risen further in the first two months of 2008, reflecting steady demand, and that the price of 1 kilogram of export wheat (US No.2 Hard) has increased by the end of March 2008 to about twice the level of the preceding year and now ranges above 0.4 USD, and given that the crop capacity for this strategic crop in Georgia has not exceeded 2 tons per hectare over the last ten years (the worst indicator for it in the European countries is only observed in Russia and Portugal, and in the Benelux countries; in the UK and Denmark the average yield exceeds 8 tons per hectare), it becomes clear, that even under the conditions of modern technology in the small areas, it will be difficult to raise a national competitive wheat. The only way out is to increase competitiveness and to optimize the demands of the inner market and export abilities accordingly. This means hat, first of all, to raise cultures whose crop capacity is related to those achieved in the developed countries, or cultures which are not raised in those countries (walnut, hazelnut, exotic fruit, tea, etc.).

A comprehensive deep free trade agreement (FTA+) is expected to be signed between Georgia and the European Union in 2009, and new efforts are needed in food and product safety and competition protection for 2008 and the next years – measures which are important for the negotiations, amongst other issues. At the same time, what is also important is the rationale of providing high-quality goods and services to the national market, as there is evidence that the absence of competition rules has already resulted in non-market pricing and an inadequate quality of supplies to domestic consumers. Georgia had adopted a law on food safety and quality in 2005, established a food safety, veterinary and plant protection service in 2006, but in 2007 decided to delay full food safety controls until the end of 2009 against the background of the service's limited capacity and the serious difficulties of institutions to comply with the rules.

Tea production was an important business for Georgia in the 1980s. With an annual production of 500,000 tons Georgia was one of the largest tea producing nations in the world, and tea accounted for around 18% percent of the agricultural sector. The collapse of the Soviet Union saw the industry take a downturn, as neglected processing plants became dilapidated, the conflict in Abkhazia decreased the total coverage of plantations, and output shrank. Reviving the tea industry is a significant challenge. Thousands of hectares of plantations were lost during the conflicts in the early 1990s and many more remain unusable due to years of neglect. Some plantations are now being used to harvest kiwi fruit and nuts instead. Most Georgian tea is exported to the Middle East. In 2006, the USAID-funded AgVANTAGE project helped set up Georgia's first "premium" green tea processing plant and is now helping tea producers find foreign markets through international exhibitions.

Table 4: Export and Import of Agricultural and Food Products (m. USD)

Year/Product	2001	2002	2003	2004	2005	2006	2007*
Total Agro import	**168.1**	**164.1**	**208.8**	**390.8**	**437.6**	**606.3**	**762.0**
O.w. Wheat and flour	40.4	28.8	37.8	123.8	90.5	99.1	139.2
Sugar	24.1	34.7	53.0	50.8	78.2	65.6	90.5
Meat	14.7	15.6	17.0	22.8	20.6	N/A	38.2
Total Agro export	**92.7**	**104.6**	**164.7**	**202.4**	**304.0**	**264.9**	**274.0**
O.w. Hazelnuts	9.8	6.7	12.6	17.7	70.3	56.6	65.1
Spirit						30.1	57.4
Wine	32.4	33.3	43.0	48.9	81.3	37.6	29.2
Mineral water	15.2	20.1	31.4	33.3	52.9	39.9	25.3
Fruits, vegetables	10.3	7.9	9.5	16.7	69.3	N/A	18.8
Citrus fruits	1.6	1.6	2.1	2.6	5.9	N/A	4.6
Tea	5.9	4.7	4.4	3.7	3.1	N/A	1.3

*Preliminary data

Source: ADB (b) (2007) and Department of Statistics (2007)

The Russian Federation's embargo on Georgian products, imposed early in 2006, and the economic sanctions imposed by Russia on Georgia have had a significant impact on agricultural exports and the economy in general (table 4), affecting the livelihoods of rural people. Air carriers and the railway are losing money because of the suspension of travel between the two countries. Wheat importers are already facing serious problems which are likely to drive bread prices high. According to the Department of Statistics, the country's trade with the Russian Federation stood at USD 574.4 million in January-November 2007 (0.4% less than in the same period of 2006). In the analyzed period, export to Russia fell by USD 22.1 million, as compared to the same period of 2006. This is directly connected with the ban imposed by Russia on the import of Georgian agricultural produce, wine and mineral waters. At the same time, the import of goods from Russia to Georgia grew by USD 19.8 million and reached USD 524.2 million (3.9% more than in January-November 2006). All in all, however, last year saw a certain enhancement of the role of Europe and primarily of the EU as a market for Georgian exports. The share of exports to the EU increased in particular, whilst that to the former Soviet Republics decreased. This was the result of the Russian embargo rather than the targeted diversification of Georgian exports. Besides, discrepancies between the countries' trade statistics make it extremely difficult to draw meaningful conclusions.

Georgia, famous throughout the former Soviet Union for its wines, was hit hard by an export ban imposed in 2006 after Russian authorities allegedly found pesticides in Georgian produce. Georgian wines and mineral water disappeared from Russian shelves, dealing a blow to Georgia's exports. Due to a sharp increase in the demand for hazelnuts and walnuts, this commodity came to be amongst the top seven main export commodities in the reviewed period. Germany, the Czech Republic, Italy and Belgium were distinguished amongst the Georgian nut consumers, due their higher shares of trade.

While the trade signals may be positive, political differences look like more of a problem. Georgia has accused Russia of fomenting separatism in the two breakaway provinces Abkhazia and South Ossetia, while Russia is against Georgia's plan to join the North Atlantic Treaty Organization, NATO. The political differences between the countries seem to be far from a resolution, but hopes have emerged that Russia's application to join the World Trade Organization could be a potential trigger to clear the trade log-jam. Russia's hopes to become a member of the WTO could be snagged by Georgia, already a member of the organization, and Russia has already backed down in trade disputes to speed up its WTO entry.

Russia's weight as a trading partner still remains important, but the history of winemaker "Tbilghvino" exemplifies recent changes in the industry. Before the ban, the company had an annual output of some 1.5 million bottles, shipping more than half of its production to Russia. After the embargo was introduced, "Tbilghvino" had to cut production by half and sell off land and buildings to survive. But one positive effect that followed the Russian embargo is that Georgia is not longer dependent on a single market, and last year, new markets such as the United Kingdom, Canada and Poland have opened up. Wine export to Ukraine increased by 48 percent, to Belarus by 160 percent and Kazakhstan by 506 percent. Meanwhile "Borjomi Mineral Water's" export to Ukraine increased by 50 percent, to Azerbaijan by 75 percent and to Kazakhstan by 60 percent. However, Georgia has a long way to go to replace the Russian market.

The embargo had another positive effect, and it was rather good news for Georgian wines; wine has been produced here since time immemorial but nobody was aware of that before the ban. The embargo also prompted the Georgian government to draft a wine law and adopt a developing strategy for the industry. Efforts to make Georgian wine better known are taking place in new markets. Also, an appellation of origin labelling system has been introduced to prevent counterfeiting, which had been a widespread problem. The unique selling point for Georgian wines is the country's rich winemaking history and country-specific white grape varieties such as Rkatsiteli, Tsitska, Tsolikauri and Mtsvane and red grapes such as Alexandrouli, Mujuretuli, Saperavi, Aladasturi and Ojaleshi. Benefiting from the climate, Georgia produces some 90 million litres of wine a year, which is extremely important for the republic. In the past few years, Georgia has liberalized and deregulated its economy, resulting in annual growth rates of more than 10% and a spike in foreign direct investment to above $1.7 billion in 2007.

The USAID supported AgVANTAGE project, in partnership with the Ministry of Agriculture, selected the popular food and wine show, the 53rd Annual Fancy Food Show in New York in 2007 to introduce Georgian food, wine and mineral waters to American consumers. The show was heavily attended, attracting over 25,000 decision-makers from supermarkets, specialty food merchants, importers, and wine and food distributors, as well as chefs and food service marketers. To more deeply penetrate the US market will go a long way to offset the impact of the Russian embargo on Georgian wine and food products, as well as introducing a new positive message about Georgia to US consumers.

The most prominent feature of the world food and feed markets has been the surge in the export prices of cereals, in particular wheat and maize, which have reached levels not seen for a decade. Poor harvests in key producing countries and a fast growing demand for biofuel production have been the main drivers of these markets, and most of the increase reflect an expansion in demand for the production of maize-based ethanol; the US and a number of other countries are in the process of establishing grain based ethanol facilities. Between 2000 and 2005, the maize-based ethanol production in the US increased by 150 percent. There are over 100 ethanol plants operating in 20 states across the country and 42 additional plants are under construction while another seven are being expanded. This development starts to have important implications for the overall supply of maize for other uses, but also for exports, considering that the US is the world's largest maize exporter. At the same time, the ocean freight rates for wheat, based on current chartering practices for vessels ready to load three to four weeks ahead, from US Gulf ports to the Black Sea ports of Georgia amounted to 30 - 42 US$/ton in the period from 1998 – 2007 for vessels of 20 000-40 000 tons, excluding vessels sailing under the flag of the United States (In comparison, the freight rates to Istanbul ranged at 16 US$/ton). For a long time, high freight rates represented a protection for the home market. However, in connection with the sharp increase of the prices for wheat on the world market, this protection appears to be insufficiently effective [FAO, 2007].

In Georgia, the practice that farmers act as independent entrepreneurs of agricultural products has only just begun. Small farms, created through land ownership reforms and privatization, are less effective from an economic viewpoint, especially if the parcelling of agricultural lands will be retained for a long time-period. At the same time, we have to remember that during the transitional period, when conditions of rapid economic recession in Georgia prevailed, farming played a decisive role in saving the largest part of the population from starvation. We have to consider the agricultural traditions of the Georgian people, as well as the mentality and attitudes of people in the rural areas. In Georgia, the land reform resulted in an agricultural sector dominated by small-scale subsistence farms. If we want to improve the performance of the agricultural sector in general, implementation measures which will assist in the commercialization of these farms are needed.

In Georgia it is practically impossible to speed up the development process through an increase in the amount of cultivated agricultural land. Increased productivity of currently farmed land is the only solution possible to guarantee a secure food supply. But it should also be noted that certain aspects of the intensification of agricultural production, such as increased application of fertilizers, pesticides, etc., creates other, even greater, problems of environment pollution. In addition, adverse changes in climate and flora take place due to global warming. The only course is to re-establish and maintain an ecological balance. Georgia must develop new agricultural approaches. The adoption of modern technologies is essential, but it is evident that the adoption of such technologies is only feasible on large farms, and practical application is only possible if modern information technologies and comprehensive technical facilities are used [Didebulidze, 2003].

Table 5: Main agricultural machinery and equipment, owned by agricultural holdings (1000 units)

Machinery/Equipment	1988	1994	1998	2006	Demand
Tractors	27.2	18.2	10.7	14.9	18.0
Combine harvesters	1.8	1.2	0.9	0.2	1.2
Loading machine	26.3	12.6	8.8	20.2	
Pump stations	1.4	0.4	0.3	0.3	0.5

Source: Department of Statistics (2007)

Private land ownership exists but, until recently, the transfer of titles to agricultural land was problematic, partly because of land fragmentation. Georgian agriculture is dominated by small family farms, cultivating 0.75 hectare on average and usually split into 2–4 plots. A related problem is that banks have been reluctant to lend to small farmers because small fragmented parcels of land are considered inadequate collateral. The weak environment for secured transactions raises credit exposure risks. As a result, the small farm and small agribusiness sectors remain underserved by the banking system and bank lending tends to be concentrated on trade and construction. The microfinance sector is served by a number of credit unions, non-government microfinance finance operators, and banks. Lack of credit information inhibits outreach and improved service quality for low-income clients. Recently, efforts were made to facilitate an exchange of credit information. As regards other financial services, the leasing market will need to develop a way to meet the demand for machinery and equipment (table 5). The insurance market is small (0.3% of GDP) but growing. The pension funds system is also underdeveloped.

In spite of price disparities, agricultural equipment and energy resources still remain the major preconditions for a successful and efficient agrarian sector (table 5). The development of the engineering and technical spheres requires extensive implementation of leasing activities. As for the energy policy strategy, it should be oriented to achieving a fuel-energy balance in the field and to a rationalization of the fuel and energy supply system. The power supply to Georgia's agricultural sector will not be able to reach the same level as the one in developed countries with similar natural conditions. Furthermore, the yield amount is not only determined by how much equipment is available, but as well by how it is used. According to these criteria, the lag in Georgia is even more apparent. A competitive agrifood sector is inconceivable without an effective machinery and equipment park implementing modern production and processing technologies for agricultural products. Mechanical engineering plants in the country should not only produce the necessary machinery and technical facilities, taking into account the export potential, but also provide a technical service package.

The rational use of energy resources in the agriculture of Georgia is not less important. It is the processing of quantitative evaluation, management and optimization methods of energy flows in the ecosystem that requires carrying out of fundamental energy analyses. The idea is to allot a share of two different types of energy – natural and artificial. Artificial energy can be divided into biological and industrial energies. Biological energy includes animal power and organic substances. Industrial energy comes from power plants, fuel, thermal, and other sources of energy. Industrial energy expenditures are divided into direct and indirect expenditures. Energy related to the activities in the agrifood sector belongs to direct expenditures, while indirect expenditures include those related to the production of equipment, mineral fertilizers, pesticides, and construction material. Clearly, improving the technical level of agriculture will be followed by an increase in industrial energy expenditures. Therefore, it is significant to determine specific values for energy used in production and to plan measures for its reduction [Didebulidze, 2003].

Agriculture is of extreme importance to the economy of Georgia. Therefore, improvement in this sector has a direct effect on the lives of the majority of common Georgian people. In 2006, agriculture, together with hunting, forestry and fishing, accounted for 11.3% of total the Gross Domestic Product (GDP) (in 1995 it was 41.7%, table 6) and 55.3% of total employment; also, the overall income of the rural population is lower than the income of the urban population (but it is necessary to consider that self-employment in agriculture is not fully included in the GDP measurement). Due to the activation of other sectors in Georgia's national economy, the share of agriculture in GDP has been reduced which implies a development of the agricultural sector. But as follows from the table[1], over 50 percent of the population have been constantly employed on the agricultural sector over the last five years; however, the sector produces less than one fifth (between 11.3 to 19.2 percent) of total GDP. This is also testified by the incomes of the agricultural population which, in comparison with the urban population, are much lower. Therefore, countrymen search for additional incomes in the cities, e.g. in construction, in trade and as transport drivers. Most rural households are trapped at the minimum subsistence level, eking out a meagre livelihood but unable to generate a surplus to invest in rebuilding their assets.

Table 6: Growth Rate of Real GDP (1996=100)

Year	Real Gross Domestic Product (GDP)	Percentage of Agriculture, Hunting, Forestry, Fishing in GDP	Agriculture, Hunting, Forestry and Fishing	Output of Agriculture in current prices, billion GEL	Employment in Agriculture as percent of total employed
1997	107,1	28,2	103,9	1,81	58,9
1998	111,7	26,2	97	1,76	48,4
1999	114,2	24,7	103,7	1,93	52,1
2000	115,8	20,6	91,3	1,63	52,1
2001	120,7	21	98,8	1,9	52,7
2002	127,3	19,2	97,3	2,02	51
2003	142,2	19,3	107,4	2,26	54,9
2004	150,2	16,4	99	2,18	54
2005	165,8	12	110,8	2,33	54,3
2006	181,7	11,3	97,9	2,09	55,3
2007	202,8	11	106,5		

Source: Department of Statistics (2007)

While agriculture has provided a lifeline to many who would otherwise have gone hungry, it is also a difficult and often unrewarding activity. Self-employment in agriculture pertains to people who predominantly work their own land plots and who, as

[1] Most persons engaged in agriculture work only 2-5 hours a day.

a rule, receive only a small monetary income. For many, agricultural activities in Georgia traditionally had an auxiliary character, because most of the urban population owned small land plots that provided a non-monetary income. Agriculture's overwhelming contribution to the number of the self-employed merely reflects the crisis in other sectors of the economy. Finally, the absence of monetary income for a large group of self-employed people contributes to the perpetuation of non-monetary economic relations in rural areas, areas, in which a significant volume of goods and services are paid for with agricultural products. This feature prevents the capitalization of individuals' businesses and limits the range of goods and services available to farmers. The importance of agriculture as a relief valve is also observed in the differences in unemployment levels between Tbilisi and the rural regions: unemployment levels are significantly higher in Tbilisi than in rural regions.

Figure 1: Real GDP (1989=100)

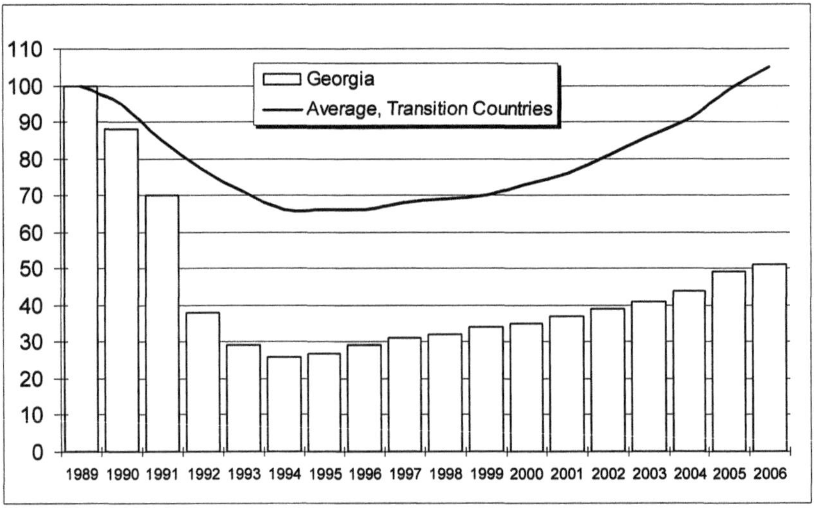

Source: EBRD (2007)

Georgia's agriculture is at such a difficult stage of transformation, because two important problems are in process of being solved. One of them is related to the final establishment of market-led relations and the other to overcoming the crisis within. Results of a more than fifteen-year-long practice of transformation have obtained different meanings. Of course, some achievements were made, namely in the sphere of Georgia's transition to market-led relations and principles, but on the other side we face unsolved problems which obstruct measures taken to lift the country out of its economic crisis. In order to realize its full potential, the agricultural sector should produce that which is most profitable, but at present Georgia is confronted with a lack of information and an uncertain political climate, both of which make it difficult for a farmer to assess profitability in the long term. However, farmers are more likely to adopt risk-averse production strategies such as low input/low output or a continuation of existing production patterns, thereby avoiding the need for significant investment.

The recent reforms did not have the desired results. About 1/3 of arable land is not used or is used for uncontrolled pasturing. Due to a poor economic policy, the level of realization of the production potential in the agrarian sector is rather low. As a result, the country has turned from a net-exporter into a net-importer of agrarian production; in addition, import exceeds export approximately 3 times. A two stage reform (stage I – January 1992 - December 1998, table 7; stage II – since July 2005, proceeding until now) was not successful. One of the main problems is the small square acreage of lands (62 % peasant farms own lands of less than 1 ha). The regime of wide reproduction is prevented, causing an intensive migration of the rural population to urban regions. An increased rate for irrigation, pasture taxation, etc. has even worsened the situation. The adoption of the laws "Concerning the Ownership of Lands of Agricultural Use" (March 22, 1996) and "On State Agricultural Land Privatization"(July 8, 2005) was not effective enough. This resulted in the decrease of agricultural lands (in particular owing to the growth of town areas), a change in the structure of land tenure and, therefore, a decrease of the main agricultural production (e.g. the areas of vineyards decreased from 112.6 thousand ha in 1990 to 63.5 thousand ha in 2006, grape production decreased from 691 thousand t in 1990 to 230 thousand t in 2006). Additionally, only about 90 thousand hectares have been privatized from 2005 until today; of those, 73 thousand hectares were rented before and only 16 thousand were sold at auctions. It is necessary to create a uniform agrarian system, to restore links between agrarian sciences and agricultural production, and, with the different consequences of the Russian economic embargo, to search for new commodity markets for the traditional agrarian production of Georgia (wine, mineral waters, fruit, greens and vegetables, beekeeping production, tea). Thus, the rational use of land resources will form a strong economic basis for the main production of Georgia as an agrarian-industrial country.

In spite of the fact that a number of government programs and projects, directed at the stimulation of agriculture and new technologies including application and new economic methods to improve the living standard of the population, have already been carried out, many problems remain unresolved; the majority of the programs is simple enough and not complex, they are easy to operate and within the limits of such programs separate private results are easily achieved. However, they do not solve the problems of the population as a whole; therefore it is necessary to urgently develop and carry out the complex programs, to steady and balance the social, economic and ecological development of rural regions, in particular taking into account that rules of steady land tenure in Georgia are regularly violated and the necessity of an improvement of land tenure and its institutional basis is obvious. Guaranteed rights to land and a legal basis supporting them are the most important factors for the achievement of a constant and sustainable development. It is also essential whether these legally grounded rights are absolute or limited by certain conditions. Only based on the knowledge of conditions, requirements, tendencies and land tenure purposes, it is possible to develop schemes on management and the development of Georgia's land resources.

Table 7: Stage I of the Privatization in Agriculture of Georgia (1998)

Type of Land	Total area	Including				State property
		Private		Rent		
	1000 ha	thsd. ha	%	1000 ha	%	percent
Arable land	785.0	431.9	55.0	255.9	32.6	12.4
Perennial plants	277.5	185.7	66.9	31.0	11.2	21.9
Meadows	140.6	47.6	33.9	28.6	20.3	45.8
Pastures	1788.0	124.5	7.0	441.4	24.7	68.4
Total	2991.1	789.7	26.4	756.9	25.3	48.3

Source: Department for Statistics (2007)

A wide range of structural measures was taken to create an enabling environment for private sector development, including, inter alias, the elimination of subsidies; liberalization of prices and trade; reduction of budgetary sector employment; a strong start on medium and large scale privatization coupled with the completion of small-scale privatization; establishment of the basic legal framework; land reform, as well as banking and energy sector reforms. In 2000, Georgia became a member of the WTO with the status of a developed country (in transition); this kind of status renders Georgia more operative than it would be if entered as a developing one. The new Customs Code, introduced in July 2006, reduced the number of tariffs to three, with a maximum tariff of 12%. Nearly 90% of imported goods are exempted from tariffs. The government has announced plans to further reduce and eliminate tariffs altogether by 2008.

Georgia's agriculture is not subsidized. Private investment in the agricultural sector is hampered by the long time required for returns. The funds designed for investment in Georgia will therefore more likely go to other sectors. In addition, previous investment in crops now without an export market such as wine and tea, have left assets which are destroyed reluctantly as there is still uncertainty over the future potential of such assets.

The Georgian agricultural sector is not saved yet. A well developed agricultural sector asks for many components to be in place, such as availability of good quality agricultural machinery and inputs, access to agricultural credits, marketing structures and modern processing facilities, but also the availability of agricultural research and applied technology, a market for agricultural land which then can be used as collateral for agricultural credits.

4 CONCLUSION

Thus, there are certain preconditions for agriculture development in Georgia. At the same time there is still a set of problems, in particular the low level of production intensification, unsatisfactory conditions in the organization of the seed growing and breeding business; there are difficulties connected to the realization of production, the volumes of crediting are unsatisfactory, the insurance question is not decided. An improvement of the situation in agriculture, the increase of incomes in the food industry, reduction of rural poverty and increase of food security based on sustainable development demand the realization of the following measures:

- Continuation of the land reform and development of the land market are of paramount importance for the sector, a new stage in the privatization of agricultural lands is under way including the plots which remain state property, and a smoothly functioning land market is a necessary precondition for the effective use of the land resources. Land reform allows farmlands to be sold in order to ensure that agricultural lands are transferred from less efficient to more efficient owners. The development of the land market is designed to stimulate foreign investment inflow into the agricultural sector;
- maintenance of agricultural biodiversity, protection of the rights of consumers and producers by ensuring that products are genuine and of good quality;
- rehabilitation and development of infrastructure and necessary structures with the purpose of granting favorable conditions for the activity of entities engaged in agriculture, the processing industry and the realization of agricultural production and processed products; improvement of access to new markets;
- technological modernization of the agricultural products processing sector, introduction of a leasing system with the purpose of updating agriculture by technical means;

- improvement of the access to financing and optimization of a stable and transparent tax system, ensuring of medium-term and long-term credits granted on accessible conditions, subsidized interest rates for young farmers, state supported disaster insurance program;
- organization of cooperation in the agricultural sector and assistance in its activities, promotion of cooperatives and farmers associations, collaboration with civil society, making the farming systems less vulnerable to climate change;
- creation of an agricultural extension system to upgrade farmer's technical and management skills, which is free of charge for small farmers, utilization of the intellectual potential of research and higher educational institutions;
- expansion of rural tourism, creation of opportunities in the rural non-farm economy that the rural poor can seize;
- ensuring of transparent governance, management of the connections among agriculture, natural resource conservation, and the environment.

5 References

Asian Development Bank (ADB) (a): Country Economic Report: Georgia. June 2007. - 57 pages.

Asian Development Bank (ADB) (b): Key Indicators of Developing Asian and Pacific Countries. Georgia., June 2007. pp. 218 – 222, http://adb.org/Docum

Department of Statistics: Agriculture in Georgia. Statistical Collection 2006., Tbilisi, 2007. – 114 pages.

Didebulidze, A. and Plachter, H.: Nature Conservation Aspects of Pastoral Farming in Georgia. In: "Pasture-landscapes and Nature Conservation", Springer-Verlag, Berlin-Heidelberg-New York, 2002, pp. 87-105.

Didebulidze, A: Sustainable Agriculture in Georgia. The South Caucasus Regional Land Policy Conference, Tbilisi, 24-26 February 2003. – 20 pages (CD-variant), http://www.aplr.org

European Bank for Reconstruction and Development (EBRD): Country factsheet, Georgia. 2007. www.ebrd.com

Felix-Henningsen, P., Urushadze, T., Narimanidze, E., Wichmann, L.-C., Steffens, D., Kalandadze, B.: Heavy Metal Pollution of Soils and Food Crops due to Mining Wastes in the Mashavera River Valley. Bulletin of the Georgian National Academy of Sciences, vol. 175, # 3, 2007. – pp. 97-106.

Food Agricultural Organization (FAO): Food Outlook. Global Market Analysis., #2, November 2007. – 91 pages.

Government of Georgia: Law "On State Budget of Georgia of 2008", Tbilisi, 28 December 2007, # 5659-rs

Gvaramia, A.: Change of Agricultural Land Parcel Category Simplified. APLR, Tbilisi, 16.01.2008. http://aplr.org/

Ministry of Economic Development of Georgia: Department of Statistics. Quarterly Bulletin. Tbilisi, 2008. – 89 pages.

Parliament of Georgia: Georgia without Poverty: Program of the government of Georgia 2008-2012 (approved in January 2008). http://www.government.gov.ge/eng /mtavroba 1172571820.php

United Nations (UN): Food Aid Statistics for Georgia, # 2, World Food Programme, 25.08.2007

United Nations Country Team (UNCT) in Georgia: Millennium Development Goals in Georgia., Tbilisi, 2004. – 58 pages.

Urushadze, T. and Urushadze, A: The problems of Land Degradation and Desertification in South Caucasus. The JRC Enlargement Action, Land Degradation, Italy, Ispra, 2003 – 5 pages.

UPTAKE OF HEAVY METALS BY FOOD CROPS FROM HIGHLY POLLUTED KASTANOZEMS IN AN IRRIGATION DISTRICT SOUTH OF TBILISI, EASTERN GEORGIA

PETER FELIX-HENNINGSEN, DIEDRICH STEFFENS, TENGIZ URUSHADZE, ELISO KING-NARIMANIDZE & BESSO KALANDADZE

1 Introduction

Because agriculture is the main economic sector of Georgia, long-term conservation of the high yield potential of the fertile soils under irrigation in SE Georgia is of common interest.

Decades of mining copper and precious metals in the mountain ranges of SE Georgia have caused severe environmental problems. Near Kazreti, at the middle reaches of the Mashavera valley, mining wastes were deposited on the mountain slopes around the opencast mine. Furthermore, deposits of waste from a flotation plant, consisting of fine ground rock debris, cover an area of about 240 ha. Mining and flotation wastes are rich in fines, which contain remains of sulphides. Due to an extremely acid environment caused by sulfuric acid, caused by oxidation of sulphides near recent surface deposits and the release of heavy metals (HM), the stockpiles are unvegetated. Sufficient measures to stabilize the slopes by planting vegetation were never taken. Therefore, runoff from the slopes leads to rill and gully erosion and severe contamination of the Kazretula river and other small creeks, which meet the Mashavera river, by fines with adsorbed heavy metals and remnants of sulphides. Presently, the Mashavera is one of the most polluted rivers of Georgia.

The use of the Mashavera waters for irrigation led to the pollution of the fertile Chernozem and Kastanozem soils by suspended fines, rich in heavy metals. The polluted soils, on the other hand, are also a potential source for the contamination of the food chain by heavy metals. Although the severe soil pollution of the Mashavera valley was known to Georgian scientists and official authorities, valid data about the extent and ecological importance of the soil pollution were not surveyed or published. In addition, basic information and analytical data about the soils in the Mashavera valley, which could be used to estimate the potential of adsorption for or mobilization of heavy metals in the root zone, were not available.

In the frame of a research project, generously funded by the German Volkswagen Foundation, the amounts and spatial distribution of heavy metals as a consequence of the deposition of mining wastes on irrigated soils of the Mashavera valley were investigated (Narimanidze et al. 2005, Felix-Henningsen et al. 2007 a, b). The soils, rich in clay, organic matter and calcium carbonate, display a high adsorption capacity for heavy metals, proven by very low amounts in NH_4NO_3 soil extracts. Therefore, the uptake of heavy metals by crops was assumed to be rather low, despite the high degree of soil pollution. In this paper we present and discuss data which indicate that a high risk for the contamination of the food chain by heavy metals nevertheless exists.

2 Study Area

The study area is situated in SE Georgia, about 60 km SW of the capital Tbilisi, in the administrative district Bolnisi, and contains the middle and lower reaches of the Mashavera valley (Figure 1). South of Bolnisi, the Mashavera valley is bordered by the volcanic high plateau of Javakheti, which passes into the Armenian highlands in the south and is the source area of the Mashavera river (Javakhishvili & Gvelesiani, 1962). The ore deposits of the Bolnisi region show a close relationship to volcano-sedimentary rocks of the Upper Cretaceous (Upper Turonian until Lower Santonian), locally penetrated by intrusive rocks. Mining of the copper sulphide deposits started in 1975. The "Madneuli" mining company produces high-grade copper concentrates by flotation, which are exported to different countries (Burnod-Requia, 2003). The open mine pit is situated 10 km SW of Bolnisi village and covers an area of about 350 ha in the summit area of a mountain range between the Mashavera valley in the west and the Poladauri valley in the east. Therefore, surface and subsurface waters from the mine and the mine spoils, which cover the slopes adjacent to the pit, form acid mine drainage (AMD) with pH values < 3 and high dissolved HM concentrations. AMD is collected in sedimentation basins by a channel and pump systems, but a lack of financial and technical possibilities in the past prevented further processing of the AMD by neutralization with lime and extraction of the dissolved copper by binding agents. Therefore, considerable amounts of untreated AMD were dumped into the adjacent rivers in most years. Especially after strong rains and power blackouts, the pump system of the mine is not able to take up the drainage water, which leads to uncontrolled runoff.

In addition, highly contaminated waste water from the Madneuli floatation plant contributes to the pollution of the river Mashavera (MSPM, 2000). Violations against existing security regulations caused a temporary closure of the mining operation in the past.

Fine grained floatation wastes are deposited on two heaps situated about 2-3.5 km NW of the open mine pit, covering a total area of 240 ha. The sulphidic rock material is subject to intensive biochemical oxidation, which causes a release of large amounts of AMD and dissolved HM. The steep slopes of the large heaps are bare and not fixed by vegetation, thus causing sheet and rill erosion and transport of fines to the Mashavera and Kazretula rivers, while AMD runoff directly contaminates the rivers with HM. In addition, the physical state of the deposits was not taken into account and resulted in slope failures (MSPM, 2000). During the dry summer months the heaps of flotation waste are the source of dust, which amounts to about 30 Mg a^{-1} (Burnod-Requia, 2003) and pollutes large areas of the surroundings with HM. Thus the non-irrigated arable soils and uncultivated soils of the mountain regions also display relatively high "natural" background concentrations of HM.

Figure 1: Location of the study area in the Mashavera valley in SE-Georgia

Source: Author's diagram

Georgia is situated at the northern boundary of the northern subtropics with a gradient of decreasing humidity and increasing continental character from west to east. The climate station of Bolnisi has a long-term average annual precipitation of 504 mm and a mean annual temperature of 12.4 °C, with 1.1 °C in January and 24.4 °C in July. The maximum precipitation occurs in the early spring time. With a potential evapotranspiration of 1,084 mm the bio-climatic index of aridity amounts to 0.47, mainly due to the severe aridity of the summer months. Therefore, the Mashavera river has a maximum discharge in May, while a minimum occurs from June until October.

The natural steppe vegetation formations (Nakhutsrishvili, 2000) of the Mashavera valley were changed by agriculture, which in this area has an 8,000 year old history. Especially potatoes were cultivated in large fields when Georgia was part of the Soviet Union (Didebulidze, 1997). After 1991 agriculture was decentralized and successively re-privatized with the consequence that the field sizes decreased. According to Libert (2003), the average cultivated area per family amounts to 0.9 ha. Vineyards and orchards with intercrops of different vegetables, as well as house gardens dominate the area of the lower terrace of the Mashavera valley in addition to fields of corn and winter wheat. The crops from gardens and fields are cultivated for subsistence and open markets in the villages of the area and along the main roads up to the capital Tbilisi 60 km away. Sufficient yields can only be obtained by irrigation during the arid summer months. While house gardens near the river are irrigated manually with river water, furrow and surface irrigation of fields, wine gardens and orchards occurs with water from a system of channels, supplied from a main channel along the lower slope of the mountain chain on the southern side of the Mashavera valley. During the Soviet period, pumps enabled furrow and sprinkler irrigation of fields above the irrigation channels.

This main channel branches off from the Mashavera river directly below the flotation plant and the large deposits of flotation waste at Kazreti. Due to the strong load of the Mashavera river water with fines, which derive from the waste deposits and the flotation plant, the irrigation channels also conduct water heavily contaminated by HM.

The fields of the northern side of the Mashavera valley are irrigated with water from a channel, which branches off from the Mashavera river upstream of Kazreti. Consequently, this water is free from contamination.

Due to the continental type of climate, the typical main soil orders belong to Kastanozems and Chernozems (Javakhishvili & Gvelesiani,1962; Purnell, 2000), which have frequently degraded to Phaeozems as a consequence of long periods of intensive irrigation. The main soils of the study area are Calcic Kastanozems, which show an accumulation of calcium carbonate in the form of concretionary mottles in the subsoil but lower humus content and a more brownish colour than a Chernozem. They are associated with Calci-Vertic Chernozems with clay contents of up to 65 mass-%. The medium to strong humic topsoils show a blocky to polyhedric structure due to the high clay content. They are weakly alkaline and have a rather high CEC.

Table 1 displays the ranges of analytical data of topsoils (n = 112), which were subject to the investigation and derived from different land use plots.

Table 1: Characteristics of topsoils (Ap horizons, n = 112) under different land use

Soil Characteristics		Land use		
		Arable Fields	House Gardens	Wine Gardens, Orchards
Organic Substance	(mass-%)	1.90 - 3.80	2.00 - 4.80	2.80 - 4.60
Calcium Carbonate [1]	(mass-%)	4.43	4.46	2.05
pH (CaCl2)		7.20 -7.50	7.10 - 7.40	7.10 - 7.40
Clay Content [2]	(mass-%)			39 - 62
CEC[2]	$cmol_c\ kg^{-1}$			31 - 53
Fe-Oxalate soluble	($g\ kg^{-1}$)	0.59 - 2.12	0.69 - 2.63	0.41 - 4.19
Fe-DCB	($g\ kg^{-1}$)	5.62 - 11.24	6.79 - 11.70	8.06 -12.43
Mn-Oxalat soluble	($g\ kg^{-1}$)	0.25 - 1.00	0.31 - 1.00	0.55 -1.00
Mn-DCB	($g\ kg^{-1}$)	0.45 - 0.82	0.39 -1.00	0.54 -1.02

1 Arithmetic Mean, 2 Samples n = 12;

Fe/Mn-DCB = dithionite-citrate-bicarbonate soluble fraction of free (pedogenic) amounts of crystalline + amorphous iron and manganese oxides. Fe/Mn-Oxalate soluble = fraction of amorphous iron and manganese oxides

The humus content decreases with the increasing intensity of soil cultivation in the sequence grape fields and wine gardens, orchards, house gardens, and arable fields (table 1). As a further reason for the high humus contents in topsoils of wine gardens, a lower mineralization rate can be assumed resulting from an accumulation of Cu due to strong irrigation (see below) as well as from the use of copper bearing fungicides (c.f. Parat et al., 2002). Arable fields, on the other hand, are subject to soil erosion during periods of rotation fallow, which diminishes the humus content in slope positions and leads to the formation of colluvium in depressions and on the flat valley floor. On slopes with severe erosion, the calcic horizon is exposed at the surface and causes the formation of Calcisols.

3 Methods

3.1 Soil investigations

Soil samples were taken in fields, house gardens, grape fields, wine gardens and orchards from the Ap horizon (0 – 30 cm) at 10 sites along double-diagonal transects. Eight volume-equivalent cores were taken with an aluminium auger within an area of 4 x 4 m at each site. The fine earth (< 2mm) of the air dried samples, ground in a porcelain mortar, was investigated in the laboratories of the Institute of Soil Science and Soil Conservation of Justus-Liebig University in Giessen, Germany.

The *pH* was determined after DIN 10390 in suspension with 0.01 M $CaCl_2$ with a pH-meter pH90 (WTW).

The amount of *carbonates* was determined by the gas-volumetric method using a calcimeter, following DIN 18129.

Total amounts of carbon (C_t) and nitrogen (N_t) were determined on fine ground samples by gas-chromatography using a C-N-S element analyzer (Heraeus). *Anorganic C* was calculated from the carbonate content by using the factor 0.1199, while the amounts of *organic carbon (C_{org})* resulted from the difference between C_t and anorganic carbon. The amounts of *organic matter* were calculated by C_{org} * 1.724.

Particle size distribution was determined by the combined sieving (and fractions) and pipette method (silt and clay) after decomposition of carbonates (HCl) and organic matter (H_2O_2) and dispersion in Na-Pyrophosphate, following DIN 19683.

Amorphous iron (Fe_{ox}) and manganese oxides (Mn_{ox}) were determined by extraction with buffered oxalic acid, pH 3.25, under dark conditions as described in Schlichting et al. (1995).

Pedogenic iron (Fe_{DCB}) and manganese (Mn_{DCB}) oxides were extracted following the procedure of Mehra & Jackson, as described in Schlichting et al. (1995).

The *mobile and exchangeable fractions of HM*, which are eco-toxicologically relevant because they are potentially plant available and easily leachable, were extracted with NH_4NO_3 according to Zeien & Brümmer (1989, 1991). They are designated in the text as HM_{AN}.

The *total amounts of subsequent deliverable HM*, which is the *supply fraction* and includes the soluble and exchangeable fractions as well as the HM strongly adsorbed to carbonates, oxides and organic substances, were extracted by EDTA. Deviating from the method described by Hornburg & Brümmer (1993), EDTA was dissolved in a buffered solution of ammonium acetate at pH 7. Elements of this fraction are designated in the text as HM_{EDTA}.

The *total amounts of HM* were extracted from finely ground samples by using Aqua Regia, following DIN ISO 11466. Elements of this fraction are designated in the text as HM_{AR}.

Element concentrations in the extracts were determined with the atomic adsorption spectrometer FAAS 4100 (Perkin Elmer). For determination of Cd in the NH_4NO_3 extracts a GFAAS SIMAA 6000 spectrometer (Perkin Elmer) was used due to the low concentrations.

3.2 FIELD AND POT EXPERIMENTS

The bioavailability of the accumulated heavy metals was analyzed in pot and field experiments. The pot experiments were carried out with 3 soils at the greenhouse experimental station in Giessen. The soil samples for the pot experiments were taken from 0.20 m depth at three experimental sites near Bolnisi village. Soil samples were air dried, ground and sieved trough a 4-mm sieve. Six small Mitscherlich pots were filled with 5 kg soil each. Before spinach (*Spinacia oleracea* L. cv. Calata) and summer wheat (Tr*iticum aestivum* L. cv. Thasos) were grown, the soil in the pots was treated with 15 g of a combined fertilizer (12% N, 5% P, 14% K as sulphate, 6% S, 2% Mg). After emergence, wheat and spinach seedlings were thinned to 20 plants per pot. Summer wheat was fertilized with 0.5 g N per pot as NH_4NO_3 at shooting stage. Spinach was grown for 41 days and wheat for 105 days.

The field experiment was carried out with growing spinach (*Spinacia oleracea* L. cv. Calata) at three irrigated sites with a different degree of HM pollution (soil data and contents of HM see table 5). Spinach was sown with 4 g seeds/m^2. At each experimental site spinach was cultivated in 8 plots (10 m^2). Before plant samples were dried at 105°C a part of the spinach shoots were washed with demineralized water in order to carry out a comparative analysis of washed and unwashed samples. The concentration of heavy metals in plants was determined by AAS after a dry ashing of the ground plant samples at 420°C and dissolution of the ash in HNO_3 (VDLUFA, 1976).

4 Concentrations and Mobility of Heavy Metals in Soils of Different Land Use

Results of the investigation of soil properties, the total amounts and mobile fractions of heavy metals in topsoils of un-irrigated and irrigated fields, meadows, house and wine gardens and orchards are presented and discussed in detail in Felix-Henningsen et al. (2007 a,b).

Screening investigations of the concentrations of diverse heavy metals in topsoils of the Mashavera valley showed that only Cu, Zn and Cd are problem elements, which accumulated in the course of irrigation and dust deposition, originating from the deposits of mining waste in the Kazreti mountainous area. Other heavy metal species such as Pb, Cr, V, Co, Ni and Hg occur in negligibly low concentrations.

The ecological importance of the soil pollution by HM has to be assessed by threshold values for contaminated soils, which are documented in bodies of laws of several European industrial countries. Up to now, such values do not exist for Georgia.

Therefore, the total amounts of HM in the aqua regia extract (HM_{AR}) and the dissolved and weakly adsorbed fraction in the NH_4NO_3 extract (HM_{AN}) will be compared with the threshold values of the Federal Soil Protection and Contaminated Sites Ordinance (BBodSchV, 1999) based on the Federal Soil Protection Act (BBodSchG, 1998). Another source of threshold values based on total amounts (AR) is the "Dutch List" (Rosenkranz et al., 1993).

Figure 2: Total amounts (box plots with medians, 25 and 75 percentiles, minimum and maximum) of heavy metals in topsoils of un-irrigated arable fields (RS = reference soils) in comparison to topsoils of irrigated sites of different land use.

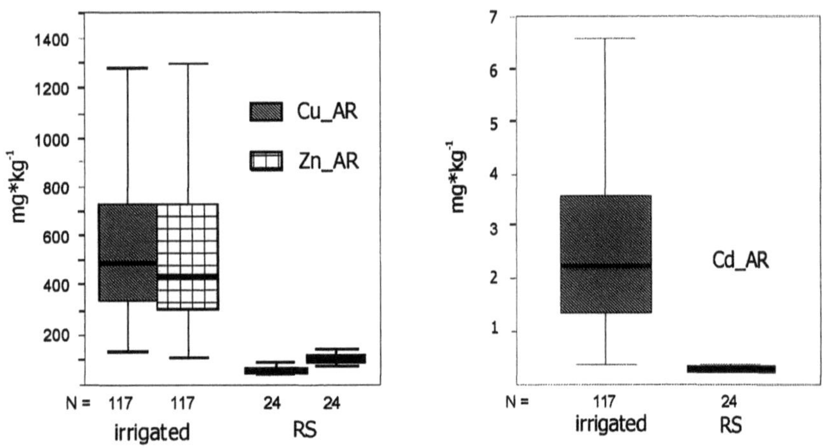

The concentrations of HM in the EDTA-extract (= HM_{EDTA}) indicate the available plus the potentially available supply fraction, more or less strongly adsorbed amounts of HM, bound on free Fe- and Mn-oxides, carbonates and organic substances (Zeien & Brümmer, 1989, 1991). They indicate the future risk potential of soil pollution (Smith et al., 1996) because the HM_{EDTA} can be mobilized when the soil conditions influencing the adsorption (e.g. pH, contents of carbonates and organic matter, redox-conditions etc.) change. Therefore this fraction was also included in this study.

The spatial distribution of the problem elements Cu, Zn and Cd shows a soil loading which depends on the kind of land use. A significant increase of the concentrations follows the sequence: Reference soils (un-irrigated) << arable fields < house gardens << orchards and vineyards.

Non-irrigated soils of arable fields used for growing cereals and maize are situated on slopes above the irrigation channels. The concentrations comply with the background values in the study area and are related with the parent material and eolian dust deposition from the mining waste deposits (Burnod-Requia, 2003). With respect to the precaution values (BBodSchG, 1998), the HM concentrations in the non-irrigated topsoils are in the uppermost range. For about 30 % of the investigated sites the soil concentration of Cu_{AR} already exceeds the precaution value of 60 mg*kg-1, while, with 0.31 mg*kg^{-1}, the maximum value of the mobile Cu_{AN} fraction lies clearly below the trigger value for the pathway soil–food crop. The Zn_{AR} and Cd_{AR} concentrations fall below all threshold values and therefore pose no concern at present for the non-irrigated soils.

Table 2: Proportion (%) of non-irrigated and irrigated sites of different land use, exceeding relevant threshold values of the German *Protection and Contaminated Sites Ordinance* (BBodSchV, 1999)

Land use (number of investigated sites)	Precaution value	Trigger value soil – food crop	Action value
arable land, non-irrigated (10)	Cu: 30%		
arable land, irrigated (42)	Cu: 100% Zn: 77 % Cd: 27 %		
permanent meadows (15)	Cu: 100 % Zn: 100 % Cd: 80 %		Cu: 60 %, in case of grazing of sheep
vegetable gardens (67)	Cu: 100 % Zn: 100 % Cd: 85 %	Cu: 50 % Zn: 15 %	Cd: 21 %, in case of non-accumulating crops; Cd: 40 %, in case of accumulating crops
orchards and grape fields (49)	Cu: 100 % Zn: 100 % Cd: 100 %	Cu: 85 % Zn: 45 %	Cd: 82 %, in case of mixed cropping with vegetables; Cd: 45 %, in case of wine grapes only

Most of the irrigated soils under different land use display a strong enrichment of HM that can be traced back to irrigation with polluted water from the Mashavera river over a period of several decades (Figure 2). The concentrations of Cu_{AR}, Zn_{AR} and Cd_{AR} are narrowly and significantly correlated (r = 0.94–0.96) due to the same source. The span of the HM concentrations in irrigated soils is by far wider than in the non-irrigated topsoils, especially towards higher concentrations above the 75th percentile. Differences in the duration, frequency and amounts of irrigation as well as changes in the type of land use and soil cultivation (e.g. depth of plowing) are the reasons for this high spatial variability. This shows that the degree of soil pollution by HM in the Mashavera valley cannot be estimated but must be investigated for each cultivated field, house garden and orchard in order to evaluate the hazard to the food chain and the population.

The medians of the EDTA fractions in topsoils of irrigated arable fields amount to 9 (Cu), 13 (Zn) and 6 (Cd) times greater than in the non-irrigated reference soils. About 33 % (Cu), 12 % (Zn) and 67 % (Cd) of the total amounts belong to this supply fraction, which indicates, especially in the case of Cd, a high potential risk of soil pollution with heavy metals.

As Cd is the most mobile heavy metal, an increasing uptake by plants already occurs under slightly acid conditions below pH 6.5. At present the arable soils also contain small amounts of calcium carbonate in the A-horizon, mostly in the range of 2–5 mass%. Because agriculture and irrigation favors decalcification and acidification, an increase of the Cd mobility must be taken into account in the future.

Permanent grassland of the alluvial floodplains along the Mashavera river, which becomes occasionally submerged by flood waters of the river, is polluted by settled fines with a high concentration of HM, distributed on the soil surface and the leaves of vegetation. Eighty percent of the samples from A-horizons of meadows exceed the precaution value for Cd_{AR}, and 100 % of the samples for Zn_{AR} und Cu_{AR}. The meadows are used for extensive grazing of cattle, sheep herds and horses, which also use the river water for drinking. Due to a sandy texture and partly low pH values, humus and carbonate contents, the alluvial soils display a higher proportion of HM in the mobile fraction as compared to irrigated arable soils.

Vegetable gardens are distributed on the lower terrace of the Mashavera valley, adjacent to villages and farms, where they are irrigated manually with water from the river. Other vegetable gardens adjoin the irrigation channels, where they concentrate around settlements. Compared to the arable land, vegetable gardens are irrigated over a longer period in the year, more intensively and with a greater amount of water brought by buckets and pumps or distributed from irrigation channels by furrow irrigation. Therefore, the concentrations of the total amounts of HM_{AR} are much higher than in topsoils of arable fields. The same trend is shown by the supply fraction (EDTA). Forty percent of Cu, 20 % of Zn and 79 % of Cd belong to the supply fraction. This indicates that with an increasing transfer of HM with the irrigation water, the supply fraction increased non-proportionally.

At nearly all sites relevant threshold values according to BBodSchV (1999) and the target values of the "Dutch List" (table 2) are exceeded. This means that especially due to the strong toxicity of Cd and a potentially high transfer into the food chain, further investigations of all irrigated vegetable gardens in the Mashavera valley are necessary. According to BBodSchG (1998), restrictions of land use would be necessary in order to protect the population.

An extreme HM load is distributed in topsoils of orchards, grape fields and wine gardens (table 2). Since the 19th century, cultivation of fruit trees and wine has a long tradition in the Mashavera valley and its tributaries. These permanent cultures cover most of the area of the Mashavera valley bottom and the lower slopes of the adjacent mountains. Traditional mixed cropping with vegetables between the grape rows requires a high intensity of irrigation and therefore causes a severe HM contamination of the soils. Background, precaution and target values of all three HM species are exceeded many times. This threatens both soil functions and growth conditions of the vegetation.

The evaluation of the Cd concentrations with respect to the relevant action values must take into account the mixed cropping. Wine grapes are not strong accumulators of heavy metals. Therefore, the higher action value of 0.1 mg Cd_{AN}*kg^{-1} (BBodSchV, 1999) is valid, which is exceeded by about 45 % of the investigated sites. On the other hand, for growing vegetables, which are partly strong accumulators of HM, the lower action value of 0.04 mg Cd_{AN}*kg^{-1} (BBodSchV, 1999) is applicable, which is exceeded by 82 % of the investigated sites.

5 Uptake of Heavy Metals by Food Crops

The high total amounts of HM in topsoils of irrigated soils used for agriculture, house gardens, orchards and grape fields as well as occasionally submerged meadows of the Mashavera floodplain, indicate a strong contamination due to the deposition of fines with sulphidic metal particles and dissolved HM in acid mine drainage waters.

The bulk of immobile, specifically adsorbed plus mobile or weakly adsorbed HM forms the supply fraction. The proportion of this fraction, soluble in EDTA, ranges from 20 % to 45 % of the total amounts (HM_{AR}). Highly significant correlations exist between amounts of the supply fraction and the amounts of adsorbents in the topsoils. The important organic substances for the adsorption of the supply fraction increase in the sequence Cd ($r = 0.43$) < Zn ($r = 0.56$) < Cu ($r = 0.58$). The highly significant correlations between the amounts of Fe and Al oxides and the supply fraction increase in the sequence Cu ($r = 0.80$ Fe_{ox}, 0.85 Al_{ox}) < Zn ($r = 0.58$ Fe_{ox}, 0.64 Al_{ox}) < Cd (no significance Fe_{ox}, 0.62 A_{ox}). This shows that the adsorption of Cu and Zn on soil particles is stronger than that of Cd, which therefore has the highest mobility and risk potential to the food chain, also indicated by the lowest threshold values displayed in the BBodSchV (1999) and the "Dutch List".

Herms and Brümmer (1984) referred to the great environmental importance of the supply fraction, due to significant correlations between supply (HM_{EDTA}) and mobile (HM_{AN}) fractions. The latter are relevant for the amount of HM uptake in food plants (Brümmer et al., 1986; Zeien & Brümmer, 1989, 1991, Hornburg & Brümmer, 1993; Allen et al., 1995). In this study, the highly significant correlations (r) between the supply fractions (*EDTA*) and the mobile fractions (NH_4NO_3-extractable) increase in the sequence Cu 0.75 < Zn 0.82 < Cd 0.86.

The supply fraction of HM in topsoils represents a high future risk potential, which is presently still increasing due to continuous irrigation with polluted water from the Mashavera and Kazretula rivers. In case of a slow change of soil properties due to land use, which influences conditions for specific HM adsorption, parts of the supply fraction will become more available. Decomposition of organic matter, decalcification, increasing acidity or intermittent reducing conditions (Charlatchka and Cambier, 2000), due to excess of irrigation water or logging of surface water, would be the main factors.

Due to the recent high adsorption capacity of the soils for HM (Tiller et al., 1984 a, b), only small amounts of HM were found in the mobile fraction with proportions of less than 1 % of the total amounts for Cu and Zn and a maximum of 1.5 % for Cd. Nevertheless, trigger values (for Cu and Zn) and action values (for Cd), which are based on the concentrations of the NH_4NO_3 soluble fraction (BBodSchV, 1999), are exceeded in most irrigated house gardens, orchards and grape fields, especially in those, which are used for growing vegetables in mixed cropping systems. As measured by the threshold values of the German Soil Protection Law, land use restrictions and remediation measures would apply to 40 % of the investigated house gardens and to about 80 % of the grape fields, wine gardens and orchards with mixed cropping of vegetables.

In order to indicate the actual risk of an HM transfer to food crops, herbs and vegetables from single house gardens were investigated.

First results from a (non-systematical) screening investigation of HM concentrations in herbs and vegetables of irrigated house gardens and grass from an occasionally submerged meadow of the Mashavera floodplain (table 3) show that in many cases the Cu and Zn concentrations exceed the tolerance threshold for plants. Therefore, garden owners in the Mashavera valley claim low yields and a bad quality of the crops, often visible in the form of growth depressions and chloroses or necroses of the leaves. In many cases, the Cd concentrations of the herbs and vegetables exceed the tolerance threshold for animals and human beings and therefore endanger the health of the population. One important source for the HM uptake into the food chain is the adhesion of fines to leaves and stems of herbs, which have direct contact to the irrigation water. Therefore, washing of the crops in clean water diminishes the HM concentrations (table 3). The opposite can be observed in the Mashavera valley: Farmers often wash fruits and vegetables directly after the harvest directly in the Mashavera river bed or with water from the river, before going to market. This behavior causes another high risk of direct contamination of market fruits and crops.

Table 3: Screening of Cd-, Cu- and Zn-contents in mg* kg^{-1} dry matter (dm) of herbs and vegetables from irrigated house gardens and of grass from an occasionally submerged meadow of the alluvial plain of the river Mashavera. Tolerance thresholds for vegetation, animals and human beings according to Sauerbeck (1985).

Plant species	Organ	Cu mg* kg^{-1} dm	Zn mg* kg^{-1} dm	Cd mg*kg^{-1} dm
Pepperoni	leaf	50.1	268.4	4.1
Pepperoni	stem	16.3	80.1	1.46
Pepperoni	fruit	20.3	52.6	1.29
Tomato	fruit	6.5	23.9	0.4
White cabbage	external leaf internal leaf	9.2 19.9	68.7 47.9	1.19 0.32
Red beet	stem leaf	15.7 4.0	52.2 6.8	0.33 0.16
Grass. washed Grass. without washing	leaf	20.3 23.1	96.4 93.5	0.73 0.72
Basil. washed Basil. without washing	leaf leaf	5.7 56.7	17.2 78.5	0.10 0.72
Parsley. washed Parsley. without washing	leaf leaf	4.3 6.7	14.6 96.4	0.09 0.13
Spinach. washed Spinach. without washing	leaf leaf	19.5 37.2	237.7 253.2	2.65 2.93
Tolerance threshold for plants		15-20	150-200	5-10
Tolerance threshold for animals and human beings		30-100	500	0.5-1.0

Furthermore, pot and field experiments with summer wheat (*Triticum aestivum*) as a non-accumulating crop and spinach (*Spinacia oleracea* L. cv. Calata) as an accumulating crop were carried out on different polluted soils from the Mashavera valley. While the pot experiment was carried out with soil material from the Mashavera valley under controlled conditions in a green house, the field experiment was influenced, during the growing season, by the natural environmental factors of the study area.

Table 4: Physical and chemical characteristics and contents of total (HM_{AR}) and soluble amounts (HM_{AN}) of heavy metals in topsoils of irrigated Kastanozems of the Mashavera valley used for pot and field experiments

Soil P = pot F = field	pH (0.01M CaCl2)	Clay mass-%	CaCO3 mass-%	OM mass-%	Cu_{AR} mg*kg-1	Cu_{AN} mg*kg-1	Zn_{AR} mg*kg-1	Zn_{AN} mg*kg-1	Cd_{AR} mg*kg-1	Cd_{AN} mg*kg-1
1 P	7.57	49.18	1.29	2.76	527	1.23	352	0.23	1.63	0.01
2 PF	7.39	50.18	0.21	3.24	761	1.43	658	0.66	4.00	0.04
3 PF	7.52	42.84	0.44	3.32	242	0.56	269	0.10	0.94	0.05
4 F	6.67	54.39	0	3.48	933	1.99	782	1.40	3.58	0.12

The topsoils of the experimental fields are rich in clay and organic matter and the low contents of calcium carbonate stabilize the pH in the weak alkaline range (table 4). Only site 4 displays weakly acid conditions due to the absence of carbonates. At all sites the concentrations of HM are exceeding the precaution values (BBodSchG, 1999) by far. Trigger values for the pathway soil – food crop are exceeded by Cu at sites 1P and 2 PF, while action values for Cd accumulating vegetables (Cd > 0,04 $mg*kg^{-1}$) are exceeded at sites 2 PF, 3PF and 4 F, and for non-accumulating crops (Cd > 0,1 $mg*kg^{-1}$) at site 4F.

Table 5: Heavy metal concentrations ($mg*kg^{-1}$ dry matter) in grains and straw of summer wheat, grown in a pot experiment. Tolerance thresholds for plants and animals see table 3, soil data see table 4.

Soil	Cu		Zn		Cd	
	Grain	Straw	Grain	Straw	Grain	Straw
1 P	12.2	11.2	60.0	69.5	0.26	0.41
2 PF	12.3	11.4	65.3	119.3	0.29	0.66
3 PF	9.5	7.0	50.9	42.7	0.13	0.26

Because summer wheat belongs to the type of vegetation with a reduced uptake or which excludes HM, tolerance thresholds (Sauerbeck, 1985) were not exceeded, except for the Cd accumulation in the straw of site 2 PF. But at all sites the Cd concentrations in the wheat grains exceed the "guideline value for Cd concentrations in bread wheat" of 0.1 mg*kg^{-1} according the German Food Ordinance (ZEBS, 1990) which means that wheat with such Cd concentrations should not be used for human nutrition. The threshold value according to the German Animal Feed Ordinance (FuttermittelVO, 1988) of 1 mg Cd*kg^{-1} for straw which is used as fodder for animals is not exceeded.

Table 6: Heavy metal concentrations (mg*kg^{-1} dry matter) in the green mass of spinach (*Spinacia oleracea* L. cv. Calata) grown in a pot experiment (P) and a field experiment (F). Spinach from the field experiment was investigated unwashed (F uw) and washed (F w). Tolerance thresholds for plants and animals see table 3, soil data see table 4.

Soil	Cu			Zn			Cd		
	F uw	F w	P	F uw	F w	P	F uw	F w	P
1 P			27.2			182.4			2.99
2 PF	22.0	10.3	33.9	206.7	182.3	209.3	3.04	2.21	4.20
3 PF	13.0	5.8	5.1	123.3	124.1	87.7	1.23	1.29	1.25
4 F	37.2	19.5		253.2	237.7		2.93	2.65	

The results of the field and pot experiments with spinach show that this plant species accumulates heavy metals, especially Zn and Cd. At site 1 P, with a Cd concentration in the soil below the action value of 0.04 mg*kg^{-1} (table 5), the transfer factors (Cd in plant / Cd_{AR} soil) are clearly above 1.

The tolerance thresholds for plants as well as for humans and animals are partly reached or exceeded, especially by Cd. In the field experiment, the spinach of field 2 and even stronger of field 4 showed severe growth depressions, possibly due to the high Cu and Zn concentrations.

Washing the spinach leaves of the field experiment leads to a marked reduction of the Cu concentrations and reduces the Zn and Cd contents to a smaller extent. Obviously, washing removes HM, which were adsorbed at the plant surface as a consequence of the direct contamination with polluted irrigation water. The high affinity of Cu to organic substances as well as the highest concentrations of Cu in the irrigation water is responsible for the fact that washing has the greatest effect on the reduction of Cu.

The threshold for leaf-vegetables of 0.2 mg Cd *kg^{-1} in fresh plant material according the EU Food Ordinance (Verordnung EG Nr. 466/2001 vom 8. März 2001) is exceeded at most sites if a dry matter content of 10 % is assumed.

6 Discussion

The majority of the irrigated soils show high total amounts of Cu, Zn and Cd and high concentrations in the supply fraction, which indicates the severe long term risk of heavy metal pollution for the food chain. In spite of the high HM_{AR} and HM_{EDTA} fractions the proportion of the mobile and weakly adsorbed HM_{AN} fraction, soluble in NH_4NO_3, is rather low and only reaches < 1% (Cu, Zn) – 1.5 % (Cd) of the total amounts. The reason for this rather low mobility is the high sorption capacity of the soils, due to high contents of clay and organic matter, and the pH values > 7 (c.f. Blume and Brümmer, 1987). Nevertheless, the concentrations of the mobile Cd fraction exceed the precaution trigger and action values of the BBodSchV (1999) and the target and action values of the "Dutch List" in topsoils of many sites. According to the German Soil Protection Law, land use restrictions and remediation measures would apply to 30 % of the investigated house gardens and to more than 50 % of the grape fields, wine gardens and orchards with mixed cropping of vegetables. Therefore, the actual risk of Cd transfer into the food chain, which affects the local population as well as people in cities up to the capital Tbilisi, where crops from the Mashavera valley are sold at the open markets, is proven.

The first results of the screening investigation of crops from house gardens as well as field and pot experiments with wheat and spinach indicate a high uptake of Cu, Zn and Cd in cereals and vegetable leafs that exceed tolerance thresholds for plants, animals and human beings and threshold values for human food and animal feed. This indicates that in spite of a high sorption capacity of the soils a plant availability of HM still exists. Although, specifically, Cu is mainly adsorbed on organic matter and therefore the least mobile element in neutral to weakly alkaline soils (indicated by the results of the NH_4NO_3 extractions), the Cu contents in vegetation indicate a rather strong uptake. Correlations between HM_{AN} and organic matter content are highly significant and increase in the sequence Cu ($r = 0.38$) < Cd ($r = 0.45$) < Zn ($r = 0.52$). According to Brümmer et al. (1986) and Welp and Brümmer (1998), the mobility of Cu and other metal ions increases in alkaline soils in the sequence Zn < Cd < Cu due to the formation of soluble organic complexes. This could be one reason for the higher HM uptake. While correlations between pH value and Zn_{AN} ($r = -0.43$) and Cd_{AN} ($r = -0.31$) are weak and negative, the correlation between pH and Cu_{AN} is weak and positive.

While the high affinity of Cu to different fractions of organic substances is well proven (Brümmer and Herms, 1983; Atanassova, 1995, 1999; Alloway, 1999), the adsorption of Zn, as reflected by the coefficient of correlation, is something special. McGrath & Cegarra (1992) showed that the high adsorption of Zn on organic matter reflects a strong loading situation. Due to a relatively low amount of Fe and Al oxides and the saturation of their adsorption capacity for HM ions, Zn is first bound to humic acids and then migrates with time into the oxide minerals where it is specifically adsorbed. Similar observations were made by Sposito et al. (1985) and Han et al. (2001).

Besides the soil reaction, the decrease of the redox potential during the irrigation periods can cause a higher HM mobility due to the dissolution of Mn and Fe oxides (e.g. Charlatchka and Cambier, 2000), which have a high binding capacity for HM at pH values > 7 .

Furthermore, it should be established that the characterization of the mobile fraction with the NH_4NO_3 extraction method according to Zeien & Brümmer (1989, 1991), which was developed for neutral to acid Middle European soils, is also valid for alkaline soils. Due to the possible volatilization of NH_3 in an alkaline environment, the concentration of NH_4 is probably too low for the exchange of weakly adsorbed HM. In a weakly alkaline milieu they could be mobilized due to direct contact with plant roots by acid root exsudates.

Further systematic investigations of soil pollution and the mobility and plant uptake of HM are necessary. Research should also focus on the development of methods suitable to characterize the mobility and risk of plant uptake of HM in alkaline soils.

7 References

Allen, H.E., Huang, C.P., Bailey, G.W. & Bowers, A.R. (1995): Metal speciation and contamination of soil. - Lewis Publ., Boca Raton, Ann Arbor, London, Tokyo.

Alloway, B.J. (1999): Schwermetalle in Böden. - Springer-Verlag Berlin Heidelberg.

Atanassova, I. (1995): Adsorption and Desorption of Cu at high equilibrium concentrations by soil and clay samples from Bulgaria. - Environmental Pollution, 87: 17-21.

BBodSchG (1998): Federal Soil Protection Act. - Federal Law Gazette I, p. 502.

BBodSchV (1999): Federal Soil Protection and Contaminated Sites Ordinance. Federal Law Gazette I, p. 1554.

Blume. H.-P. & Brümmer, G. (1987): Prognose des Verhaltens der Schwermetalle in Böden mit einfachen Feldmethoden. Mittlg. Dtsch. Bodenk. Ges., 53: 111 - 117

Brümmer, G.W. & Herms, U. (1983): Influence of soil reaction and organic matter on the solubility of heavy metals in soils. - In: Ulrich, B. u. Pankrath, J. (eds.): Effects of accumulation of air pollutants in forest ecosystems: 233 – 243, Reidel, Dordrecht.

Brümmer, G.W., Gerth, J. &. Herms, U. (1986): Heavy metal species, mobility and availability in soils. - J. Plant Nutr. Soil Sci., 149: 382 -398.

Burnod-Requia, K. (2003): Mining, industry and environment. - Environmental Performance Reviews, Georgia, UNECE: 107-118.

Charlatchka, R. & Cambier, P. (2000): Influence of reducing conditions on solubility of trace metals in contaminated soils. - Water, Air and Soil Pollution, 118:143-167.

Didebulidze, A. (1997): Problems and prospects of agriculture and rural development in Georgia. - UNDP Discussion Paper Series 5, Tbilisi.

Felix-Henningsen, P. , Urushadze, T. F., Narimanidze, E. I., Wichmann, L, Steffens, D. Kalandadze, B. B. (2007): Heavy metal pollution of soils and food crops due to mining wastes in an irrigation district south of Tbilisi, eastern Georgia. – Annales of Agrarian Science, Vol. 5, No. 3: 11 – 27.

Felix-Henningsen, P. , Urushadze, T. F., Narimanidze, E. I., Wichmann, L, Steffens, D. Kalandadze, B. B. (2007): Heavy metal pollution of soils and food crops due to mining wastes in the Mashavera river valley. – Bull. Georgian Nat. Acad. Sci, Vol. 175, No. 3: 97 – 106.

Futtermittelverordnung (1988): Verordnung zur Änderung der Futtermittelverordnung vom 22. Juni, 1988 BGBl.I, S.881.

Han, F.X., Banin, A. & Triplett, G.B. (2001): Redistribution of heavy metals in arid-zone soils under a wetting-drying cycle soil moisture regime. - Soil Science, 166: 18-28.

Herms, U. & Brümmer, G. (1984): Einflussgrößen der Schwermetalllöslichkeit und –bindung in Böden. - J. Plant Nutr. Soil Sci.,147: 400-424.

Hornburg, V. & Brümmer, G.W. (1993): Verhalten von Schwermetallen in Böden. I. Untersuchungen zur Schwermetallmobilität. - J. Plant Nutr. Soil Sci.,156: 467-477.

Javakhishvili, A. & Gvelesiani, G. (1962): Soviet Georgia. - Academy of Sciences of the Georgian S.S.R., Vakhushti Institute of Geography. Progress Publishers, Moscow.

Libert, B. (2003): Agriculture and environment. - Environmental Performance Reviews, Georgia, UNECE, 129-137.

McGrath, S.P. & Cegarra, J. (1992): Chemical extractability of heavy metals during and after longterm application of sewage sludge to soil. - J. Soil Sci., 43: 313-321.

MSPM (2000): JSC Mdneuli - Technical Highlights.

Nakhutsrishvili, G. (2000): Georgia's biomes. - In: Beroutchavili, N., Kushlin, A. & Zazanashvili, N. (eds.): Biological and landscape diversity of Georgia. WWF Georgia Country Office, Tbilisi.

Narimanidze et al. (2005): Bergbaubedingte Schwermetallbelastungen von Böden und Nutzpflanzen in einem Bewässerungsgebiet südlich von Tiflis/Georgien - Ausmaß, ökologische Bedeutung, Sanierungsstrategien. ZEU Discussion Paper 21, 92 pp, Giessen.

Purnell. M.F. (2000): World Reference Base for Soils. - Report of the correlation tour in Georgia, 20. July to 1 August 2000.

Rosenkranz, D., Bachmann, G., Einsele, G. & Harress, H.-M. (Eds., 1993 ff.): Bodenschutz. - Ergänzbares Handbuch der Maßnahmen und Empfehlungen für Schutz, Pflege und Sanierung von Böden, Landschaft und Grundwasser, Erich Schmidt Verlag, Berlin.

Sauerbeck, D. (1985): Funktonen, Güte und Belastbarkeit aus agrikulturchemischer Sicht. - Materialien zur Umweltforschung 10, Kohlhammer, Stuttgart.

Schlichting, E., Blume H.-P. & Stahr, K. (1995): Bodenkundliches Praktikum. - 2. Aufl.. Pareys Studientexte 81, Blackwell, Berlin.

Smith, C. J., Homans, P. & Cook, F.J. (1996): Accumulation of Cr, Pb, Cu, Ni, Zn and Cd in soil following irrigation with treated urban effluent in Australia. – Environm. Pollution, 94: 317 – 323.

Sposito, G. (1985): The distribution of potentially hazardous trace metals. - In: Siegal, H. (ed.): Concepts in metal toxicology, Metal ions in biological systems, Dekker, New York.

Tiller, K. G.; Gerth, J. & Brümmer, G. (1984 a): The sorption of Cd, Zn and Ni by soil clay fractions: procedures for partition of bound forms and their interpretations. - Geoderma, 34: 1 - 16.

Tiller, K. G.; Gerth, J. & Brümmer, G. (1984 b): The relative affinities of Cd, Ni and Zn for different soil clay fractions and geothite. - Geoderma, 34: 17 – 36.

Welp, G. & Brümmer, G.W. (1998): Adsorption and solubility of ten metals in soil samples of different composition. - J. Plant Nutr. Soil Sci., 162: 155-161.

VDLUFA (1976): Methodenbuch, Bd. III. Die chemische Untersuchung von Futtermitteln. Verlag J. Neumann-Neudamm, Melsungen, Berlin, Basel, Wien.

ZEBS (Zentrale Erfassungs- und Bewertungsstelle für Umweltchemikalien des Bundesgesundheitsamtes) (1990): Richtwerte für Schadstoffe in Lebensmitteln, Bundesgesundhbl. 5: 224-226.

Zeien, H. & Brümmer, G.W. (1989): Chemische Extraktion zur Bestimmung von Schwermetallbindungsformen in Böden. - Mitt. Deutsche Bodenkundl. Ges., 59/I: 505 - 510.

Zeien, H. & Brümmer, G.W. (1991): Chemische Extraktionen zur Bestimmung der Bindungsformen von Schwermetallen in Böden. - In: Sauerbeck, D. u. Lübben, S. (Eds.): Auswirkungen von Siedlungsabfällen auf Böden, Bodenorganismen und Pflanzen. Ber. z. ökol. Forsch. 6: 62 – 91, Jülich.

CLIMATE CHANGE IN GEORGIA DURING THE 20TH CENTURY

MICHAEL SCHAEFER, MARIAM ELIZBARASHVILI, LORENZ KING & RAMAZ MESKHIA

1 Introduction

Climate change is an important topic worldwide. In its Fourth Assessment Report of Climate Change 2007 the Intergovernmental Panel on Climate Change (IPCC) writes that the 100-year linear air temperature trend worldwide (1906-2005) is + 0.74°C. Eleven of the twelve hottest years since starting the observation of temperature in 1850 have been between 1995 and 2006. Warming of the climate is unequivocal and world temperatures could rise between 1.1 and 6.4°C during the 21st century. Precipitation has increased between 1900 and 2005 in North and South America, northern Europe and northern and Central Asia, whereas precipitation declined in the Sahel, the Mediterranean, southern Africa and parts of southern Asia (IPCC, 2007).

In Georgia, a country with an area of 69,000 km² surrounded by the Black Sea, the Greater and the Lesser Caucasus Mountains, the landscape is very heterogenic. The relief is mountainous, large orographic raisings alternate with intermountain hollows. The Greater Caucasus Mountain Range stretches across the northern part of the territory from northwest towards southeast. Mt Elbrus (5642 m) on the Russian side and Mt Shkhara (5200 m) on the Georgian side are the highest peaks in this high mountain area and covered by glaciers. Neighbouring countries in the north and east are Russia and Azerbaijan. In the south, the Lesser Caucasus Mountain Range forms a border towards Turkey and Armenia. Between the Greater and Lesser Caucasus Mountains is an intermountain depression with lowlands, plains and plateaus.

Various climate regions can be found within short distances. The chief-climate forming factors in the Caucasus are the relief, atmospheric circulation and radiational peculiarities (Borisov, 1965). The landscape in Georgia varies from a subtropical Black Sea coast in the west to a semi-arid steppe in the east. The Surami Mountains, which connect the Greater Caucasus Mountains with the Lesser Caucasus Mountains, separate the maritime west from the continental east (Schaefer, 2005).

Precipitation and temperature are decisive factors for a productive agriculture and their distribution causes different conditions for land use. Therefore, the knowledge of climate trends is important for a country that depends on agriculture as the main source of income and for self-subsistence of the population. Which climate trends can be observed in the mountainous country of Georgia? How have temperature and precipitation primarily changed in the last decades of the last century? What about local and seasonal differences in climate, is there a change in summer- and winter-temperature? These are the questions this article deals with.

2 Data, Methods and Working Area

Time series of NCEP/NCAR Reanalysis 1 data (Kistler et al., 2001) provided by the NOAA/OAR/ESRL PSD, Boulder, Colorado, USA from their Web site at http://www.cdc.noaa.gov and time series of station data for six climate stations in Georgia are used for the climate analyses. One time series of grid data covers western Georgia and one time series covers eastern Georgia and western Azerbaijan for the years 1951 to 2007. Furthermore, the data sets of six stations with monthly precipitation and temperature data have been analysed. The monthly data have been provided by the State Department of Hydrometeorology of Georgia and contain data from January 1968 to December 1990.

Figure 1: Map of Georgia with the six analysed climate stations and the two analysed grids

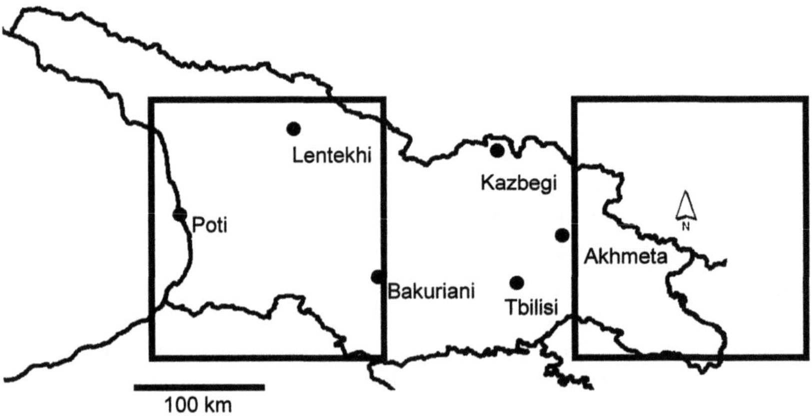

In Soviet times about 100 climate stations in Georgia were in use. Civil war and a difficult political situation at the beginning of the 1990s caused a reduction of climate stations to only ten climate stations. In the year 2008 15 climate stations are in use.

Data gaps are a challenge in analysing Georgian climate data. Therefore, time series for six stations with fewer gaps are used for this analysis. Werbelow (2006) compared different sources of climate data for Georgia and underlined the difficult situation.

To get an impression of trends in temperature and precipitation, grid data for the years 1951 to 2007 is analysed by a linear trend analysis. For more detailed information station data will also be analysed by a linear trend analysis. Monthly temperature and precipitation data is analysed for the six stations Poti, Lentekhi, Bakuriani, Kazbegi, Tbilisi and Akhmeta (Figure 1). These stations are in different climate zones. The elevation, average precipitation and mean air temperature are presented in table 1. The mean values are calculated from the time series between 1968 and 1990 (Schaefer, 2005).

Poti is a city at the Black Sea coast situated in the delta of the river Rioni, the largest river of western Georgia. In the last years, floods have often occurred in this humid climate zone. Lentekhi is a village in the low-mountain zone of the Greater Caucasus Mountains. Bakuriani is situated in the Lesser Caucasus Mountains near Borjomi where the famous mineral water comes from. Because of data gaps, only precipitation data is analysed for this climate station. Kazbegi, today called Stepantsminda, near Mt Kazbeg and located near the border between Georgia and Russia, is the third highest mountain in Georgia and covered by glaciers. Tbilisi is the capital of Georgia with about 1.25 million inhabitants. The first climate observation station was founded there in 1845. A climate graph of Tbilisi is presented in Figure 2. Akhmeta in eastern Georgia is situated in the Alazani valley in Kacheti.

Table 1: Elevation, mean air temperature and mean precipitation for six climate stations in Georgia based on the time series 1961-1990

	Elevation	Mean air temperature	Mean precipitation
Poti	3 m	14.1 °C	1891 mm
Lentekhi	760 m	9.4 °C	1312 mm
Bakuriani	1665 m	4.3 °C	859 mm
Kazbegi	3653 m	-5.8 °C	973 mm
Tbilisi	403 m	13.1 °C	499 mm
Akhmeta	567 m	12.3 °C	199 mm

Figure 2: Climate graph for Tbilisi, the capital of Georgia

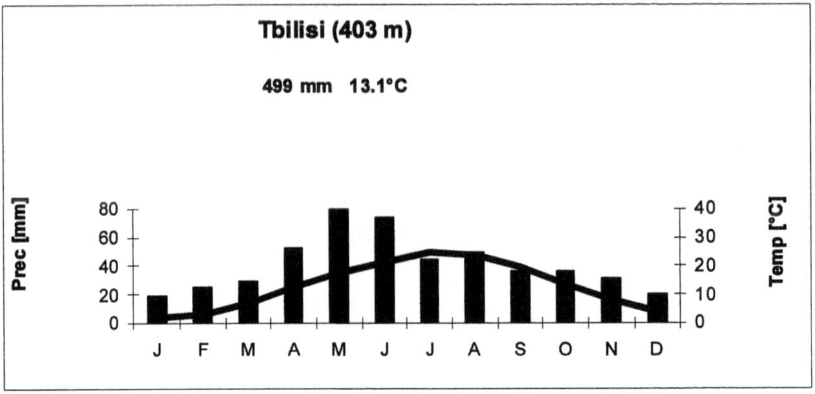

Tbilisi is located on the same latitude as Rome in Italy and has a maximum in precipitation in May and June. The summers are hot with a mean air temperature of 25°C in July and the winters are cold with a mean temperature of 2°C in January.

3 Temperature

The temperature distribution in Georgia is very heterogenic because of the heterogenic landscape. The subtropical western part of Georgia is characterized by a mild winter with mean temperatures of about 7.5 - 10°C in January at the southern tip of Georgia's Black Sea. The January temperatures in the Kolchis lowlands range between 2.5°C and 7.5°C. The mean January temperatures in the Greater Caucasus Mountains are not more than 0 - 2°C. In Tbilisi temperature is about 2 - 4°C and in the continental western Georgia 5 - 7.5°C. The mean July temperatures in western Georgia reach 29°C in Kutaisi, 3°C on the Kashbek and 20°C in Tbilisi. In Georgia the yearly mean temperature for the period 1961-1991 is 14°C. A complex investigation of the temperature field of the territory of Georgia has been carried out by Elizbarashvili (1999a, 1999b) and Tavartkiladze (1999), who considers geographical factors and mathematical models. Interpolated temperature and precipitation maps of Georgia based on the time series from 1961 to 1990 and a calculated climate zoning map of Georgia are published in Schaefer et al (2007).

To get a first impression of the temperature development, grid data from the NCEP/NCAR Reanalysis Project is analysed for western and for eastern Georgia / western Azerbaijan. The results are presented in Figure 3 and Figure 4.

Figure 3: Yearly mean temperature and decreasing temperature trend for western Georgia using data from the NCEP/NCAR Reanalysis Project

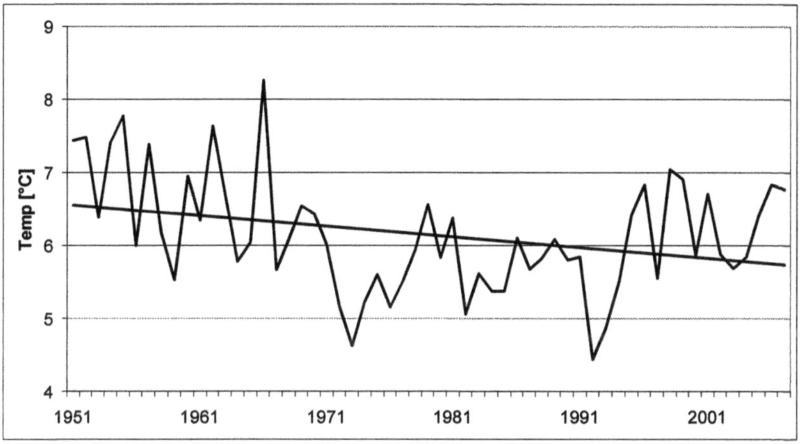

Figure 4: Yearly mean temperature and increasing temperature trend for eastern Georgia/Azerbaijan using data from the NCEP/NCAR Reanalysis Project

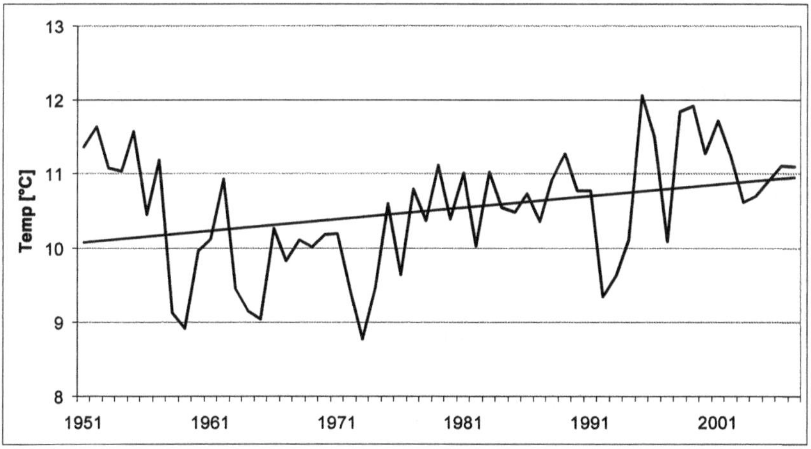

Figure 5: Summer and winter temperature in western Georgia using data from the NCEP/NCAR Reanalysis Project

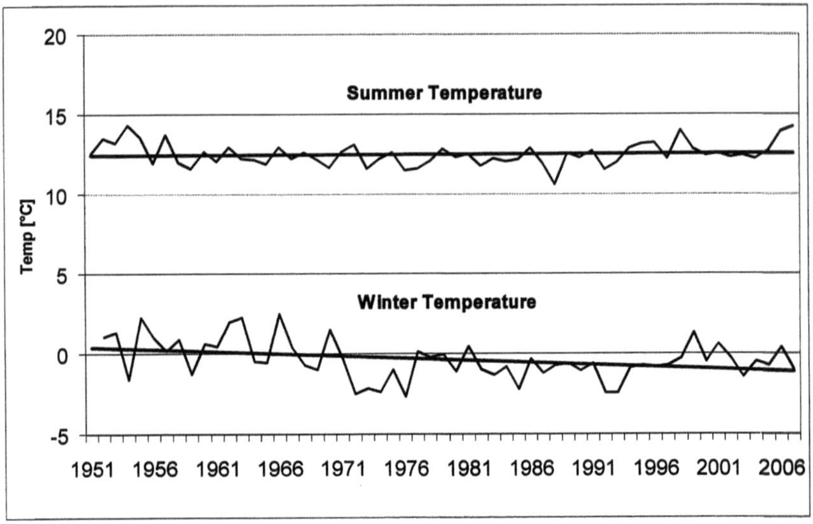

Figure 6: Summer and winter temperature in eastern Georgia using data from the NCEP/NCAR Reanalysis Project

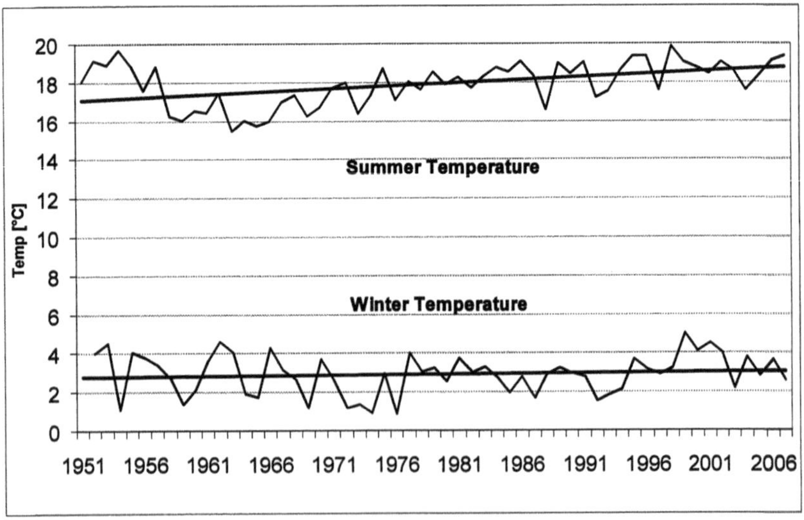

Figure 7: A weak decreasing trend in temperature in Tbilisi in western Georgia using station data from Hydromet Tbilisi

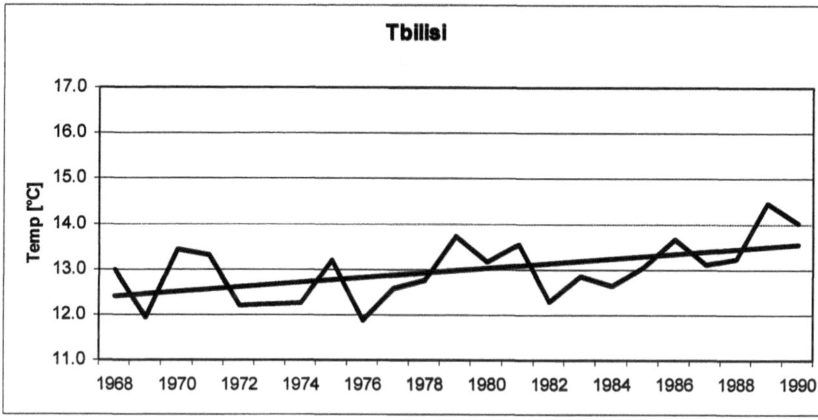

Figure 8: An increasing trend in temperature in Poti in eastern Georgia using station data from Hydromet Tbilisi

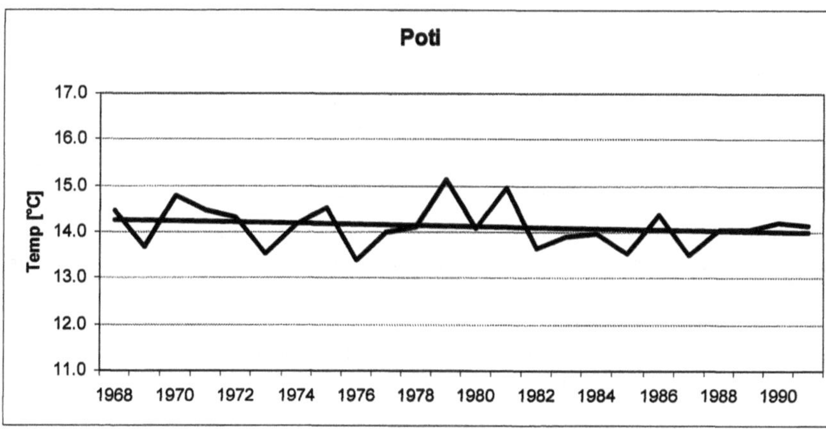

Arid landscapes became warm more quickly than humid ones, which were even cooled in many cases. Warming is mainly observed in eastern Georgia and Azerbaijan, which is characterized by a continental climate. In western Georgia, which is under the influence of the Black Sea, the temperature trend decreases.

In a next step, the trends in summer and winter temperature are analysed, cf Figure 5 and 6. The summer period contains the months May until October, the winter period the months November until April. The decreasing trend in mean temperature in western Georgia is mainly caused by a decreasing trend in winter temperature, whereas the increasing trend in eastern Georgia is mainly due to an increasing trend in summer temperature.

A trend analysis of the station data shows similar results. Poti (Figure 7) in western Georgia has a decreasing trend in temperature, similar to Lentekhi. Tbilisi (Figure 8) in eastern Georgia has an increasing trend, also Akhmeta and Kazbegi in the Greater Caucasus Mountains. Temperature increase in the higher mountain areas seems to have stronger trends than in the lowland areas.

4 Precipitation

Georgia shows the most diverse distribution of precipitation in comparison to all former Soviet states (Lydolph, 1977). In western Georgia there are climate stations with more than 2700 mm precipitation a year, concerning the mean values of the period 1968 to 1990, and in eastern Georgia there are climate stations with less than 200 mm precipitation a year (Schaefer et al, 2007).

The maximum annual precipitation is recorded along the Black-Sea coast, increasing from north to south. While the annual precipitation in Sochumi (Abchasia) is 1634 mm, it amounts to 1829 mm in Zugdidi and even to 2704 mm in Batumi. However, in Abchasia total annual rainfall reaches more than 2200 mm due to heavy rain eastwards in front of the massif of the Greater Caucasus Mountains.

Lydolph (1977) states that in the Southern Caucasus rainfall rapidly decreases as one heads eastwards. The reduction of rainfall can be attributed to the distance from the Black Sea and the leeward location to offshore mountain ranges like the Surami Mountains, which connect the Greater with the Lesser Caucasus Mountains and form a watershed between the Black and the Caspian Sea. Minimum mean precipitation is recorded in Akhmeta with 199 mm.

The trends in precipitation are weak, as shown for the climate stations Bakuriani (Figure 9), with a weak increasing trend in western Georgia. In Poti and in Lentekhi the trends in precipitation are also increasing.

Figure 9: A weak increasing trend in precipitation in Bakuriani in western Georgia

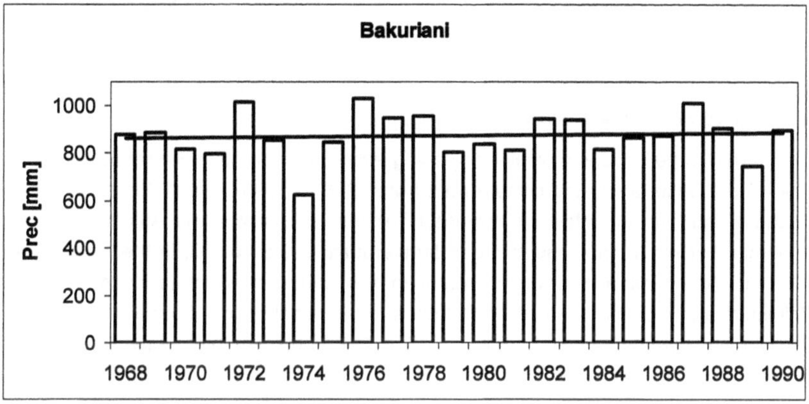

In Kazbegi there is no trend in precipitation and for Tbilisi (Figure 10) there is a weak decreasing trend in eastern Georgia. With small absolute totals in Akhmeta, less than half of the totals in Tbilisi, the trend in precipitation in Akhmeta is increasing.

Figure 10: A weak decreasing trend in precipitation in Tbilisi in eastern Georgia

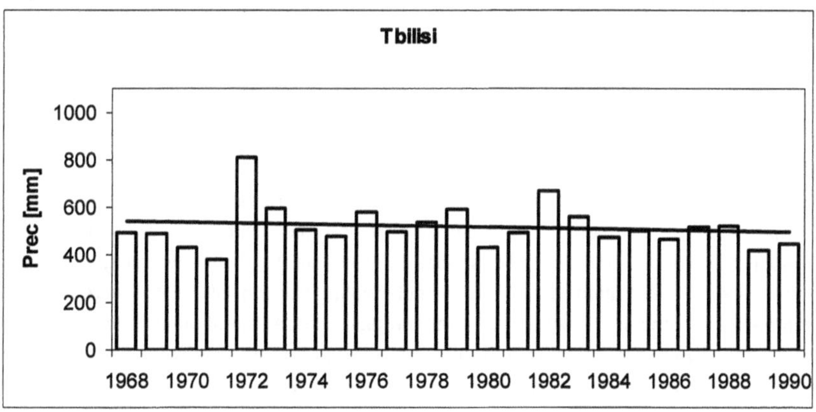

5 Climate Change in Georgia - Review of Results

The topic of climate variability and climate change in Georgia has been considered in a number of Georgian researches. Among them it is necessary to point out the works within the National Program on Climate Change according to the United Nations Frame Convention that are published in the Bulletin of the National Climate Research Center (Beritashvili et al 1999, Gzirishvili et al 1999).

Begalishvili (2006) found a weak decreasing temperature trend in western Georgia with 0.2 - 0.3°C while analysing climate data from 1906 to 1995. A high decrease is to be found in the high mountainous regions of Abchasia, Svaneta and Adjara. In eastern Georgia there is an increase in mean temperature of 0.5 - 0.6°C, in the region of the Surami Mountains and Southern Kartli in the centre of Georgia even by 0.7°C. Minimum and maximum temperature trends are increasing, too. Concerning the Tbilisi temperature trend (Figure 8) Begalishvili (2006) obtained the same results using climate data starting in 1845 till 1995. Beritashvili et al (1999) published a map with temperature trends for Georgia. The Black Sea coast is characterized by a decrease of -0.3 to -0.1°C, the eastern part by an increase of 0.3 to 0.7°C in South East Georgia. The higher mountain areas in the northwest have decreasing temperature trends, in the northeastern part increasing temperature trends.

These results are conform with IPCC (2007). According to IPCC, the Black Sea area is characterized by temperature decrease and the Caspian Sea area and Central Asia are characterized by temperature increase. Davitashvili (2008) underlines the climatic diversity of Georgia and climate cooling in western Georgia and warming in eastern Georgia. The specific properties of the regional climate cooling process in western Georgia are studied by mathematical modelling in Daviatishvili & Khantadze (2008). VASClimO-data analysed by Werbelow (2006) pointed out an additional temperature increase in the middle of the 1990s. Also, the climate data from the NCEP/NCAR reanalysis project in Figure 3 and 4 affirms this.

Elizbarashvili et al (2002a, 2002b, 2002c, 2005a) and Kolomyts (1985) analysed the climate change in Georgia, the impacts of climate change and especially the respondings of the different landscapes of Georgia to global warming. In humid landscapes, evaporation consumes a substantial fraction of heat. Therefore, the temperature trend is negligible or even negative there. Non-humid landscapes are more sensitive to the warming and the temperature there increases faster. The rate of temperature change greatly depends on the type of landscape. In Elizbarashvili et al (2006) the climate conditions for a formation of certain natural landscapes are evaluated and a scheme of possible transformations of natural landscapes is worked out on the basis of landscape zoning and climate change scenarios. Modelling of temperature (Kheladze, 1990) and of climate and runoff (Hagg et al, 2008) is an important challenge, also the effect of urbanization on climate variability, especially in winter (Elizbarashvili et al, 1999).

The danger of desertification in eastern Georgia is analysed in Elizbarashvili (2002b) and in Tourmanidze (2003). A severe drought took place in Georgia in the year 2000 (Red Cross report, 2001). Begalishvili (2006) expects a reduction in crop production especially in southern and eastern Georgia. Transformation of the alpine vegetation in the Caucasus Mountains due to climate change is analysed by Nakhutsrishvili et al (1999, 2006). It is expected that further climate warming can significantly change the vegetation and the landscapes in the mountain regions of Georgia.

The impacts of climate change can be observed by regarding the area that is covered by glaciers (table 2). Svanidze et al (1999) summarized that the area covered by glaciers was reduced from 2319 km^2 in the year 1850 to 511 km^2 in the year 1999. This means the area covered by glaciers is reduced by 5 to 10 km^2 a year. This has a great impact on the run-off in Georgia. The firn line is about 3400 m high and will rise about 160 m by an increase of air temperature by 1°C. To quantify recent and future regional climate, Hagg et al (2008) employed climate models in the Caucasus Mountains and concluded that the calculated severe loss of glaciated areas by the end of this century will be the reason for spring floods and summer droughts.

Tab 2: Retreat of glaciated area in Georgia, Svanidze et al (1999).

Year	Area covered by glacier
1850	2319 km^2
1910	1986 km^2
1965	1424 km^2
1999	511 km^2

6 Conclusion and Outlook

Changes in air temperature, precipitation, glacier retreat and related changes in surface water run-off have affected Georgia for several decades. Especially changes in air temperature and precipitation were analysed in this article. There is a change of climate in the different climate zones of Georgia. Eastern Georgia is getting warmer with a weak decreasing trend in precipitation, western Georgia is getting colder with a weak increasing trend in precipitation.

Altogether climate change has a great impact on Georgia and there is a need for further analyses. For example, the influence on the occurrence of natural disasters has to be analysed. Floods and droughts receive increased attention because they have great impacts on the life of people and on infrastructure because the number of climatic extreme events seems to increase. Heavy rain events can trigger debris flows which are a great danger in a mountainous country like Georgia and the number and dimension of debris flows seems to increase (Gavardashvili et al 2007). Therefore, preparations have to be made to reduce the deficiencies and uncertainties in Georgian climate data especially after 1991.

The maintenance of as many of the climate stations in the different climatic zones as possible is of high importance, as the understanding of climate change in the past is the base for projecting the climate of the future. However, not only the analysis of climatic trends is important. It is furthermore relevant to examine if the time series show - in addition to trends - quasi-periodic behaviour (Hartmann et al. 2008a, b). This in turn is the background for a possible adaptation strategie to climate change (King et al 2006, 2007).

7 References

Begalishvili NA, Beradze NI, Beritashvili BSh, Kordzakhiya GI, Lazriev GL, Turmanidze TI (2006): Problem of climate change in mountainous countries (Case study for Georgia), Russian Meteorology and Hydrology 2006 No 3, 46-54, in Russian.

Beritashvili B, Janelidze P (1999): Georgia's Initial National Communication under the United Nations Framework Convention on Climate Change. Tbilisi. Georgia.

Borisov AA (1965): Climates of the U.S.S.R. Oliver and Boyd LTD, Edinburgh.

Davitashvili T (2008): Research of conditions of the climate change in Georgia in view of regional and antropogenic factors. EMS8/ECAC7 Abstracts, Vol. 5. 8th Annual Meeting of the EMS / 7th ECAC.

Davitashvili T, Khantadze A (2008): On climate, desertification and water pollution problems for the territory of Georgia. In: Simulation and assessment of chemical processes in a multiphase environment. Springer 2008.

Elizbarashvili M (1999a): The temperature field of the territory of Georgia. Monograph 87 p. Tbilisi, in Georgian.

Elizbarashvili M (1999b): Response of the temperature field in Georgia to the current global warming. Bulletin of the Georgian Academy of Sciences, Tbilisi, 159, 2: 254-156, in Georgian.

Elizbarashvili E, Elizbarashvili M (2002a): Reaction of some types of landscapes of Transcaucasia to the global warming. Transaction of the Russian Academy of Science, Part Geography 5, Russ. Moscow: 52-56, in Russian.

Elizbarashvili E, Elizbarashvili M (2002b): The estimation of probability of desertification of the steppe and semi-desert landscapes on the basis of the concept of the periodic principle of geographical zoning. Meteorol. Hydrol. 11: 44-48. Russ. Moscow, in Russian.

Elizbarashvili E, Elizbarashvili M (2002c): The reaction of various types of Trans-Caucasus landscapes on global warming. Bulletin of the Russian Academy of Sciences, Ser.Geogr. 5, Moscow: 52-56 in Russian.

Elizbarashvili E, Meskhia R, Elizbarashvili M (2005a): Climate change of the western Transcaucasia. Transaction of the Russian Academy of Sciences, Part Geography 4: 35-38. Russ. Moscow, in Russian.

Elizbarashvili E, Elizbarashvili M (2005b): Possible transformations of the Caucasus natural landscapes in connection with global warming. Meteorology and Hydrology 10: 53-58. Russ. Moscow, in Russian.

Elizbarashvili E, Elizbarashvili M (2006): Main problems of landscape climatology. Inst. of Hydrometeorology. Monograph 118 p. Tbilisi, Georgia, in Georgian.

Elizbarashvili E, Elizbarashvili M (1999): Effect of urbanization on the modern climate variability of Georgia, Bulletin of the National Climate Research Center 8.

Gavardashvili G, Schaefer M, King L (2007): Debris flows at the river Mletis Khevi (Greater Caucasus Mountains, Georgia) and its assessment methods, Discussion Paper 32, Center for international Development and Environmental Research, Giessen, Germany.

Gzirishvili T, Shvangiradze M (1999): Bulletin of the National Climate Research Center 8, Georgia.

Hagg W, Shahgedanova M, Zacios M, Mayer C, Lambrecht A, Popovin V (2008): Climate and runoff modelling in the Caucasus mountains. Geophysical Research Abstracts, Vol. 10, EGU2008-A-09575.

Hartmann, H; Becker, S; King, L (2008a): Quasi-periodicities in Chinese precipitation time series. Theoretical and Applied Climatology 92: 155-163.

Hartmann, H; King, L; Jiang, T; Becker, S (2008b): Quasi-cycles in Chinese precipitation time series and in their potential influencing factors. In press at Quaternary International. Published online November 12, 2008.

IPCC (2007): Fourth Assessment Report of Climate Change 2007: Impacts, adaptation and vulnerability. Cambridge University Press.

Kheladze TV, Lortkipanidze GP, Elizbarashvili ME (1990): Mathematical modelling of a ground field of temperature for territory of Georgia. Reports of the Academy of Sciences of the GSSR Vol. 138, 3, Georgia: 67-72, in Russian.

King L, Nasr Z, Yaalaoui E & Lahache Gafrej R (2006): Climat et Hydrologie. Chapitre 1 de « Changements climatiques ; effets sur l'économie tunisienne et stratégie d'adaptation pour le secteur agricole et les ressources naturelles » Rapport 1ère Etape, Projet GTZ / MARH, Eschborn and Tunis, in French.

King L, Nasr Z, Almohamad H, Maag CC (2007): Climat – Cahier 2 de la « Stratégie nationale d'adaptation de l'agriculture tunisienne et des écosystèmes aux changements climatiques ». Rapport 2ème Etape, Projet GTZ / MARH, Eschborn and Tunis, in French.

Kistler R, Kalnay E, Collins W, Saha S, White G, Woollen J, Chelliah M., Ebisuzaki W, Kanamitsu M, Kousky V, van den Dool H, Jenne R, Fiorino M (2001): The NCEP_NCAR 50-Year Reanalysis: Monthly Means CD-ROM and Documentation. Bulletin of the American Meteorological Society 82: 247-267.

Kolomyts EG (1985): Forecast of influence of global climate fluctuations on the landscape structure of highland. Bulletin of the Academy of Sciences of the USSR, Ser, Geogr. 1, Moscow: 14-30, in Russian.

Lydolph PE (1977): Climates of the Soviet Union. Elsevier Scientific Publishing Company, Amsterdam.

Nakhutsrishvili G, Abdaladze O, Zazanashvili N, Gamkrelidze T, Chiboshvili M (1999): Expected transformation of high-mountain vegetation of the Caucasus in connection with possible climate change. Bulletin of the National Climate Research Center 8, Tbilisi, Georgia.

Nakhutsrishvili G, Akhalkatsi M, Abdaladze O (2006): First signs of transformation of alpine vegetation in the Caucasus as a result of climate warming in the Kazbegi region. J. Cramer Publ. Berlin, Stuttgart.

NCEP/NCAR Reanalysis Project: www.cdc.noaa.gov/cdc/reanalysis

RedCross Report (2001): Georgia - Drought Appeal No. 31/00 Situation Report.

Schaefer M (2005): Klimazonierung von Georgien - Dokumentation und Kritik bestehender Klimaklassifikationen anhand von langjährigen Klimadaten und mittels eines Geographischen Informationssystems. Published in Giessener Elektronische Bibliothek (GEB), University of Giessen, in German.

Schaefer M, Narimanidze E, King L (2007): Climate zoning of Georgia - a GIS-based regionalisation of longstanding precipitation and temperature data. Annals of Agrarian Sciences, Vol. 5, No. 1, 53 - 56, Tbilisi, Georgia.

Svanidze G, Tsomaia V (1999): Assessment of the vulnerability of Georgian glaciers to expected climate change. Bulletin of the National Climate Research Center 8, Tbilisi, Georgia.

Tavartkiladze KA, Elizbarashvili ES, Mumladze DG, Vachnadze DI (1999): Empirical model of change ground temperature field of Georgia. Tbilisi, Georgia.

Tourmanidze T (2003): Drought classification and assessment of drought caused losses. Annals of Agrarian Sciences. Tbilisi, Georgia.

Werbelow F (2006): Folgen des Klimawandels für die Bewässerungslandwirtschaft an den Beispielen Tunesien und Georgien. Diploma thesis, Institute of Geography, University of Giessen, in German.

INVESTIGATION OF THE ECOLOGICAL STATE OF GEORGIA'S MAIN RIVER KURA IN THE FRAMEWORK OF THE NATO - OSCE PROJECT: SOUTH CAUCASUS RIVER MONITORING

NODAR KEKELIDZE, TEIMURAZ JAKHUTASHVILI, EREMIA TULASHVILI, DAVID KEKELIDZE & LELA MTSARIASHVILI

1 Introduction

Thorough investigation of the hydrosphere as one component of the environment is a priority of modern ecological science. Water is an irreplaceable resource of the country and the basis of its development and well-being. Water discharge strongly depends on the level of development of the country. Recently, owing to demographic outburst and an intensification of manufacture, water discharge has increased. This creates a serious danger to human beings, as a considerable part of the used water is polluted and reaches environmental compartments.

On the other hand, the hydrosphere is one of the sources of biological diversity. Accordingly, the preservation of biological variety essentially depends on water quality; insignificant changes of water chemistry cause the reduction of the typical variety of organisms living in water. Furthermore, rivers have special significance because they are a source of potable water and are used for irrigation, the fish industry and other purposes.

The river Kura – the largest river of South Caucasus - is among those rivers, being the basis for economies not only in Georgia, but also for its neighbor – Azerbaijan. The river Kura is 1515 km long. It begins in the Armenian highland (Turkey), approximately at an altitude of 2740 m above sea level, and runs into the Caspian Sea. The river Kura is subject to strong influences of anthropogenic factors. The main reasons of pollution are chemical fertilizers, pesticides, industrial and domestic sewage. Water is polluted with various salts, heavy metals, radionuclides, and persistent organic pollutants (pesticides, polychlorinated biphenyls). Many of them are toxic and dangerous for ecosystems and human beings. The river Kura and its inflows have frequently drawn the attention of scientists.

A detailed physical and chemical study of the river Kura and its inflows was carried out in the 90s[1]. As a result, concentrations and behavior of the main ions, some biogenic substances, and biochemical consumption of oxygen are available.

Owing to heavy economic reasons, regular and complex research has practically stopped for the river Kura and its inflows during the last 15 years. In the meantime, measuring equipment is outdated. Researches had incidental character and had rather small scales. So, for example, some physical and chemical parameters and inorganic components of the river Kura (only 22 components) were studied within the limits of Tbilisi during several months of 2002[2].

During the last years certain works were carried out within the limits of several projects financed by different international funds, however, generally, they all touched the legislative issues and management of water resources. For this reason, the carrying out of regular water quality monitoring in South Caucasus - even within the limited scales on the basis of modern international physical methods - and the elaborated use of Standard Operational Procedures (SOP) and up-to-date equipment have become the primary aims of the international project "South Caucasus River Monitoring" that was prepared by the initiative of leading scientific organizations of Georgia, Azerbaijan, and Armenia (as well as the USA, Norway and Belgium). This project was approved in 2002 under the NATO program "Science for Peace".

The main feature of the NATO project that started under the initiative of local experts was the creation of a modern methodological and material-technical infrastructure and a restoration within the limited scale of regular monitoring of key parameters of water quality at thirty points of the Kura-Araks river basin in three countries of the South Caucasus. There are no similar projects in the South Caucasus region.

The above mentioned initiative is very important from the regional safety point of view as Armenia and Azerbaijan have considerable political issues. All this has led OSCE to take an interest and support the project. Today there is a tight collaboration between the scientists from South Caucasus countries and the USA, Belgium and Norway. This will allow us to achieve very important results regarding economic development and environmental protection in these countries.

[1] See G. Supatashvili, Hydrochemistry of Georgia, TSU Publishing House, 2003; Bulletins of the Hydrometeorological Department of Georgia until 1990.

[2] See N.Buachidze et al., Chemical-ecological research of Kura (within the limits of Tbilisi), Soros Educational magazine, #1, pp. 33 – 41, 2002).

1.1 WORK-SPECIFIC OBJECTIVES

- Increase of technical capabilities (analytical chemistry and its application in water systems, sampling and standards for good laboratory practice [GLP], quality assurance [QA] and quality control monitoring, database management and communications) among the partner countries.
- Cooperative establishment of standardised operating procedures (SOP) for common sampling, analysis and data management techniques for all partner countries. Implementation of s (QC).

1.2 MAIN OUTCOME OF THE PROJECT

SOPs were developed on the basis of international methods ISO, EPA and programs of quality assurance (QA, QC) of the water control parameters.

Within four years a regular monitoring of water quality parameters was carried out at more than thirty points of the Kura-Aras river basin in Azerbaijan, Armenia and Georgia, in particular:

1. physicochemical characteristics: discharge, specific conductivity, total dissolved solids, salinity (mineralization), pH, temperature, dissolved oxygen and redox potential;
2. main ions and biogenic substances: Na+, K+, Ca^{2+}, Mg^{2+}, SO_4^{2-}, Cl^-, HCO_3^-, CO_3^{2-}, Total P, Total N;
3. heavy metals: Cu, Mo, Pb, Zn, Ag, Co, Cr, Ni, Mn, Cd, etc.;

1.3 TO BE INVESTIGATED IN SUBSEQUENT STUDIES

- radionuclides: ^{40}K, ^{208}Tl, ^{210}Pb, ^{212}Pb, ^{214}Pb, etc.;
- radon and thoron: ^{222}Rn, ^{220}Rn;
- persistent organic pollutants (POPs):
- pesticides: aldrin, chlordane, DDT, dieldrin, endrin, heptachlor, hexachlorobenzene, mirex and toxaphene (9 pesticides);
- industrial chemicals: PCBs (polychlorinated biphenyls);
- unpremeditated waste products: dioxins, furans;

2 Methodology

Sampling and measurements were carried out according to the standard operating procedures (SOP) developed and coordinated between the countries, in particular:

- with regard to selection, 2 SOPs have been developed;
- for the control of physical-chemical parameters - 8 SOPs
- for the control of the main ions and biogenic substances - 8 SOPs
- for the control of heavy metals - 2 SOPs;
- for the control of radionuclides - 2 SOPs are in development.

2.1 Equipment

The main means used in the project was the newest equipment received within the framework of the project, in particular:

- Atomic-absorptive spectrophotometer AAnalyst 800 (Perkin-Elmer Instruments, 2003);
- Semi-conductor gamma-spectrometer GC2020 (Canberra Industries, 2005);
- Gaschromatograph-mass spectrometer TRACE DSQ (Thermo Electron Corp., 2006);
- Water Quality Checker U-10 (2M) (Horiba, 2003);
- Complete Water Quality Laboratory DREL/2400 (Hach Company, 2004);
- Electronic Radon Detector RAD7 (Durridge Company Inc., 2005).

2.2 Field expeditions

According to plan, 47 expeditions to 10 control points have been carried out. Of those, 7 control points are located directly on the river Kura, and three on its inflows – Suramula, Lekhura, and Khrami, not far from potentially large sources of pollution (industrial and municipal objects).

Within the framework of these expeditions 360 measurements of physical and chemical parameters have been carried out and selected:

- 260 samples for the analysis of the main ions and biogenic substances;
- 357 samples for the analysis of metals;
- 33 samples for the analysis of radionuclides (since the 26th expedition);

According to the Program Quality Assurance Program Plan (QAPrP) Iso measurements were carried out on control samples. Regular measurements were carried out on physical and chemical parameters, the main ions and biogenic elements, and also on heavy metals. This was done in cooperation with the Ministry of Environment Protection and Natural Resources of Georgia. Radionuclides were analysed in some water samples selected from several control points along the river Kura and its inflows, and also in several samples of bottom sediments and coastal ground. The methodology of researches and results are at a stage of improvement. The method now in use is based on a three-day study of the radiation spectrum. Furthermore, together with colleagues from Armenia and Azerbaijan, it is planned to master a concentration method of water samples with the use of special resins.

Work on persistent organic pollutants is at an initial stage. The analytical equipment was received several months ago. Analyses require gases of high cleanliness and plenty of reagents, which are lacking in the country. Purchase and reception from abroad is now in process. The methodology of sample preparation and analysis is based on methods of SW846. Initial variants of SOPs are prepared.

3 Summary of results

The data on three main groups of parameters:

- physical-chemical parameters,
- main ions and biogenic substances,
- heavy metals

are given in 6 tables (separately for the river Kura and for inflows) in the annex. The tables show the minimal and maximal values of corresponding parameters for every three-year period (2004, 2005 and 2006) for each control point. For comparison of the same place, the corresponding historical data for the period of 1980-1989 are also depicted.

3.1 Physical-chemical parameters[3]

Discharge of water. For corresponding stations the certain tendency of an increase of both minimal and maximal values is observed in comparison with the historical data. This tendency is very interesting and is apparently connected with the degree of common global-climatic changes. In fact, over the last years, intensive precipitation as

[3] See tables1 and 1a.

snow and rains was observed in the country, which frequently caused strong flooding that did not have any precedent in the past and represents a big threat for nearby settlements. This tendency is characteristic for all regions of the South Caucasus and deserves separate independent study.

Specific conductivity. Comparison of the received data with corresponding historical values shows that they basically lie within the same limits. Some excess is observed only for point Khertvisi (which is the first control point on the territory of Georgia after Turkey). As will be apparent later, it is necessary to note that such a tendency is also observed for some other parameters. Now we practically do not have any information on similar data and corresponding geographic-economic forces, influencing parameters of water quality on the territory of Turkey. Thus, in the future, a joint monitoring work with the participation of researchers from Turkey would be expedient. It would facilitate things if the influence of territorial factors of Turkey on the parameters of water quality were taken into account.

The study of the correlation of specific conductivity with values of ions is of special interest and is supposed to be carried out after all data for the current year have been received.

Total dissolved solids (TDS). Values of TDS lie within the limits of the corresponding historical data, except for point Khertvisi where all maximal values appreciably exceed the corresponding historical data. Furthermore, an essential reduction takes place at the following point (Borjomi), after which a monotonous increase is observed.

Salinity. All values of salinity lie below the sensitivity of the device used and correspond to the data on small salinity of waters of a river basin of the Kura.

pH. Approximately all values lie within the limits of the historical data. Any essential dependence on an arrangement of control points was not observed.

Temperature. The minimal values of temperatures appreciably exceed the minimal values of the historical data. To a certain degree, this is also true for the maximal values, especially for the stations located in a flat part of the basin. Such a tendency, as well the discharge of water, implies effects of climate-change. The last years were really characterized by rather soft winters and hot summer periods. This circumstance once again emphasizes the need for special studies concerning this phenomenon.

Dissolved oxygen (DO). The maximal values are close to corresponding with the historical data, minimal values are appreciable higher. Special dependences on the location of control points are not observed.

3.2 MAIN IONS AND BIOGENIC SUBSTANCES[4]

Na+. K+. Ca2+. Mg2+. In some cases the minimal values of some ions are lower than the corresponding historical minimal values (for example, K^+ at Borjomi and other points, Ca^{2+} for Tbilisi). This circumstance is probably connected with the reduction of anthropogenic impact in these regions. Insignificant reduction is observed after Khertvisi down to Borjomi or Gori, and then the concentration of ions grows monotonously. The same phenomenon is observed for the historical data as well.

SO_4^{2-}, Cl-. Their maximal values are appreciably lower than the corresponding historical data. From Khertvisi down to Gori their concentrations practically do not vary, and, at the sampling point of Tbilisi, some growth is observed. This may also testify to a decrease of the factor of anthropogenous pollution, however, it is necessary to note, that their minimal value was practically never below the corresponding historical minimal values. This circumstance may mean that though the general anthropogenous loading has decreased, its influence still has a more stable character.

HCO_3^- CO_3^{2-}. The character of changes of concentration of these ions differs a little from the ions SO_4^{2-}*and* Cl^-. In this case, both maximal values and minimal values are close to historical minimal values. It is difficult to explain their behavior, which demands special analysis and, apparently, researches. For these ions, special dependences on the location of control points is not observed, a fact which can apparently testify to an insignificant influence of anthropogenous factors on the concentration of these ions.

Biogenic elements. (Total P, Total N) Basically, their maximal values are appreciable smaller, and the minimal values are higher than the corresponding historical data. Depending on the arrangement of the station a certain increase is observed in the maximal values along a watercourse, however, the minimal values practically do not vary. This circumstance demands more detailed consideration, and further investigations are planned.

3.3 HEAVY METALS: CU, MO, PB, ZN, AG, CO, CR, NI, MN, CD[5]

Methods used for the historical data were rather insensitive (high MDL) and had rather big errors. Therefore, the zero results of concentration fixed in many cases should be

[4] See tables 2 and 2a.

[5] See tables1 and 1a.

counted as uncertain. Proceeding from this, the comparison of the received minimal values with the corresponding historical data loses significance and a comparison is only carried out concerning the maximal values.

Cu. In many cases the maximal values of *Cu* concentration appreciably exceed the corresponding historical data. The range of their changes was quite significant in 2004 and 2005 and appreciably decreased in 2006. The probable reason can be connected to differing methodological features: from the second half of 2005 a more effective method of filtration was applied that could result in a reduction of influence of the suspension particles on the received values. Given this circumstance, the influence of this step on the analytical method should be further investigated. On the other hand, this methodical aspect has not affected all results of the analysis of metals.

Influence of the geographical factor on *Cu* concentration is practically not observed.

Mo. The maximal values of *Mo* concentration are quite close to the corresponding historical data and the same is true in the case of *Cu.* Little dependency on geographical arrangement of control points was observed.

Pb. The maximal values of *Pb* concentration differ appreciably from historical data for the majority of points (only for point Tbilisi – Gachiani - they are approximately comparable). Similar to the case of *Cu,* the range of their changes was quite significant in 2004 and 2005 and appreciably decreased in 2006 (this is also, apparently, connected to questions of methodology). Little dependency on the geographical arrangement of control points was observed.

Zn. The maximal values of *Zn* concentrations are basically comparable to the corresponding historical data (appreciably below only for point Rustavi). Any dependency on the arrangement of stations is not observed.

Ag. Concentrations of Ag in the waters of the Kura are quite low. Insignificant increase of values is observed downstream the river Kura.

Co. For practically all stations the maximal values of *Co* concentrations are appreciably lower than the corresponding historical data. The range of their changes is insignificant, any dependence on an arrangement of control points is also not observed.

Cr. The corresponding historical data of regular measurements are unavailable. The data received in the project are actually the first of such measurements. As is true for *Co*, the range of their changes is insignificant, any dependence on an arrangement of control points is also not observed.

Ni. The maximal values of concentration are appreciably lower than the corresponding historical data. Dependence of the maintenance on the location of control points does not practically exist.

Mn. The maximal values of *Mn* concentration are comparable to the similar historical data. Dependence of the maintenance on the location of control points does also not practically exist.

Cd. The corresponding historical data of regular measurements are not available. As was true for *Cr*, the data received in the project were actually the first of such measurements. The range of changes of concentrations of *Cd* is rather significant, especially in 2004 and 2005, and appreciably decreased in 2006 (this is apparently also connected to questions of a methodological character). The problem of Cd is very interesting, and next year it is planned to pay special attention to this question.

4 Conclusion and Outlook

In this work all results received within four years of sampling and measurement on three basic groups of parameters are briefly compiled for the first time. A comparative analysis with historical data is carried out and basic features are marked.

The determination of natural and anthropogenic influences on parameters of quality of surface waters is one of the major subjects of further research. The carrying out of corresponding experiments is a challenge, and, as a rule, the received results reflect a total influence of these factors. It seems to us that, with some reliability, such a rating can be carried out on the basis of statistical processing available on historical long-term data, given monthly measurements of corresponding parameters.

The carrying out of the all-round analysis of ground deposits and the weighed particles also represents a special interest. Such researches will give important information on mechanisms of origin, distribution and transformation of polluting substances in the river system.

Table 1 **Generalized data (the minimal and maximal values) on physical-chemical parameters for 2004, 2005, 2006, 2007 and corresponding historical (1980-1989) data for 7 control points on river Kura (Georgia)**

№	Point	Year	**Physical-chemical parameters**												
			D, m^3/sec		**Ec,** mS/cm		**TDS**, mg/l		**Salinity** %	**pH**		T, ^{0}C		**DO,** mg/l	
			min	max	min	max	min	max		min	max	min	max	min	max
1	Khertvisi	2004	-	-	0.12	0.36	116	490	<0.1	7.41	9.08	3.0	23.9	8.3	13.4
		2005	-	-	0.12	0.30	120	430	<0.1	6.87	8.60	2.5	24.8	8.3	12.3
		2006	-	-	0.14	0.33	198	950	<0.1	7.68	8.53	3.8	23.5	7.3	10.8
		2007	13.6	436	0.25	0.33	210	210	<0.1	7.95	8.02	2.1	29.7	7.3	13.0
		history	***4.8***	***300***	***0.08***	***0.32***	***71.6***	***296.7***	-	***6.8***	***8.9***	***0.0***	***24.0***	***6.3***	***11.6***
2	Borjomi (down)	2004	-	-	0.15	0.29	110	220	<0.1	7.52	8.90	3.1	22.5	9.2	13.4
		2005	-	-	0.15	0.35	123	340	<0.1	7.22	8.35	2.0	23.1	8.7	12.5
		2006	-	-	0.16	0.32	107	250	<0.1	7.71	8.51	3.5	22.1	7.5	11.0
		2007	55	920	0.24	0.31	180	180	<0.1	7.07	8.08	2.0	29.0	7.2	11.3
		history	***13.5***	***750***	***0.12***	***0.41***	***100.1***	***293.3***	-	***7.1***	***9.1***	***0.0***	***26.0***	***6.6***	***8.9***
3	Gori (down)	2004	-	-	0.20	0.41	51	390	<0.1	7.3	9.06	5.5	24.3	9.2	13.4
		2005	-	-	0.23	0.41	138	380	<0.1	7.40	8.51	4.7	26.1	8.0	12.5
		2006	-	-	0.27	0.41	147	220	<0.1	7.83	8.48	3.8	22.3	7.5	12.8
		2007	-	-	0.29	0.39	240	240	<0.1	7.97	8.85	3.5	28.0	7.4	12.2
		history	***5.2***	***5912***	***0.19***	***0.73***	***148.9***	***498.6***	-	***6.2***	***9.9***	***0.0***	***21.0***	***5.4***	***13.2***

Table 1 (continued)

4	Tbilisi (Zahesi)	2004	65	662	0.21	0.44	120	380	<0.1	7.25	8.67	6.5	23.6	8.3	14.3
		2005	75	1050	0.27	0.50	123	460	<0.1	7.17	8.35	4.1	24.7	8.3	13.3
		2006	130	900	0.23	0.47	154	340	<0.1	7.91	8.42	5.0	20.5	7.6	12.6
		2007	70	1300	0.18	0.43	240	240	<0.1	7.88	8.06	3.8	29.5	7.1	12.0
		history	***53.0***	***682***	***0.10***	***0.82***	***145.1***	***512.3***	-	***7.2***	***9.1***	***0.0***	***23.0***	***5.0***	***13.0***
5	Tbilisi (V.B. Bridge)	2004	79	678	0.22	0.45	134	400	<0.1	7.55	8.5	6.6	23.9	8.0	13.6
		2005	90	1100	0.26	0.57	124	470	<0.1	7.21	8.50	4.3	26.2	8.1	12.3
		2006	113	932	0.22	0.58	133	210	<0.1	7.40	8.44	5.1	20.8	7.6	12.5
		2007	86	1390	0.28	0.45	230	230	<0.1	7.90	8.04	3.8	29.2	5.8	12.1
		history	***7.9***	***798***	***0.20***	***0.60***	***107.4***	***644.6***	-	***7.2***	***9.2***	***0.0***	***25.0***	***5.3***	***12.9***
6	Tbilisi (Gachiani)	2004	-	-	0.24	0.47	160	540	<0.1	6.5	8.36	6.7	22.7	8.8	13.6
		2005	-	-	0.27	0.48	90	430	<0.1	7.48	8.38	4.7	25.7	8.1	12.1
		2006	-	-	0.24	0.50	197	500	<0.1	7.45	8.45	5.2	23.0	7.4	12.5
		2007	-	-	0.38	0.47	250	250	<0.1	7.95	8.08	4.0	29.7	6.1	12.0
		history	***11.0***	***810***	***0.10***	***0.63***	***195.5***	***663.0***	-	***7.1***	***8.9***	***0.0***	***25.4***	***4.4***	***13.8***
7	Rustavi (down)	2004	-	-	0.27	0.53	66	1130	<0.1	7.5	8.37	7.6	24.8	9.1	13.8
		2005	-	-	0.28	0.61	97	390	<0.1	7.59	8.41	5.3	26.5	8.2	13.7
		2006	-	-	0.31	0.52	200	320	<0.1	7.65	8.47	5.3	25.3	7.0	12.3
		2007	-	-	0.41	0.54	320	320	<0.1	7.59	8.05	4.2	29.2	7.5	11.9
		history	***38.7***	***856.0***	***0.07***	***1.2***	***156.9***	***1130.0***	-	***6.6***	***8.8***	***0.0***	***29.4***	***4.8***	***12.1***

Table 1a: Generalized data (the minimal and maximal values) on physical-chemical parameters for 2004, 2005, 2006, 2007 and corresponding historical (1980-1989) data for 3 control points on inflows of river Kura (Georgia)

№	Point	Year	Physical-chemical parameters													
			D, m^3/sec		**Ec,** mS/cm		**TDS**, mg/l		**Salinity**, %	**pH**		**T**, ^{0}C		**DO,** mg/l	
			min	max	min	max	min	max		min	max	min	max	min	max
1	Suramula-Khashuri (down)	2004	-	-	0.12	0.40	85	370	<0.1	7.25	8.93	5.9	25.0	8.2	13.1
		2005	-	-	0.13	0.40	116	430	<0.1	6.88	8.21	3.6	25.0	8.1	13.6
		2006	-	-	0.18	0.41	152	250	<0.1	7.51	8.61	3.6	23.7	7.1	12.9
		2007	-	-	0.26	0.58	310	310	<0.1	7.80	8.04	2.3	28.0	5.0	12.3
		history	***0.35***	***8.2***	***0.10***	***0.56***	***47***	***462***	**-**	***6.0***	***9.48***	***0.0***	***24.0***	***5.7***	***11.8***
2	Lekhura-Kaspi (down)	2004	-	-	0.38	0.75	50	300	<0.1	7.48	8.54	9.7	25.9	8.7	12.0
		2005	-	-	0.33	0.68	195	400	<0.1	7.25	8.34	4.3	25.8	8.1	13.2
		2006	-	-	0.29	0.49	170	280	<0.1	7.53	8.45	4.0	24.0	7.1	12.8
		2007	-	-	0.51	0.72	360	360	<0.1	7.94	8.08	3.6	28.9	7.1	12.1
		history	***0.27***	***23.9***	***0.24***	***0.92***	***111***	***1141***	**-**	***7.4***	***10.24***	***-2.4***	***25.0***	***5.8***	***14.5***
3	Khrami (Red Bridge)	2004	-	-	0.27	0.59	200	510	<0.1	6.90	8.59	7.3	24.5	9.1	13.0
		2005	-	-	0.24	0.70	147	410	<0.1	7.40	8.42	5.3	27.0	7.9	13.1
		2006	-	-	0.25	0.48	196	320	<0.1	7.84	8.42	5.4	25.5	7.0	12.3
		2007	-	-	0.50	0.63	420	420	<0.1	7.88	8.04	4.9	29.1	8.0	11.9
		history	***0.34***	***169.0***	***0.16***	***1.00***	***232***	***932***	**-**	***7.0***	***8.4***	***0.9***	***27.0***	***4.3***	***14.2***

Table 2: Generalized data (the minimal and maximal values)on main ions for 2004, 2005, 2006, 2007 and corresponding historical (1980-1989) data for 7 control points on river Kura (Georgia)

№	Point	Year	*Main ions (mg/l)*																*Biogenic elements (mg/l)*			
			Na^+		K^+		Ca^{2+}		Mg^{2+}		SO_4^{2-}		Cl^-		HCO_3^-		CO_3^{2-}		Total P		Total N	
			min	max	min	max	min	max	min	max	min	max	min	max	min	max	min	max	min	max	min	max
1	Khertvisi	2004	12.5	32.0	1.9	4.6	13.9	22.0	4.1	11.3	7.5	20	4.8	10.9	87.8	200	<1.5	12.0	0.020	0.050	0.71	4.45
		2005	10.0	24.0	0.8	2.0	15.9	27.3	4.8	7.8	7.0	18.0	5.0	9.2	95.2	161	<1.5	12.0	0.005	0.050	0.62	3.37
		2006	9.0	19.0	0.8	1.7	16.0	31.3	3.3	5.8	8.0	18.0	6.4	8.5	90.3	107	<1.5	6.0	0.005	0.020	1.14	5.02
		2007	10.0	22	1.00	2.7	16.2	22.7	5.1	8.9	4.1	65	4.3	8.8	88	146	4.5	6.6	0.020	0.226	1.3	4.1
		history	***5.0***	***30.0***	***0.8***	***5.5***	***7.9***	***41.6***	***1.0***	***18.9***	***3.0***	***36.0***	***2.0***	***38.9***	***37.2***	***192.0***	***<1.5***	***12.0***	***0.0***	***0.133***	***0.19***	***5.24***
2	Borjomi (down)	2004	12.0	26.0	1.0	4.7	18.5	25.0	3.6	10.5	11.0	25	4.1	7.8	92.7	175	<1.5	13.5	0.025	0.100	1.36	5.88
		2005	8.0	27.0	0.7	1.9	18.0	26.0	3.9	9.5	9.0	20.0	4.3	8.4	100.0	193	<1.5	12.8	0.010	0.070	1.46	4.95
		2006	9.0	20.0	0.9	1.7	17.6	30.8	4.8	7.6	8.5	20.0	4.9	7.4	107.3	124	2.7	5.7	0.010	0.050	1.15	5.19
		2007	8.0	21	1.00	1.50	19.2	26	4.4	6.6	8.0	24	5.4	6.5	100	159	4.6	6.3	0.047	0.227	1.4	5.8
		history	***6.0***	***28.0***	***1.0***	***4.4***	***10.9***	***35.6***	***1.2***	***22.9***	***2.5***	***220.0***	***1.6***	***12.4***	***26.8***	***191.0***	***<1.5***	***13.9***	***0.005***	***0.167***	***0.17***	***9.75***
3	Gori (down)	2004	10.2	41.0	1.1	8.6	25.5	40.0	3.9	11.8	15.0	30	4.1	7.7	107.3	234.2	<1.5	7.5	0.020	0.060	0.68	5.25
		2005	11.0	23.0	0.9	1.8	24.3	35.9	3.6	10.3	12.0	24	4.0	9.4	116.6	242.0	<1.5	13.5	0.010	0.040	0.88	3.58
		2006	10.0	20.0	0.9	2.2	19.4	36.9	3.6	9.8	9.0	19.5	5.1	7.8	109.8	136.6	3.0	5.7	0.012	0.025	1.89	4.01
		2007	9.0	21	0.90	1.50	26.6	35	5.4	9.5	9.0	24	5.1	6.8	110	193	6.0	7.8	0.025	0.118	1.6	5.6
		history	***6.9***	***48.0***	***0.5***	***2.9***	***10.3***	***95.4***	***2.8***	***115.4***	***6.5***	***138.0***	***1.4***	***27.2***	***40.0***	***216.6***	***<1.5***	***21.0***	***0.003***	***0.130***	***0.01***	***15.08***

Table 2 (continued)

	Tbilisi (V.B. Bridge)	2004	12.5	28.0	1.4	2.4	25.8	47.0	5.4	12.4	17.0	50	4.0	9.1	115.0	217	<1.5	9.0	0.020	0.060	1.10	5.71
		2005	11.0	31.0	0.6	1.8	25.7	46.7	4.9	10.6	16.0	30	4.9	10.6	90.3	225	<1.5	13.5	0.015	0.050	1.85	7.21
		2006	11.5	22.0	0.9	1.8	33.5	37.4	5.6	11.5	14.5	30.5	5.7	7.9	141.5	168	<1.5	6.0	0.011	0.020	2.59	5.32
		2007	11.0	22	0.90	1.60	20.8	51	6.2	11.7	20	44	5.9	7.5	149	215	4.8	7.8	0.035	0.164	1.8	6.2
		history	***7.0***	***60.0***	***0.8***	***3.8***	***32.1***	***70.3***	***2.1***	***31.5***	***5.0***	***131.0***	***3.2***	***30.2***	***110.4***	***225.7***	***<1.5***	***21.0***	***0.005***	***0.300***	***0.64***	***19.43***
6	Tbilisi (Gachiani)	2004	10.0	27.0	1.5	3.1	26.5	50.0	6.3	12.6	24	51	4.1	8.9	120.0	207	<1.5	6.0	0.025	0.075	1.44	6.60
		2005	13.5	33.5	0.6	2.0	26.9	53.9	6.8	13.9	23	42	5.9	14.3	112.2	234	<1.5	13.8	0.015	0.060	1.91	7.82
		2006	14.5	23.0	0.9	1.8	36.8	56.1	5.8	11.9	18.0	80	6.7	8.5	143.9	180	<1.5	6.6	0.015	0.050	3.00	6.24
		2007	16.0	24	0.90	1.50	23.7	55	5.1	10.1	20	60	5.7	7.1	154	181	2.7	6.6	0.025	0.155	2.9	6.2
		history	***7.0***	***62.0***	***0.8***	***4.6***	***30.5***	***104.9***	***4.4***	***31.6***	***19.0***	***270.0***	***3.2***	***43.6***	***105.5***	***232.4***	***2.7***	***21.0***	***0.010***	***0.310***	***0.33***	***19.83***
7	Rustavi (down)	2004	14.0	35.0	1.5	3.0	32.0	56.0	6.7	13.6	32.5	70	5.1	11.9	124.0	210	<1.5	5.0	0.029	0.075	1.57	7.52
		2005	18.0	38.0	0.6	2.1	33.7	56.1	6.1	14.2	16.0	50	6.4	12.8	163.5	239	<1.5	12.0	0.010	0.075	2.01	6.32
		2006	16.0	35.0	0.9	2.1	36.6	55.3	5.8	10.9	28	68	4.3	8.5	151.3	185	<1.5	6.9	0.020	0.060	3.18	6.74
		2007	22.0	32	1.10	2.0	36.4	51	9.1	10.8	28	103	6.2	9.6	151	182	4.2	7.8	0.024	0.404	2.6	7.7
		history	***10.0***	***83.0***	***0.7***	***4.5***	***31.3***	***105.0***	***4.9***	***37.9***	***6.0***	***490.0***	***4.7***	***49.8***	***109.2***	***263.5***	***<1.5***	***9.3***	***0.012***	***0.550***	***0.30***	***32.64***

Table 2a: Generalized data (the minimal and maximal values) on main ions for 2004, 2005, 2006, 2007 and corresponding historical (1980-1989) data for 3 control points on inflows of river Kura (Georgia)

№	Point	Year	*Main ions (mg/l)*																*Biogenic elements (mg/l)*			
			Na^+		K^+		Ca^{2+}		Mg^{2+}		SO_4^{2-}		Cl^-		HCO_3^-		CO_3^{2-}		Total P		Total N	
			min	max	min	max	min	max	min	max	min	max	min	max	min	max	min	max	min	max	min	max
1	Suramula-Khashuri (down)	2004	10.0	26.0	1.2	3.8	16.0	35.3	4.9	12.3	13.0	24	4.6	13.5	102.0	230	<1.5	4.5	0.030	0.140	1.19	8.11
		2005	16.0	30.0	0.6	1.8	25.7	48.1	3.3	12.3	14.0	34	7.1	12.9	161.0	256	<1.5	10.5	0.020	0.060	2.19	7.84
		2006	12.0	30.0	1.1	2.3	30.3	57.7	4.4	16.5	22	36	7.1	12.8	119.6	219	<1.5	5.1	0.005	0.125	2.36	6.95
		2007	15.0	25	1.10	3.7	24.1	45	5.8	9.8	26	38	8.8	14.9	151	240	3.0	7.2	0.020	0.656	2.2	12.1
		history	***4.0***	***39.5***	***0.3***	***4.8***	***6.4***	***71.2***	***1.2***	***27.5***	***5.9***	***57.5***	***1.6***	***150.5***	***18.3***	***337.0***	***<1.5***	***13.5***	***0.003***	***0.552***	***0.15***	***18.72***
2	Lekhura-Kaspi (down)	2004	20.0	46.0	1.6	3.1	35.1	60.9	8.7	15.7	22	84	6.8	20.0	22.6	273	<1.5	8.7	0.025	0.080	1.26	8.12
		2005	13.0	48.0	0.7	2.9	32.9	64.2	9.2	13.9	22	50	4.5	11.2	143.9	281	<1.5	13.5	0.020	0.060	2.19	3.80
		2006	13.0	35.0	0.9	2.2	30.5	47.9	7.6	13.9	34	40	5.7	12.7	180.6	240	1.5	6.8	0.015	0.035	2.90	8.93
		2007	13.0	42	1.10	2.4	27.3	45	5.1	10.9	24	104	7.8	16.8	171	268	1.8	7.8	0.040	0.040	2.3	10.3
		history	***6.4***	***125.0***	***0.4***	***3.5***	***11.1***	***104.0***	***2.1***	***51.9***	***14.5***	***470.0***	***3.4***	***111.6***	***41.5***	***390.0***	***1.2***	***32.4***	***0.005***	***0.084***	***0.10***	***10.26***
3	Khrami (Red Bridge)	2004	21.0	52.0	0.9	3.1	39.0	55.3	5.8	18.5	18.0	60	7.1	21.4	127.0	230	<1.5	9.0	0.025	0.080	1.65	7.45
		2005	18.0	40.0	0.6	2.0	28.9	47.2	5.8	13.6	22	52	7.8	18.4	122.0	195	<1.5	13.8	0.015	0.050	1.60	4.95
		2006	15.0	28.0	1.0	1.7	28.9	47.3	9.2	9.7	18.0	56	5.7	10.5	148.8	210	<1.5	6.6	0.020	0.040	3.25	7.27
		2007	20.0	34	1.00	2.1	28.8	50	9.0	9.7	28.0	92	9.6	14.0	156	205	6.6	9.3	0.025	0.096	3.6	11.7
		history	***10.0***	***125.0***	***1.0***	***7.5***	***13.7***	***130.0***	***4.6***	***63.8***	***16.0***	***335.5***	***4.2***	***134.0***	***60.9***	***266.0***	***3.2***	***7.2***	***0.00***	***0.200***	***0.49***	***25.19***

Table 3 **Generalized data (the minimal and maximal values) on metals for 2004, 2005, 2006, 2007 and corresponding historical (1980-1989) data for 7 control points on river Kura (Georgia)**

№	Point	Year	Metals (ug/l)																			
			Cu		Mo		Pb		Zn		Ag		Co		Cr		Ni		Mn		Cd	
			min	max	min	max	min	max	min	max	min	max	min	max	min	max	min	max	min	max	min	max
1	Khertvisi	2004	1,56	87	0,102	4,9	0,46	9,5	1,93	16,4	0,023	0,68	0,25	1,36	0,64	5,2	0,48	6,5	12,2	48	0,057	7,4
		2005	7,2	105	0,94	4,4	1,04	7,8	2,7	28	0,031	0,59	0,150	0,99	0,67	8,5	1,03	8,7	1,97	35	0,004	1,70
		2006	6,2	14,3	1,15	3,5	0,70	1,35	4,3	18,3	0,100	0,130	0,17	0,72	0,67	3,2	1,24	2,4	7,4	18,1	0,050	0,180
		2007	5.2	9.4	1.7	3.3	-	-	11.4	13.3	0.100	0.100	<0.70	<0.70	1.56	2.5	1.05	2.1	11.1	16.0	0.060	0.060
		history	***0***	***22***	***0***	***5***	***0***	***39***	-	-	***0***	***3***	***0***	***8***	-	-	***0***	***31***	***0***	***40***	-	-
2	Borjomi (down)	2004	7,0	152	0,40	5,3	1,00	8,6	2,6	70	0,013	0,128	0,170	1,85	0,93	8,0	0,95	22	7,8	61	0,047	9,3
		2005	8,1	203	0,97	5,0	1,08	8,8	2,1	39	0,026	0,50	0,150	0,88	0,52	9,8	2,4	9,5	2,5	30	0,002	1,25
		2006	5,6	13,3	1,08	3,9	0,70	2,4	7,9	19,4	0,100	0,150	0,150	0,82	0,95	5,3	1,29	4,4	8,3	21	0,050	0,83
		2007	4.3	10.1	2.4	3.1	-	-	11.4	14.2	0.100	0.100	0.77	0.77	2.1	3.6	1.33	2.7	12.3	16.3	0.550	0.550
		history	***0***	***21***	***0***	***4***	***0***	***41***	-	-	***0***	***3***	***0***	***6***	-	-	***0***	***17***	***0***	***34***	-	-
3	Gori (down)	2004	2,4	98	0,30	4,8	2,7	8,5	2,0	90	0,009	0,33	0,174	1,39	0,81	3,9	0,83	14,8	18,0	60	0,025	15,2
		2005	7,1	167	0,85	4,0	1,03	7,9	2,0	42	0,010	1,37	0,149	0,97	1,08	8,8	1,98	7,1	2,0	53	0,050	1,16
		2006	5,4	11,4	1,02	4,4	0,70	1,34	8,4	19,7	0,100	0,190	0,34	0,80	0,65	2,5	1,08	3,6	8,4	18,8	0,050	0,24
		2007	4.8	10.7	2.1	5.0	-	-	13.4	14.7	0.100	0.100	<0.70	<0.70	2.3	3.2	2.1	3.0	11.0	16.3	0.120	0.120
		history	***0***	***24***	***0***	***5***	***0***	***10***	***0***	***175***	***0***	***0***	***0***	***6***	-	-	***0***	***25***	***0***	***72***	-	-

Table 3 (continued)

4	Tbilisi (Zahesi)	2004	4,1	206	0,20	6,8	0,090	8,8	2,5	26	0,006	0,49	0,194	2,5	0,37	4,6	0,83	6,7	23	52	0,022	12,8
		2005	5,3	105	1,06	4,0	0,94	8,1	2,0	34	0,032	0,81	0,150	0,79	0,68	8,3	1,01	6,8	1,05	26	0,012	1,15
		2006	6,2	10,9	0,91	5,8	0,70	1,27	8,1	23	0,100	0,82	0,150	0,70	0,61	4,3	1,13	3,9	7,8	19,3	0,050	0,30
		2007	5.1	8.2	2.1	4.3	-	-	11.1	16.2	0.100	0.100	<0.70	<0.70	2.2	2.7	1.24	2.6	9.9	15.4	0.090	0.090
		history	***0***	***65***	***0***	***5***	***0***	***16***	***0***	***65***	***0***	***0***	***0***	***7***	-	-	***0***	***13***	***0***	***25***	-	-
5	Tbilisi (V.B. Bridge)	2004	6,6	189	0,32	5,3	0,15	9,6	2,0	225	0,016	0,94	0,188	2,0	0,090	2,8	0,75	10,2	16,1	60	0,61	302
		2005	6,9	100	0,99	3,6	1,29	8,4	2,9	31	0,047	0,45	0,150	0,86	1,06	8,1	2,2	5,3	3,8	42	0,014	1,00
		2006	7,1	12,4	1,12	4,2	0,70	1,39	8,4	19,2	0,100	0,21	0,70	0,97	1,28	3,7	1,08	3,7	9,4	19,6	0,050	0,170
		2007	5.2	10.7	2.6	4.2	-	-	11.3	13.4	0.60	0.60	0.70	0.70	1.09	3.4	2.0	3.1	12.0	14.1	0.050	0.050
		history	***0***	***42***	***0***	***13***	***0***	***13***	***0***	***190***	***0***	***0***	***0***	***3***	-	-	***0***	***20***	***0***	***32***	-	-
6	Tbilisi (Gachiani)	2004	1,62	348	0,25	5,6	1,95	19,3	2,0	143	0,019	0,70	0,34	2,2	0,010	3,8	0,98	36	18,6	102	0,008	14,2
		2005	7,0	144	0,67	4,0	1,57	8,9	3,0	35	0,024	0,39	0,150	0,88	1,29	7,3	1,36	8,3	2,0	38	0,008	0,94
		2006	7,1	11,6	1,05	4,7	0,70	2,8	9,6	19,7	0,100	0,130	0,180	0,91	0,77	4,2	0,94	3,6	7,3	19,0	0,050	0,26
		2007	6.1	12.0	2.5	6.9	-	-	11.8	14.9	0.41	0.41	<0.70	<0.70	2.0	3.1	1.99	2.7	11.2	15.5	0.070	0.180
		history	***0***	***30***	***0***	***4***	***0***	***7***	***0***	***98***	***0***	***0***	***0***	***11***	-	-	***0***	***36***	***0***	***52***	-	-
7	Rustavi (down)	2004	1,66	329	0,51	6,7	1,41	10,3	3,3	59	0,005	0,46	0,22	1,64	0,82	3,5	0,071	7,3	0,83	83	0,008	8,1
		2005	5,2	106	1,07	5,9	0,76	8,4	2,0	38	0,029	0,53	0,137	0,83	1,45	9,8	2,4	7,4	3,2	59	0,006	1,54
		2006	8,5	13,9	1,72	4,4	0,70	1,54	9,1	30	0,100	0,180	0,130	0,93	0,68	3,2	1,08	4,1	9,9	28	0,050	0,60
		2007	6.9	10.5	2.2	4.8	-	-	11.1	15.7	0.12	0.41	0.77	0.77	2.3	2.6	1.82	3.3	12.1	16.6	0.050	0.050
		history	***0***	***42***	***0***	***8***	***0***	***41***	***0***	***450***	***0***	***0***	***0***	***7***	-	-	***0***	***20***	***0***	***38***	-	-

Table 3a: Generalized data (the minimal and maximal values) on metals for 2004, 2005, 2006, 2007 and corresponding historical (1980-1989) data for 3 control points on inflows of river Kura (Georgia)

№	Point	Year	***Metals (ug/l)***																			
			Cu		**Mo**		**Pb**		**Zn**		**Ag**		**Co**		**Cr**		**Ni**		**Mn**		**Cd**	
			min	max	min	max	min	max	min	max	min	max	min	max	min	max	min	max	min	max	min	max
1	Suramula-Khashuri (down)	2004	3,1	130	0,190	5,3	0,080	9,2	4,0	47	0,031	1,85	0,152	1,46	0,53	6,5	0,87	8,1	10,4	81	0,175	10,3
		2005	8,5	145	0,87	5,0	0,97	8,4	4,0	29	0,047	0,61	0,027	0,75	1,00	9,4	1,85	6,8	2,1	32	0,016	0,85
		2006	7,2	14,4	1,34	2,9	0,70	1,97	6,1	19,3	0,100	0,26	0,25	0,75	0,86	4,1	1,08	4,0	5,3	15,4	0,000	0,090
		2007	6.9	10.2	2.6	3.9	-	-	11.4	15.0	0.20	0.24	<0.70	<0.70	1.94	2.8	1.37	2.4	11.3	15.3	0.090	0.090
		history	***0***	***32***	***0***	***6***	***0***	***60***	-	-	***0***	***0***	***0***	***3***	-	-	***0***	***20***	***0***	***40***	-	-
2	Lekhura-Kaspi (down)	2004	4,4	203	0,36	6,6	0,92	15,8	2,5	42	0,009	0,62	0,28	1,67	1,19	4,4	0,63	8,1	14,0	65	0,46	8,6
		2005	8,2	136	0,93	3,1	0,95	8,7	2,8	57	0,022	0,34	0,117	0,50	0,31	8,0	0,54	5,9	3,0	42	0,011	1,78
		2006	6,1	13,8	0,95	4,4	0,70	2,7	9,0	21	0,100	0,26	0,100	1,21	0,68	2,7	1,27	2,9	7,5	17,9	0,050	0,28
		2007	5.1	9.6	3.1	3.7	-	-	11.3	13.6	0.16	0.34	0.75	1.55	1.33	2.9	1.88	3.6	11.3	14.2	<0.050	<0.050
		history	***0***	***26***	***0***	***5***	***0***	***16***	-	-	***0***	***0***	***0***	***2***	-	-	***0***	***7***	***0***	***35***	-	-
3	Khrami (Red Bridge)	2004	8,1	193	1,10	8,0	1,55	14,8	6,9	425	0,007	0,26	0,192	2,0	1,59	4,8	1,10	5,9	20	44	0,130	32,6
		2005	7,1	98	1,42	4,2	1,18	9,4	4,0	49	0,027	0,63	0,113	0,94	1,25	7,3	1,78	8,6	2,9	36	0,019	1,05
		2006	8,3	12,3	1,23	4,6	0,70	1,78	4,3	30	0,100	0,35	0,28	0,90	1,06	2,6	1,09	3,0	8,1	16,4	0,050	0,58
		2007	7.2	10.8	2.4	4.5	-	-	12.1	14.3	0.11	0.11	0.82	0.82	2.4	4.3	2.5	3.7	13.3	15.1	0.070	0.070
		history	***0***	***63***	***0***	***10***	***0***	***32***	***35***	***63***	***0***	***0***	***0***	***3***	-	-	***0***	***28***	***0***	***56***	-	-

List of the Authors in Alphabetical Order

Teimuraz Beridze
International Black Sea University
David Agmashenebeli Alley 13th km #2
0131 Tbilisi, Georgia

Lado Chanturia
Professor für Bürgerliches Recht,
Gesellschaftsrecht,
Rechtsvergleichung
Universität Bremen
Fachbereich Rechtswissenschaft
28353 Bremen

Nino Chkoidze
Justus-Liebig-Universität Gießen
Zentrum für internationale Entwicklungs- und Umweltforschung (ZEU)
Senckenbergstraße 3
35390 Gießen

Alexander Didebulidze
Georgia State Agricultural University,
13 km D. Agmashenebeli Ave., Digomi,
0131 Tbilisi, Georgia

Mariam Elizbarashvili
Ivane Javakhishvili Tbilisi State University
Faculty of Exact and Natural Sciences
1, Chavchavadze Ave.
0128 Tbilisi, Georgia

Marie-Carin von Gumppenberg
Election/Democratisation Officer,
OSZE Mission to Georgia
Schleissheimer Str. 43,
80797 München

Peter Felix-Henningsen
Justus-Liebig-Universität Gießen
Institut für Bodenkunde, IFZ
Heinrich-Buff-Ring 26-32
35392 Gießen

Teimuraz Jakhutashvili
Ivane Javakhishvili Tbilisi State University
Faculty of Exact and Natural Sciences
3, Av.Chavchavadze,
0128, Tbilisi, Georgia

Beso Kalandadze
Ivane Javakhishvili Tbilisi State University
Faculty of Exact and Natural Sciences
3, Chavchavadze Ave.
0128 Tbilisi, Georgia

David Kekelidze
Ivane Javakhishvili Tbilisi State University
Faculty of Exact and Natural Sciences
3, Av.Chavchavadze,
0128, Tbilisi, Georgia

Nodar Kekelidze
Ivane Javakhishvili Tbilisi State University
Faculty of Exact and Natural Sciences
3, Av.Chavchavadze,
0128, Tbilisi, Georgia

Elguja Khokrishvili
Universität Potsdam
Lehrstuhl Finanzwissenschaft
August-Bebel-Str. 89
14482 Potsdam

Lorenz King
Justus-Liebig-Universität Gießen
Zentrum für internationale Entwicklungs- und Umweltforschung (ZEU) und
Institut für Geographie
Senckenbergstraße 3
35390 Gießen

Eliso King-Narimanidze
Justus-Liebig-Universität Gießen
Zentrum für internationale Entwicklungs- und Umweltforschung (ZEU)
Senckenbergstraße 3
35390 Gießen

Tanja Kirn
Universität Potsdam
Lehrstuhl Finanzwissenschaft
August-Bebel-Str. 89
14482 Potsdam

Mamuka Komakhia
BTKK - Policy Research Group
Tbilisi Postleitzahl?
Georgia

Marietta S. König
Institut für Friedensforschung und Sicherheitspolitik
Beim Schlump 83
20144 Hamburg

Kerstin Kötschau
Justus-Liebig-Universität Gießen
Zentrum für internationale Entwicklungs- und Umweltforschung (ZEU)
Senckenbergstraße 3
35390 Gießen

Giorgi Khubua
Ivane Javakhishvili Tbilisi State University
1, Chavchavadze Ave.
0128 Tbilisi, Georgia

Eric Livny
International School of Economics (ISET), Director
Zandukeli St. 16
0108 Tbilisi, Georgia

Ramaz Meskhia
The National Environmental Agency
Department of registration and analysis of natural disasters
150, David Agmashenebeli Ave.
0112 Tbilisi, Georgia

Lela Mtsariashvili
Ivane Javakhishvili Tbilisi State University
Faculty of Exact and Natural Sciences
3, Av.Chavchavadze,
0128, Tbilisi, Georgia

Nino Narimanidze
Ivane Javakhishvili Tbilisi State University
Faculty of Economics and Business
1, Chavchavadze Ave.
0128 Tbilisi, Georgia

David Narmania
Georgian Institut of public affairs (GIPA)
Marie Brosset St. #2
0108 Tbilisi, Georgia

List of the Authors in Alphabetical Order

Mack Ott
Gravitas International LLC, Director
904 Plum Street SW
Vienna, VA 22180
Vladimer Papava
Georgian Foundation for Strategic and International Studies (GFSIS)
Shio Chitadze Street 3a
0108 Tbilisi, Georgia

Michael Schäfer
Justus-Liebig-Universität Gießen
Zentrum für internationale Entwicklungs- und Umweltforschung (ZEU)
Senckenbergstraße 3
35390 Gießen

Eka Sepashvili
Ivane Javakhishvili Tbilisi State University
University st. 11
Tbilisi, Georiga

Diedrich Steffens
Justus-Liebig-Universität Gießen
Institut für Pflanzenernährung, IFZ
Heinrich-Buff-Ring 26-32
35392 Gießen

George Tomaradze
Dr. Hans Kapfinger Str. 13 A123
94032 Passau

Karine Torosyan
International School of Economics (ISET), Research Assistant
Zandukeli St. 16
0108 Tbilisi, Georgia)

Eremia Tulashvili
Ivane Javakhishvili Tbilisi State University
Faculty of Exact and Natural Sciences
3, Av.Chavchavadze,
0128, Tbilisi, Georgia
Tengiz Urushadze
Ivane Javakhishvili Tbilisi State University
Faculty of Exact and Natural Sciences
3, Chavchavadze Ave.
0128 Tbilisi, Georgia

Gerd Winter
Professor für Öffentliches Recht
Universität Bremen
Fachbereich Rechtswissenschaft
28353 Bremen

Schriften zur Internationalen Entwicklungs- und Umweltforschung

Herausgegeben vom

Zentrum für internationale
Entwicklungs- und
Umweltforschung
der Justus-Liebig-Universität Gießen

Band 1 Hans-Rimbert Hemmer / Rainer Wilhelm: Fighting Poverty in Developing Countries. Principles for Economic Policy. 2000.

Band 2 Lorenz King / Martin Metzler / Tong Jiang (eds.): Flood Risks and Land Use Conflicts in the Yangtze Catchment, China and at the Rhine River, Germany. 2001.

Band 3 Ingrid-Ute Leonhäuser (ed.): Women in the Context of International Development and Cooperation. Review and Perspectives. Selected Papers and Abstracts presented at the Justus-Liebig-University Gießen 26.–28. October 2000. 2002

Band 4 Margit Schratzenstaller: Internationale Mobilität von und internationaler fiskalischer Wettbewerb um Direktinvestitionen. 2002.

Band 5 Armin Bohnet u.a.: Theoretische Grundlagen und praktische Gestaltungsmöglichkeiten eines Finanzausgleichssystems für die VR China. Unter Mitwirkung von Chen Biyan, Chen Shixin, Ge Licheng, Ge Naixu, Ge Zhuying, Ma Shuanyou, Markus Peplau, Yang Zhigang, Zhu Qiuxia. 2003.

Band 6 Armin Bohnet / Matthias Höher (eds.): The Role of Minorities in the Development Process. 2004.

Band 7 Thi Phuong Hoa Nguyen: Foreign Direct Investment and its Contributions to Economic Growth and Poverty Reduction in Vietnam (1986–2001). 2004.

Band 8 Andreas Böcker / Roland Herrmann / Michael Gast / Jana Seidemann: Qualität von Nahrungsmitteln. Grundkonzepte, Kriterien, Handlungsmöglichkeiten. 2004.

Band 9 Christina Mönnich: Tariff Rate Quotas and Their Administration. Theory, Practice and an Econometric Model for the EU. 2004.

Band 10 Reimund Seidelmann / Ernst Giese (eds.): Cooperation and Conflict Management in Central Asia. 2004.

Band 11 Claudia Ohly: Das Steuersystem im ungarischen Transformationsprozess. Ein Beitrag zur Transformationstheorie. 2004.

Band 12 Nicole Mau: Umweltzertifikate. Der Einsatz von Umweltzertifikaten in der Landwirtschaft am Beispiel klimarelevanter Gase. 2005.

Band 13 P. Michael Schmitz (Hrsg.): Water and Sustainable Development. 2005.

Band 14 Ira Pawlowski: Die Wettbewerbsfähigkeit der ukrainischen Milchwirtschaft. Messung von Marktverzerrung und Politikeinfluß im Transformationsprozeß. 2005.

Band 15 Kirsten Westphal (ed.): A Focus on EU-Russian Relations. Towards a close partnership on defined road maps? 2005.

Band 16 Andreas Langenohl / Kirsten Westphal (eds.): Conflicts in a Transnational World. Lessons from Nations and States in Transformation. 2006.

Band 17 Rosemarie von Schweitzer: Home Economics Science and Arts. Managing Sustainable Everyday Life. 2006.

Band 18 Dörthe List: Regionale Kooperation in Zentralasien. Hindernisse und Möglichkeiten. 2006.

Band 19 Michael Gast: Determinanten ausländischer Direktinvestitionen. OECD-Länder als Investoren und besonderer Aspekte der Ernährungswissenschaft. 2007.

Band 20 Kim Schmitz: Die Bewertung von Multifunktionalität der Landschaft mit diskreten Choice Experimenten. 2008.

Band 21 Kerstin Kötschau / Thilo Marauhn (eds.): Good Governance and Developing Countries. Interdisciplinary Perspectives. 2008.

Band 22 Armin Bohnet (ed.): Poland on its Way to a Federal State? 2008.

Band 23 Johannes Harsche: Regionale Inzidenz und ökonomische Bestimmungsgrößen der Gemeinsamen Europäischen Agrarpolitik. 2009.

Band 24 Aikaterini Kavallari: Agricultural Trade Policy Reforms and Trade Liberalisation in the Mediterranean Basin. A Partial Equilibrium Analysis of Regional Effects on the EU-27 and on the Mediterranean Partner Countries. 2009.

Band 25 Marc Christopher Kramb: Sanitäre und phytosanitäre Handelsbeschränkungen unter dem Einfluss des WTO-Abkommens. Ein Gravitationsansatz unter besonderer Berücksichtigung des EU-Rindfleischsektors. 2009.

Band 26 Lorenz King / Giorgi Khubua (eds.): Georgia in Transition. Experiences and Perspectives. 2009.

www.peterlang.de